Global Conference on Polymer and Composite Materials

(PCM 2014)

IOP Conference Series: Materials Science and Engineering Volume 62

Ningbo, China
27 – 29 May 2014

Editors:

E. Broitman

ISBN: 978-1-63439-873-2
ISSN: 1757-8981

Printed from e-media with permission by:

Curran Associates, Inc.
57 Morehouse Lane
Red Hook, NY 12571

Some format issues inherent in the e-media version may also appear in this print version.

Printed by Curran Associates, Inc. (2015)

For permission requests, please contact the Institute of Physics
at the address below.

Institute of Physics
Dirac House, Temple Back
Bristol BS1 6BE UK

Phone: 44 1 17 929 7481
Fax: 44 1 17 920 0979

techtracking@iop.org

Additional copies of this publication are available from:

Curran Associates, Inc.
57 Morehouse Lane
Red Hook, NY 12571 USA
Phone: 845-758-0400
Fax: 845-758-2634
Email: curran@proceedings.com
Web: www.proceedings.com

TABLE OF CONTENTS

IOP Conf. Series: Materials Science and Engineering **62** (2014) 011001 doi:10.1088/1757-899X/62/1/011001

Preface

The *2014 Global Conference on Polymer and Composite Materials (PCM 2014)* sponsored by Ningbo Adhesives and Products Industry Association, Shanghai Bonding Technology Association, Zhejiang Bonding Technology Association, Wuhan Bonding Technology Association, Hebei Bonding and Coatings Association and Polyurethane Industry Association was held from May 27 to May 29 2014 in Ningbo, China. The technical program consisted of 8 international keynote speakers, oral presentations, and a poster session. The conference also included an industrial exhibition where more than 50 companies displayed in their booths their most recent advanced products and services.

The present issue of IOP Conference Series: Materials Science and Engineering (MSE) records the proceedings of PCM 2014 and contains 37 specially selected manuscripts submitted to PCM2014 conference. The electronic submission and handling of manuscripts via the conference website, including the selection of reviewers and evaluation of manuscripts, were identical to the procedures applied to manuscripts submitted as regular contributions for publication.

The organization of this conference and the preparation of proceedings volumes would have been impossible without the tremendous efforts and dedication of many individuals, especially from Ms. Yin Pan, who oversaw the organization of the conference and the program; and a large team of reviewers with their timely submission of quality reports. We express our sincere thanks to all authors and presenters for their contributions. We also thank very much our sponsors for their generous support.

The 2015 Global Conference on Polymer and Composite Materials (PCM2015) will be held in Beijing, China on May 16–18, 2015. Beijing, the capital of the People's Republic of China and one of the most populous cities in the world, will welcome to all participants for a renewed and vibrant conference.

Prof. Dr. Esteban Broitman
Linköping University, Sweden
Editor in Chief – PCM2014

IOP Conf. Series: Materials Science and Engineering **62** (2014) 011002 doi:10.1088/1757-899X/62/1/011002

Peer review statement

All papers published in this volume of *Journal of Physics: Conference Series* have been peer reviewed through processes administered by the proceedings Editors. Reviews were conducted by expert referees to the professional and scientific standards expected of a proceedings journal published by IOP Publishing.

Vibrational and chemical kinetics in plasma of CO containing gases

G.M.Grigorian, A.Cenian*

St.Petersburg State University, St.Petersburg, Russia,
*The Szewalski Institute of Fluid-Flow Machinery, Gdansk, Poland

e-mail: galina_grigorian@yahoo.com

Abstract. This paper discusses the experimental results pointing to the efficient channel of the CO vibrational to the C_2 electron energy transfer. The radiation spectra of $e^3\Pi$, $C^1\Pi$, $D^1\Sigma$ states of C_2 molecules were investigated and the relation of their kinetics to vibrational excitation of CO ($X^1\Sigma$, v) molecules in the He-CO-O_2 gas discharge plasma was discussed. The rate constant for VE- process CO ($X^1\Sigma$, v$\geq$ 25) + $C_2 \rightarrow$ CO($X^1\Sigma$) + $C_2(D^1\Sigma_u)$, was estimated, and $k_{VE} \sim 10^{-14}$ cm^3/c.

1. Introduction

The low temperature plasma initiated in He-CO-O_2 mixture can be characterized by high degree of CO ($X^1\Sigma$) vibrational excitation. In addition, the steady-state vibrational distributions, CO($X^1\Sigma$,v), significantly differ from the standard Boltzmann type distribution, due to the fast vibration-vibration (VV) energy exchange processes between vibrationaly excited (anharmonic) CO molecules [1]. The difference becomes more pronounced with increasing mismatch between the gas temperature, T_g and the vibrational temperature given by the expression ($T_v = E_1/\ln(N_0/N_1)$, where E_1 is the energy of first vibrational level and N_0 (N_1) denote populations of the ground (the first excited) vibrational states. For a high enough mismatch between T_g and T_1, the characteristic plateau on vibrational distribution for levels $v > T_g /2T_1 x_e$ (x_e - anharmonicity factor) appears, and the distribution changes slowly, proportionally to 1/(v+1). Therefore cryogenic cooling of working gas mixtures was used to increase the overpopulation of high vibrational levels of CO($X^1\Sigma$).

This leads, in the case of He-CO-O_2 discharge plasma with cryogenic cooling, to relatively high values of populations ([CO(v)] ~ 10^{-2} - 10^{-3} [CO(v = 0)]) even for levels with v ~10-15; examples can be found in [2] and [3]. Such high concentration of excited vibrational molecules significantly influences plasma-chemistry of the considered gas mixture, e.g. various VE-processes (energy transfer from vibrational to electronic states) can be observed. For example, the C_2 ($d^3\Pi$, $e^3\Pi$ and $C^1\Pi$) excited states were observed, with excitation imposed by energy transfer processes from vibrational to electronic states [4].

Figure 1 presents electronic C_2 states and the first 30 vibrational states of CO($X^1\Sigma$) molecules coinciding energetically. This coincidence and the resulting resonances enable efficient energy transfer from vibrational CO states to electronic states of C_2 (V-E process):

$$CO(v) + C_2 > CO(v - n) + C_2^* \, , \quad (1)$$

where C^*_2 - is an excited state (see Fig. 1).

Here, we continue our investigations of the influence of vibrationally ecited CO molecules on electronically excited C_2 clusters [4], including highly excitations $e^3\Pi$, $C^1\Pi$, $D^1\Sigma$ (E>4 eV). The vibrational energy transfer to these electronic states has not yet been reported.

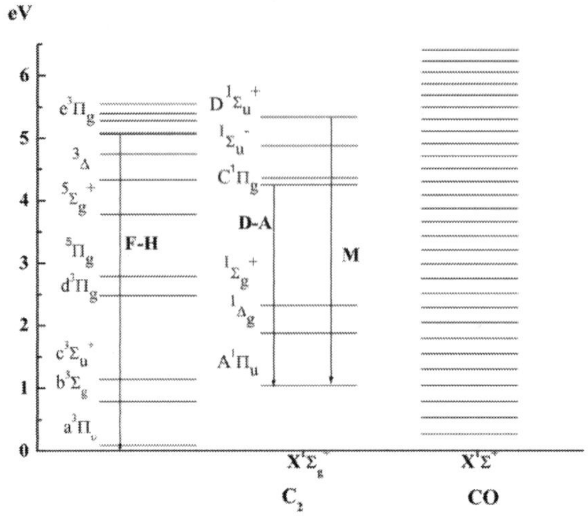

Fig.1 Scheme of the energetically-low levels of the electronic excited states of C_2 cluster and the vibrational states of CO molecule. Arrows shows the transitions related to so called Fox-Herzberg (F-H), Deslandres-d'Azambuja (D-A) and Mulliken (M) systems.

2. Experimental set-up and methodology

A molybdenum-glass discharge tube that is 20 mm in radius and 50 cm in length, nitrogen- or water-cooled, was used in the experiments (see details in [4, 6]). The hollow cylindrical electrodes made from tantalum were installed in the side branches of the discharge tube at a distance of 4 cm from its axis. A discharge was excited using DC currents of 20 - 100 mA. Gas flowing set-up (v ~2-5 m/s) was applied. The relative CO concentrations in the applied He-CO-O_2 mixtures were varied in the range of 3-15 %. The electric field in the discharge was determined by measuring the voltage drops across the discharge gap and the electrode sheaths. The electrode sheath voltages were found from the voltage measurements in experiments with different discharge lengths. A typical electrode sheath voltage was 350-400 V. The estimated value of the reduced electric field E/N was (1.5 - 2.5) 10^{-16} V cm^2.

The spectra in the range of 200 - 3000 nm were registered using the Czerny-Turner monochromator with a diffraction grating of 300 and 600 lines/mm. The recorded spectra allowed determination of the concentrations of electronic excited species, gas temperature (through the analyses of the vibrational-rotational spectra) as well as vibrational populations of CO molecules (by analyzing the radiation in the first and second vibrational harmonics). Radiation from the discharge plasma was collected from near axis volume 3 mm in radius for visible and UV and 5 mm in radius for IR spectra. HgCdTl photoresistors cooled by liquid nitrogen and InSb photoresistors were used as light detectors in IR region. The sensitivity of this technique allowed us to measured the population of CO($X^1\Sigma$) vibration levels with vibration level numbers up to v = 29.

The influence of CO vibrational-excitation on the radiation intensities from electronic excited states of C_2 molecules was studied in a discharge tube with an appropriate resonator (enabling CO-laser generation at levels v > 6-10 - depending on experimental conditions) installed at both ends of the discharge tube. Laser generation led to a rapid decrease of vibrational levels CO(X, v) population, for levels close to those participating in laser transition. This means that, by covering one mirror of the laser resonator, it was possible to change significantly the vibrational population of CO. It should be mentioned that the laser output was not very high, usually a few Watts. The plasma parameters, besides vibrational temperatures (i.e. electron and gas temperature, voltage, pressure, etc.) did not vary significantly during the modulation of the laser radiation. For example, it was found that the gas temperature decreased by only 10 K, when the laser was operating. At the same time, E/N was practically constant.

The dependence of the $C_2(X^1\Sigma)$ concentration on experimental conditions (i.e. discharge current) was determined from absorption spectra. For $C_2(X^1\Sigma)$ population the Phillips spectral systems ($A^1\Pi$- $X^1\Sigma$ transition) was used. The spectrum was observed using the CW deuterium lamp. The $C_2(D^1\Sigma)$ population was determined from emission spectra (Mulliken bands).

The laser generation was accomplished using selective (enabling tuning) and non-selective resonators. The non-selective resonator was formed by a highly reflective gold mirror with the curvature radius of 5 m and a plane mirror having transmission of ~ 30%. The tunable laser cavity consisted of 5% transmitting output mirror with radius of curvature 5 m and a metal grating blazed at 5.4 mkm.

3. Results and discussion

The results of experiments [4] showed that changes of populations of high CO vibrational states enforce changes of radiation intensities from the studied $e^3\Pi_g$ and $C^1\Pi_g$ electronic states of the C_2 molecule. This can result from a proposed VE process (1). To clarify the contribution of different vibrational states CO(w) to the process (1), the consecutive changes of CO vibrational energy distribution were imposed by an application of a tunable laser resonator to the discharge tube.

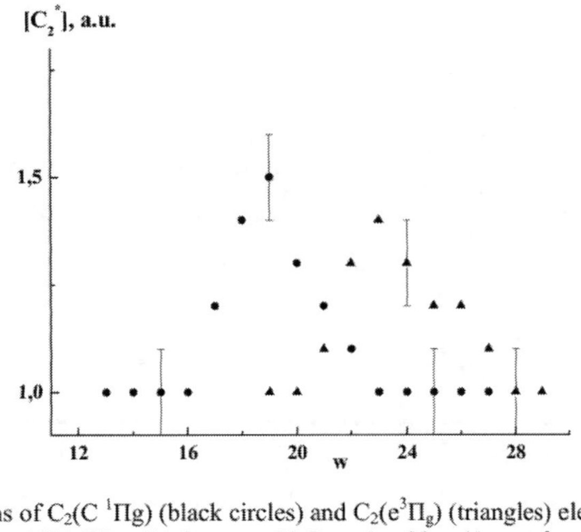

Fig. 2 Relative concentrations of $C_2(C\ ^1\Pi g)$ (black circles) and $C_2(e^3\Pi_g)$ (triangles) electronic states (in arb. units) in the discharge plasma (He + 12% CO +1% O_2, p = 13 Torr, I = 30 mA) as a function of the upper vibration level number for lasing transition (CO(w) → CO(w-1)); the number tuned using a selective resonator

Figure 2 shows the measured relative concentrations of $C_2(C\ ^1\Pi g)$ and $C_2(e^3\Pi_g)$ in the discharge plasma as a function of the upper vibration-level number 'w' for lasing transition CO(w) → CO(w-1). The concentration $[C_2^*] = 1$ corresponds to the value of $[C_2^*]_0$ with lasing absent. It can be seen (from Fig.2) that the CO vibrational levels, which energetically coincide with considered C_2states affect the concentration of these states mostly. For the C $^1\Pi g$ state they are vibrational levels v ~ 17 – 21 and for the $e^3\Pi_g$ state, vibrational levels v ~ 21 - 27.

The intensity of $D^1\Sigma_u \rightarrow X^1\Sigma$ radiation was to low and its investigation using set-up equipped with selective resonator was not possible. Therefore non-selective resonator was used to study this radiation. Moreover, Mulliken bands ($D^1\Sigma_u \rightarrow X^1\Sigma$) were not observed in the water-cooled discharge tube, under all considered here conditions. The $D^1\Sigma$ Mulliken band was registered only in the case of cryogenic cooled discharge in He-CO-O_2 (with CO $\geq$ 9%) mixture without laser generation. The radiation vanishes when the laser operates. Figure 3 presents the spectrum of Mulliken band ($\Delta v = 0$) without laser generation (a) and the same spectral region (without band) with laser turned-on (b).

Figure 4 presents the dependence of Mulliken-band intensity and population of CO($X^1\Sigma$, v = 25) vibrational state (the vibrational state energetically close to $C_2(D^1\Sigma)$ electronic state). It is clearly seen that the populations of CO($X^1\Sigma$, v = 25) and C_2 ($D^1\Sigma$) states depend on discharge current likewise

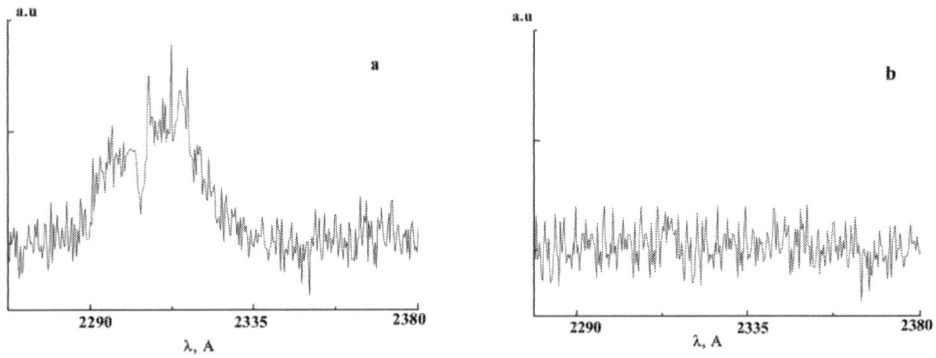

Fig.3 Spectrum of Mulliken band (0-0) in cryogenic-cooled discharge, He:CO:O_2 = 88:11:1tur p = 10 Torr (a) without laser generation and (b) with laser generation (P = 5W).

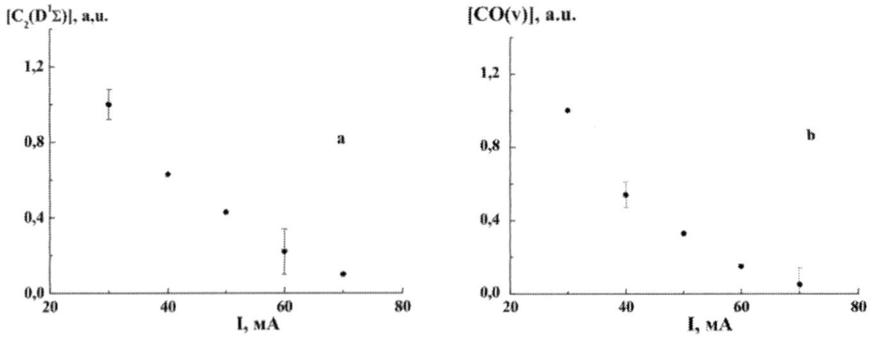

Fig.4 Intensity of (a) Mulliken band radiation ($D^1\Sigma_u \rightarrow X^1\Sigma$) and (b) population of CO($X^1\Sigma$, v = 25) state as function of discharge current in He:CO:O_2 = 91.5:8.5:0.03 mixture, p = 12 Torr

It should be mentioned that the $C_2(D^1\Sigma$) state can be populated by VE process with the participation of CO($X^1\Sigma$, $v \geq 25$) states or by a process of direct electron collision. In the second case when $C_2(D^1\Sigma)$ state would be excited in direct collisional process:

$$C_2(X) + e \rightarrow C_2(D^1\Sigma) + e, \tag{2}$$

the concentration of the $C_2(D^1\Sigma)$ state should grow with the discharge current. However, from the observed absorption of Phillips spectral system ($A^1\Pi \rightarrow X^1\Sigma$) it was found that for mixtures with [CO] $\geq$ 9% the [$C_2(D^1\Sigma)$] concentration decrease with the current (Fig. 4). At the same time, there wasn't any change of [$C_2(X)$] concentration observed under various discharge currents applied, at least with the achieved precision of 25%.

The observed Mulliken band in cryogenically cooled discharges in He-CO-O_2 mixtures, which vanish when laser operates, can be explained by VE processes of type (1) with the participation of CO($X^1\Sigma$, v

≥ 25) states. This was confirmed by the correlation of $CO(X^1\Sigma, v \geq 25)$ and C_2 $(D^1\Sigma)$ states populations presented in Fig. 4.

In addition, as mentioned above, radiation from Mulliken band was observed only for systems with cryogenic cooling. In the system with water cooling the radiation was not observed. Figure 5 shows that in the case of water cooling vibrational states with v, which can participate in VE process leading to the excitation of C_2 $(D^1\Sigma)$ state, are significantly less populated (5 times or more) than in the case with cryogenic cooling. It was also seen that laser generation lead to one order of magnitude lower populations of these states. These low populations of $CO(X^1\Sigma$, $v \geq 25)$ states explain why the Mulliken band was not observed when laser was operating and in water cooled systems.

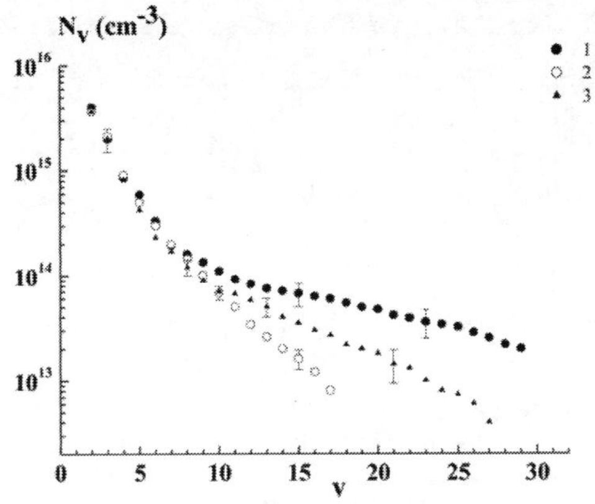

Fig.5 Populations of CO vibrational states in discharge ($He:CO:O_2$ = 91.5:8.5:0.03 mixture, p = 12 Torr, I = 30 mA) in (1,2) cryogenic and (3) water cooled set-up; with (2) laser generation and (1,3) laser turned-off

Let us estimate the rate for the process (1) in the plasma of discharge in $He:CO:O_2$ = 91.5:8.5:0.03 mixture, p = 12 Torr, I = 30 mA. Assuming that VE process constitutes the only channel for $C_2(D^1\Sigma)$-state excitation, the steady-state balance equation assumes the following form

$$\Gamma = [C_2]\cdot[CO(v \geq 25)]\cdot k_{VE} = [C_2(D^1\Sigma)]\cdot A(D^1\Sigma)\,, \tag{3}$$

where $A(D^1\Sigma)$ is the probability of spontaneous emission for $C_2(D^1\Sigma)$-state [5]. The rate Γ determined from the band intensity was equal to 10^{12} cm^{-3}s^{-1}. The population of $[CO(v = 25)]$ state was determined from IR intensities of CO(v) spontaneous emission and under considered conditions was ~ $3.4\cdot10^{13}$ cm^{-3}; $[C_2] \sim 3\cdot10^{12}$ cm^{-3}. Finally, the rate constant for VE process is obtained as

$$CO(v \geq 25) + C_2 \rightarrow CO(v\text{-}25) + C_2(D^1\Sigma) \tag{4}$$

$k_{VE} \sim 10^{-14}$ cm^3/s.

Acknowledgments

This study was partially supported by grant № 07-02-01114 of Russian Foundation for Basic Research.

References

[1] Treanor C E, Rich J W and Rehm R G 1968 *J.Chem.Phys.* **48** 1798
[2] Fisher E R and Lightman A J 1978 *J.Appl.Phys.* **49** 530
[3] Grigorian G M, Ionikh Y Z, Kochetov I V and Pevgov V G 1992 *J.Phys.D.* **25** 1064
[4] Grigorian G M and Cenian A 2009 *Chem. Phys. Lett.* **469** 247
[5] Chabalowski C F and Peyerimhoff S D 1983 *Chem.Phys.* **81** 57
[6] Grigorian G M and Kochetov I V 2004 *Plasma Phys.Reports*, **30** 788

Study of the thermal treatment of SiO_2 aggregate

N Tahiri[1, 2], L Khouchaf[1,*], M Elaatmani[2], G Louarn[3], A Zegzouti[2] and M Daoud[2]

[1] Univ-Lille Nord de France, Research Center of Ecole des Mines de Douai, Rue Charles Bourseul BP10838 59808, Douai, France.
[2] Cadi Ayyad University, Chemical Department, Faculty of Sciences Semlalia, Bd Prince My Abdellah, 40000, Marrakech, Morocco.
[3] Institut des Matériaux Jean Rouxel, UMR 6502, Université de Nantes-CNRS, 2, 14 Rue de la Houssinière 44322, Nantes Cedex, France.

*Corresponding author : lahcen.khouchaf@mines-douai.fr

Abstract: Crystalline quality and the silanoles defects (Si–OH) present within the structure of natural SiO_2 play an important role in its reactivity. In this work, the relationship between the loss of silanoles and the crystallinity improvement upon heating between 450 °C and 650 °C was shown using X-Ray Diffraction (XRD) and Fourier Transform mid Infrared Spectroscopy in Attenuated Total Reflection (ATR). A shift of the principal band Si-O-Si from 1078 cm^{-1} to 1082 cm^{-1} and the decrease of the intensity of the Si–OH bands at 555$cm^{-1}$and at 950cm^{-1} upon heating between 450 °C and 650 °C were shown. The reduction of the band is correlated to the loss of silanoles. In agreement with FT-IR results, the decrease of the FWHM of the XRD peaks shows that the crystalline quality is improved upon heating. This result leads to a decrease of the reactivity of SiO2 aggregate under chemical attacks.

Keywords: SiO_2; Reactivity; Thermal treatment; Crystallinity

1. Introduction

To improve the properties, such as strength and durability, of some materials, SiO_2 compounds may be used [1,2]. The reactivity of SiO_2 compounds depends on the chemical process that occurs in amorphous or poorly crystallized SiO_2 present in the mineral aggregates [3, 4, 5]. SiO_2 compound is a natural SiO_2 used as an aggregate in composite materials such as concrete. The degradation of concrete depends on the crystalline quality of the aggregate [6].

The presence of defects in the structure of the silica increases its reactivity [7, 8 ,9] with respect to alkali-silica reaction (ASR) which causes the dissolution of the silica network, the formation of expansive gels and hence the damage of the concrete [10,11,12,13]. Previous studies have confirmed the presence of silanoles groups in natural silica used directly in the initiation step of the ASR [14, 15, 16].

P. Schmidt et al. [17, 18] showed that this type of defects can be reduced by heat treatment. According to them, a heat treatment of natural silica results in the formation of siloxane bridges Si-O-Si according to reaction below,

$$\text{Si-OH} + \text{HO-Si} \rightarrow \text{Si-O-Si} + H_2O.$$

According to the equation above, consumption of silanoles Si-OH leads to siloxane Si-O-Si bonds formation, which improves the recrystallization of the material due to an increase of crystallite size [18, 19]. However, the choice of temperature, speed and duration of the heat treatment are important parameters for the maximum silanoles evaporation and at the same time to avoid the fracturing aggregate phenomenon [17, 18, 19].

The loss of silanoles begins at 250 °C, resulting in the formation of siloxane and water molecules trapped in small pores isolated within the grains. Most silanoles are removed between 550-600 °C [17]. Increasing the temperature above 500 °C leads to a rise in the pressure level, which causes the isolated fracture appearance, allowing water molecules to drain [17] from pores.

Another factor that affects the temperature of fracturing is the speed of the heat treatment; if this speed is high enough, it leads to the occurrence of a temperature gradient between the interior and exterior of the grain in heat treatment, resulting in a heterogeneous thermal expansion and thus fracturing [18].

The duration of heat treatment plays an important role in the process of dehydration of flint [20]. P. Schmidt et al. [17] found that there is no difference between samples heat treated for 3 h and 6 h regarding the loss of silanoles, and a heating time of 50 minutes is largely sufficient [20].

The aim of this work was to follow the structural changes of a natural flint aggregate subjected to different temperatures. The objective was to minimize defects (silanoles type) and improve the crystallinity of the aggregate to optimize its use as a concrete skeleton. The study was conducted using the X-ray diffraction, infrared spectroscopy, and the transmission electron microscope.

2. Materials and methods

The aggregate used in this study is a natural flint with 99.1% of SiO_2; it has a disordered crystallographic structure compared to alpha quartz. The sample underwent experimental heat treatment at two temperatures (table 1) and was analyzed by means of different techniques.

The procedure of heat treatment is summarized in figure 1.

Table 1. Annealing temperatures

Sample	T_3	T_3t_1	T_3t_2
Annealing temperature	25°C	450°C	650°C

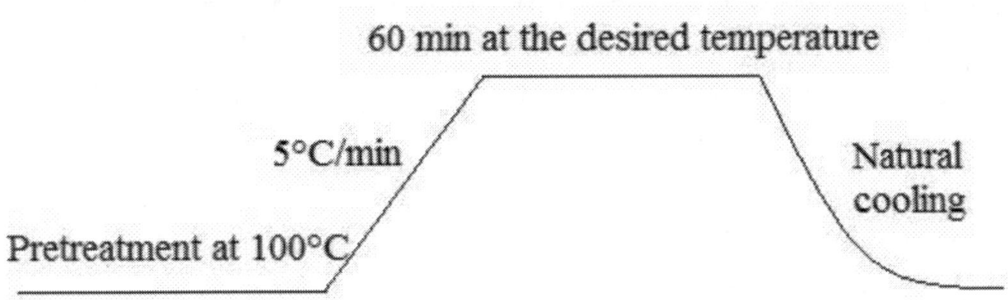

Figure 1. Procedure of heat treatment

XRD measurements were performed in the reflection mode using a Shimadzu Bruker D8 Advance diffractometer (Cu-K λ radiation = 1.5418 A°), which operates at 40 kV and 40 mA. Data were collected in the range of 25°–75° in the 2θ scale with a step size of 0.02° and a counting time of 0.5 s/step.

Spectra were acquired with a Brucker VERTEX 70 FTIR spectrometer in reflection mode. They were recorded by collecting 100 scans at 4 cm^{-1} resolution in the range of 400–5000 cm^{-1}.

The TEM investigations were performed with a FEI Tecnai G2-20 instrument operating at 200 kV accelerating potential. It is equipped with a filament of lanthanum hexaboride LaB_6, a double-tilt holder and Gatan digital camera.

3. Results and discussion

3.1. Transmission Electron Microscope

Figure 2 shows an image of sample T_3. Zones of different sizes ranging from ten nanometers to a few hundred micrometers, with angular sides characterizing angles of quartz can be seen. Moreover, other areas of a rather disordered or amorphous appearance are observed.

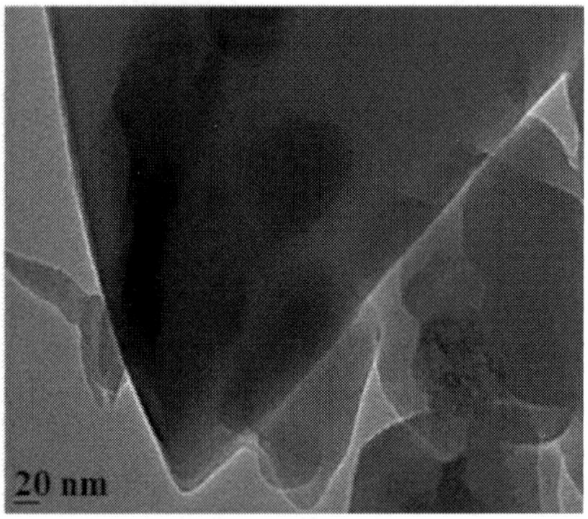

Figure 2.TEM micrograph of T_3

3.2. X-ray diffraction

Figure 5 shows the XRD patterns of sample T_3 and quartz, which were analyzed in the same conditions indicating the crystallized structure of both compounds. By comparing the two diagrams of samples, it can be seen that they have the same peak positions, while T_3 has broader and less intense ones compared to the quartz peak. It can be deduced that sample T_3 is mainly composed of a crystalline-phase quartz, with the presence of structural defects and disordered fraction.

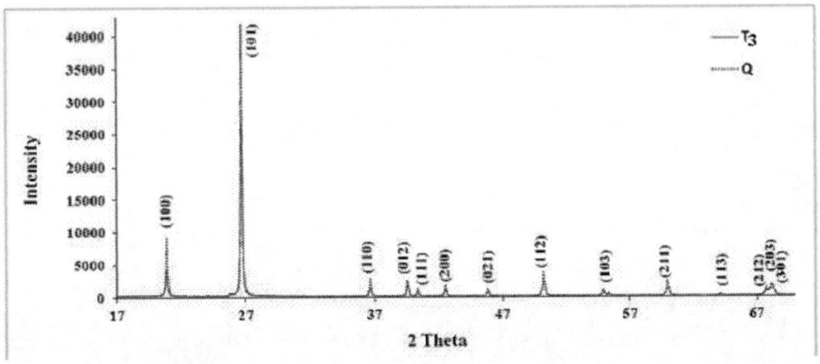

Figure 3. XRD pattern of T_3 compared to α-quartz (Q)

To evaluate changes in the crystallinity of the samples as a function of temperature, the width at half height of the diffraction peaks of the raw flint and after heat treatment was studied.

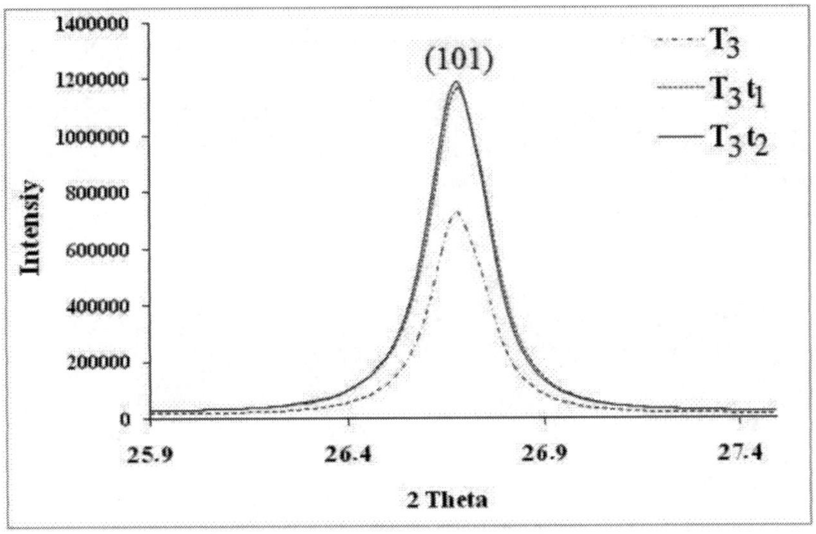

Figure 4. Peak (101) for unheated (T_3) and heated (T_3t_1, T_3t_2) flints

According to figures 4 and 5, the intensity of the peak (101) increases and the Full Width at Half Maximum (FWHM) decreases with temperature. This result indicates an improvement in the crystallinity of heat treated T_3. It can also be seen that the FWHM is lower at the temperature of 650 °C.

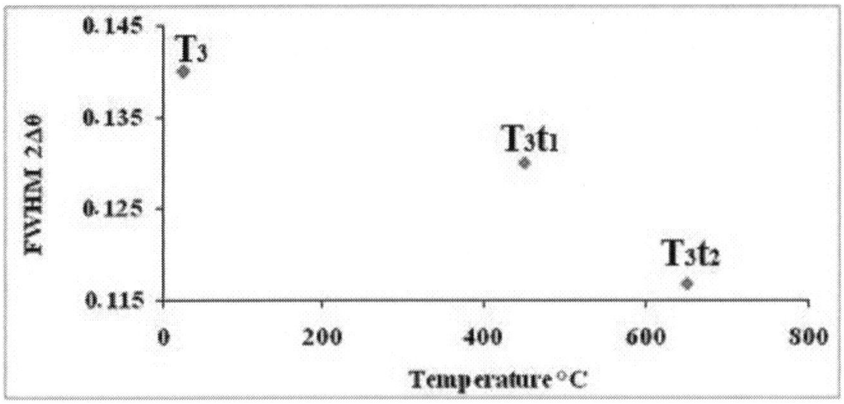

Figure 5. FWHM of the peak (101) for unheated (T_3) and heated (T_3t_1, T_3t_2) flints

3.3. Infrared spectroscopy

Figure 6 shows the infrared IR spectra between 400 cm^{-1} and 1400 cm^{-1} of the unheated (T_3) and heated (T_3t_1, T_3t_2) flints.

The infrared spectrum of unheated flint T_3 has a broad band between 1000 and 1300 cm^{-1}. This band consists of two strong peaks located at 1078 cm^{-1} and 1163 cm^{-1}. These peaks are associated with the stretching vibration of Si-O-Si [21, 22, 23, 24, 25]. In addition, bending vibrations attributed to Si-O-Si [18, 20, 21, 23, 25, 26] located at 455 cm^{-1}, 509 cm^{-1}, 778 cm^{-1} and 800 cm^{-1} are observed in the spectrum.

The bands located at 555 cm^{-1} and 950 cm^{-1} are associated with Si-O bending vibrations of no bridging Si-OH bonds and stretching vibration respectively [20, 21, 22, 23, 26, 27]. Heated flint T_3t_1 did not show significant changes in the infrared IR spectrum.

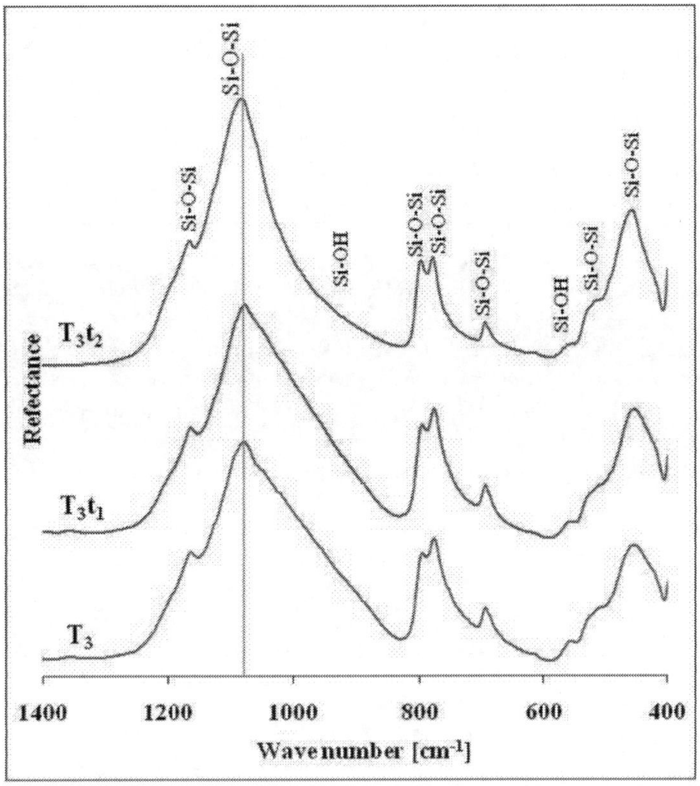

Figure 6. FT-IR spectra of the unheated (T_3) and heated (T_3t_1, T_3t_2) flints

The spectrum of the sample heated at 650 °C (T_3t_2) shows a reduction of bands located at 555 cm^{-1} and 950 cm^{-1}, an increased intensity of the bands located at 455 cm^{-1} and 800 cm^{-1} and a shift of the main band from 1078 cm^{-1} to 1082 cm^{-1}. The position of this band is connected to the structure in the order of SiO_2. The shift of the observed position of the band from 1078 cm^{-1} to 1082 cm^{-1} may be attributed to an improvement in structural order with the increase in temperature in accordance with previous studies [21]. The more the structure is ordered, the more this band is located at the large wave numbers. Similarly, the reduction bands at 555 cm^{-1} and 950 cm^{-1} at 650 °C can be attributed to the departure of some silanoles groups. These results are in agreement with the XRD changes with temperature.

4. Conclusion

Using XRD in combination with FT, the structural change of SiO_2 natural flint with temperature was successfully investigated. A shift of the principal Si-O-Si band from 1078 cm^{-1} towards 1082 cm^{-1} and decrease of Si–OH band intensity at 555cm^{-1} and at 950 cm^{-1} upon heating between 450 °C and 650 °C were observed. XRD patterns showed a decrease of the FWHM of the peaks (101) upon heating between 450 °C and 650 °C.

According to existing studies, the observation implies the crystalline quality improvement and the decrease of the silanoles Si-OH defects, which lead to a decrease of the reactivity of SiO_2 aggregate under chemical attacks typically by ASR.

References

[1] Najigivi A, Khaloo A, Iraji zad A and Abdul Rashid S 2013 Investigating the effects of using different types of SiO_2 nanoparticles on the mechanical properties of binary blended concrete *J. Composites* B **54** 52-

58

[2] Hanus M J and Harris A T 2013 Nanotechnology innovations for the construction industry *J. Progress in Materials Science* **58** 1056-1102

[3] Verstraete J, Khouchaf L and Tuilier M H 2004 Contributions of the environmental scanning electron microscope and X-ray diffraction in investigating the structural evolution of a SiO_2 aggregate attacked by alkali-silica reaction *J. Materials Science* **39** 6221-6226

[4] Khouchaf L and Boinski F 2007 Environmental scanning electron microscope study of SiO_2 heterogeneous material with helium and water vapor *J. Vacuum* **81** 599

[5] Khouchaf L, Hamoudi A and Cordier P 2009 Evidence of depolymerisation of amorphous silica at medium- and short-range order: XANES, NMR and CP-SEM contributions *J. Hazardous Materials* **168** 1188–1191

[6] Yurtdas I, Chen D, Hu D W and Shao J F 2013 Influence of alkali silica reaction (ASR) on mechanical properties of mortar *J. Construction and Building Materials* **47** 165–174

[7] Boinski F, Khouchaf L and Tuilier M H 2010 Study of the mechanisms involved in reactive silica *J. Materials Chemistry and Physics* **122** 311–315

[8] Ichikawa T 2009 Alkali–silica reaction, pessimum effects and pozzolanic effect *J. Cement and Concrete Research* **39** 716–726

[9] Khouchaf L, Boinski F, Tuilier M H and Flank A M 2006 Characterization of heterogeneous SiO_2 materials by scanning electron microscope and micro fluorescence XAS techniques *J. Nuclear Instruments and Methods in Physics Research* B **252** 333–338

[10] Ben Haha M, Gallucci E, Guidoum A and Scrivener K L 2007 Relation of expansion due to alkali silica reaction to the degree of reaction measured by SEM image analysis *J. Cement and Concrete Research* **37** 1206–1214

[11] Prezzi M, Monteiro P J M and Sposito G 1997 The alkali-silica reaction, part I: use of the double layer theory to explain the behavior of reaction-product gels *J. ACI Materials* **94** 10-17

[12] Gao X X, Stéphane Multon S, Cyr M and Sellier A 2013 Alkali–silica reaction (ASR) expansion: pessimum effect versus scale effect *J. Cement and Concrete Research* **44** 25–33

[13] Rivard P, Ollivier J P and Ballivy G 2002 Characterization of the ASR rim application to the potsdam sandstone *J. Cement and Concrete Research* **32** 1259–1267

[14] Verstraete J, Khouchaf L, Bulteel D, Garcia-Diaz E, Flank A M and Tuilier M H 2004 Amorphisation mechanism of a flint aggregate during the alkali–silica reaction: X-ray diffraction and X-ray absorption XANES contributions *J. Cement and Concrete Research* **34** 581–586

[15] Khouchaf L and Verstraete J 2007 Multi-technique and multi-scale approach applied to study the structural behavior of heterogeneous materials: natural SiO_2 case *J. Materials Science* **42** 2455–2462

[16] Hamoudi A, Khouchaf L, Depecker C, Revel B, Montagne L and Cordier P 2008 Microstructural evolution of amorphous silica following alkali–silica reaction *J. Non-Crystalline Solids* **354** 5074–5078

[17] Schmidt P, Badou A and Fröhlich F 2011 Detailed FT near-infrared study of the behaviour of water and hydroxyl in sedimentary length-fast chalcedony, SiO_2, upon heat treatment *J. Spectrochimica Acta* A **81** 552-559

[18] Schmidt P, Masse S, Laurent G, Slodczyk A, Le Bourhis E, Perrenoud C, Livage J and Fröhlich F 2012 Crystallographic and structural transformations of sedimentary chalcedony in flint upon heat treatment *J.Archaeological Science* **39** 135-144

[19] Domanski M and Webb J A 1992 Heat treatment of Polish flints *J.Archaeological Science* **19** 601-614

[20] Schmidt P and Fröhlich F 2011 Temperature dependent crystallographic transformations in chalcedony, SiO_2, assessed in mid infrared spectroscopy *J. Spectrochimica Acta* A **78** 1476-1481

[21] Garnica-Romo M G, Yanez-Limon J M, Villicana M, Perez-Robles J F, Zamorano-Ulloa R and Gonzalez-Hernandez J 2004 Structural evolution of sol–gel SiO_2 heated glasses containing silver particles *J. Phys. Chem. Solids* **65** 1045-1052

[22] Barbosa G N and Oliveira H P 2006 Synthesis and characterization of V_2O_5–SiO_2 xerogel composites prepared by base catalysed sol–gel method *J. Non-Cryst Solids* **352** 3009-3014

[23] Kim H U and Rhee S W 2000 Electrical properties of bulk silicon dioxide and SiO_2/Si interface formed by

tetraethylorthosilicate-ozone chemical vapor deposition *J. Electrochemical Society* **147** 1473-1476

[24] Chen Z L and Shen P 2012 Thermally activated sintering–coarsening–coalescence-polymerization of amorphous silica nanoparticles *J. Ceramics International* **39** 2365-2373

[25] Jung K T, Chu Y H Chu, Haam S and Shul Y G 2002 Synthesis of mesoporous silica fiber using spinning method *J. Non-Cryst Solids* **298** 193–201

[26] El Rassy H and Pierre A C 2005 NMR and IR spectroscopy of silica aerogels with different hydrophobic characteristics *J. Non-Cryst Solids* **351** 1603-1610

[27] Domanski M and Webb J A 1992 Effect of heat treatment on siliceous rocks used in prehistoric lithic technology *J. Archaeological Science* **19** 601-614

IOP Conf. Series: Materials Science and Engineering **62** (2014) 012003 doi:10.1088/1757-899X/62/1/012003

Grain refinement and Lattice Imperfections in Commercial Aluminum Alloy Processed by Severe Plastic Deformation

Saifeddine Charfeddine[1,2], Karim Zehani[2], Lotfi Besais[2], Atef Korchef[1,*]

[1] Centre National de Recherches en Sciences des Matériaux, Technopole de Borj-Cédria, BP 73 Soliman 8027, Tunisia.
[2] CMTR, ICMPE, UMR7182, CNRS, Université Paris 12, 2-8 rue Henry Dunant BP 94320 Thiais, France.

*Corresponding author: e-mail address atef.korchef@inrst.rnrt.tn

Abstract. In the present work, investigations on the microstructure of an aluminum alloy that had been subjected to severe plastic deformation (SPD) by equal channel angular pressing (ECAP), filing and ball milling, were carried out using X-ray diffraction and scanning electron microscopy. SPD leads to lattice distortions, increased dislocation density and an intensive refinement of the microstructure. The refinement and lattice imperfections of the material are greatly affected by the deformation modes and loading performance occurring during SPD. During the milling, the dislocation annihilation increases at higher strains thereby resulting in a smaller crystallite size. After ECAP, the material manifests a strong shear texture and anisotropy of the deformation behavior. Strain anisotropy is less pronounced in filed and ball milled powder particles.

1. Introduction

In recent years, various techniques of severe plastic deformation (SPD) are being developed to achieve high strains at a relatively low temperature. High strains leads to enhanced grain refinement to the submicrometer (100-1000 nm) or nanometer (<100 nm) levels. Ultrafine grained materials (UFG) obtained by SPD have attracted considerable scientific interest in the last decade because of their unusual physical, chemical and mechanical properties. For example, the intensive grain refinement is responsible for the enhancement of physical and mechanical properties of the processed materials, such as highly enhanced flow stress [1] and Vickers microhardness [2,3].

SPD techniques include, among others, torsion under high pressure [4], accumulative roll bonding (ARB) [5], high energy milling [6] and equal channel angular pressing (ECAP) [7,8]. The latter procedure, introduced almost 30 years ago to achieve enhanced plastic strains, is especially attractive for many near superplastic applications. Since the cross section of the sample remains unchanged during ECAP, repetitive pressings on the same sample are possible to achieve the required strains and to invoke different slip systems by rotating differently the sample between consecutive passes.

In high energy milling, powder particles are subjected to SPD through repetitive compressive loads arising from the impacts between the balls and the powder particles, which lead to their fracturing and welding. As a result, new crystalline and amorphous materials could be produced with crystallite sizes at the nanometer scale [9]. Furthermore, the mechanical milling kinetics depends on the energy transferred to the powder from the balls during milling [10]. The energy transfer is governed by many parameters such as milling speed, size and size distribution of the balls, dry or wet milling and temperature and duration of milling [11]. High energy milling has advantages of being simple, relatively inexpensive to produce [12], applicable to any class of materials and easily scaled up to large quantities [13].

Several techniques of analysis are used to examine the microstructure of ultrafine grained (UFG) materials obtained by SPD. Transmission electron microscopy (TEM), scanning electron microscopy (SEM) and X-ray diffraction (XRD), considered as complementary tools, are the most powerful methods used. TEM and SEM provide a direct image of the grain sizes and distribution and XRD gives quantitative estimations on the lattice imperfections, i.e. lattice distortions and dislocation density, lattice parameter, texture, etc.

The present work aims, essentially, to examine the microstructure of aluminum alloy processed by various SPD techniques, that are, ECAP, filing and ball milling using XRD and SEM. Also, mechanical tests were performed using tension and Vickers microhardness measurements. From XRD line profile analysis, the coherent domain size, lattice distortions, dislocation density and lattice parameter were calculated. The estimated values obtained from the different severe plastic deformation techniques used were compared.

2. Material and experimental procedures

2.1. Material and severe plastic deformation processes

The experiments were carried out on an aluminum (99.1%) material received in the form of cast ingots. The chemical composition, determined by inductively coupled plasma optical emission spectrometry (ICPOES), is given in Table 1. The material contains a low volume fraction of α-AlFeSi phase [2]. The material was annealed for 24 hours at 500°C and quenched in iced water. After annealing, the alloy was formed by equiaxed grains of about 85 μm in size.

Samples with dimensions ~$10\times10\times70$ mm^3 were cut for ECA pressing. The die used for ECA pressing has two square channels of 10×10 mm^2 cross sectional area intersecting at an angle $\phi = 90°$. The arc of curvature at the outer point of the intersection of the two channels delineated an angle $\Psi = 16°$ (figure 1). According to the equation given by Iwahashi et al. [14], the total equivalent strain, ε_N, accumulated by N passes through the die was equal to 1.07N. In the present work, repetitive pressings of the same sample were conducted to achieve an equivalent strain up to 4.28 (N=4) without rotation of the sample between two successive passes (the so called route A [15]). ECAP was performed at room temperature.

Table 1. Chemical composition (weight %) of the investigated aluminium alloy.

Ca	0.0014	Ni	0.0036
Co	0.0002	Pb	<0.0001
Cr	0.0056	Si	0.1071
Cu	0.0558	Ti	0.0060
Fe	0.2500	Zn	0.1660
Mg	0.2190	Al	balance
Mn	0.0600		

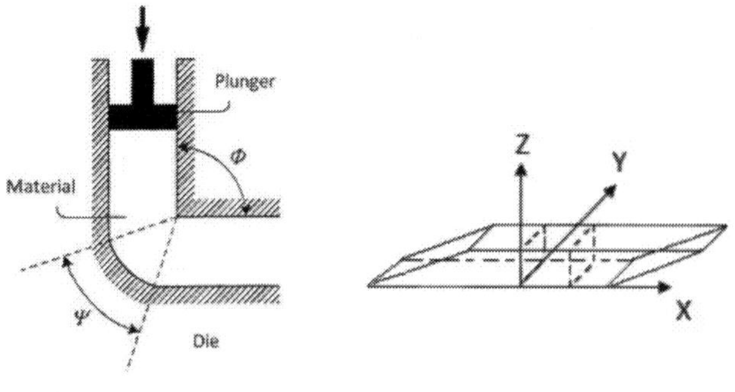

Figure 1. Schematic illustration of the die used for ECAP showing the three directions X, Y and Z.

The same material was also filed and ball milled up to 20 h. The hand filing was carried out using a steel file. The collected powders were then milled in P7 planetary mill up to 20 h. Ball to powder ratio was about 15:1 and the mill speed maintained at 300 rpm. Milling proceeded with a stationary speed of rotation. The duration of milling was automatically fixed by an electronic regulator and fixed on 15 min milling 15 min breather. Ethanol (2 weight %) was added to retard excessive welding. The milling atmosphere was argon that was purged into cups before milling to prevent oxidation.

For XRD and SEM investigations, specimens of ~5×5×10 mm3 were cut from the central part of each ECA pressed sample. Their surfaces were mechanically ground and electrolytically polished for 30 min at a temperature below 10°C under an operating voltage of 30V, using a mixture of 66% $(CH_3CO)_2O$ and 34% $HClO_4$. The scanned area was 80 × 80 μm^2 with a step size of 0.1 µm for each processed sample. XRD analysis of the samples deformed by ECAP and the hand filed and ball milled powders was carried out at room temperature with a diffractometer in step scanning mode using Cu-Kα radiation. The XRD patterns were recorded with a scan rate of 2×10^{-3} deg./s in the scanning range $2\theta = 20\text{-}120°$.

2.2. *Mechanical tests*

Mechanical tests were performed using tension and Vickers microhardness measurements. The Vickers microhardness, H_V, was measured using a DURIMET microhardness tester with a diamond pyramidal indenter. At least eight measurements were taken at randomly selected points with a load of 100 g applied for 15 s. For compression tests, cylindrical specimens of ~6 mm in diameter and ~9 mm in length were machined from ECA pressed samples, so that the compression direction was parallel to the X-axis (Figure1). A maximum compression stress of 15 kN was applied to the specimens. Compression tests were performed at room temperature, using an Instron testing machine, at a cross-head displacement speed of 0.05 mm/min corresponding to initial strain rate of 5.5×10^{-5} s^{-1}.

2.3. *X-ray diffraction analysis*

The broadening of the experimental reflection is caused by the small size of crystallites (or coherent length), internal lattice distortions (or microstrains) and the instrumental broadening [16]. In principle, the experimental broadening of reflections is the convolution of the instrumental broadening and the intrinsic broadening. The instrumental effects vary with each diffractometer and are caused by imperfect measurement capabilities, such as unfocused beams and unresolved α1 and α2 reflections. It is necessary to quantify the instrumental broadening of the diffractometer so that the effects of crystallite size and lattice strain can be individually identified in the diffraction patterns of experimental samples. A correction for the contribution from the instrumental effects to the experimental full-width at half maximum (FWHM), βexp, for each peak was obtained using the following expression:

$$ß = ([ß_{exp}^2 - ß_{ins}^2]^{\frac{1}{2}} \times [ß_{exp} - ß_{ins}])^{\frac{1}{2}} \quad (1)$$

where βins is the instrumental FWHM, which was determined using the undeformed aluminium as a standard reference material. However, subtracting the instrumental broadening effect from the diffraction pattern of a material sample, the remaining widths of the diffraction peaks are attributed to the small crystallite sizes and lattice strains. A decrease in crystallite size or enhanced microstrains leads to much broader reflections [16,17]. The Williamson–Hall (WH) and Halder–Wagner (HW) methods [16], among others, are used extensively for determining the crystallite size and the lattice microstrains of SPD materials. The WH equation used in the present work, in which the FWHM, β, due to lattice imperfections is related to the crystallite size, D, and the lattice distortions, ε, is expressed as follows:

$$\beta^* = d^*\varepsilon + \frac{1}{D} \quad (2)$$

where $\beta^* = \beta \cos\theta/\lambda$ and $d^* = 2 \sin\theta/\lambda$; θ the Bragg angle and λ is the wavelength used. From Eq. (2), the intercept and the slope of the plot of β* against d* give the crystallite size (1/D) and the microstrains ε, respectively. The WH plot shows whether the breadth depends on d* and the nature of any hkl dependence. In the HW method, which assumes a Voigt peak shape, the crystallite size, D, and the strain ε, are related to β* by the following equation:

$$(\frac{\beta^*}{d^*})^2 = \frac{1}{D}\left[\frac{\beta^*}{(d^*)^2}\right] + (\frac{\varepsilon}{2})^2 \quad (3)$$

From Eq. (3), the intercept of the plot of $(\beta^*/d^*)^2$ versus $\beta^*/(d^*)^2$ provides the value of the microstrains and the slope gives the crystallite size. In the present work, the FWHM and the position of each peak were obtained using the X'Pert HighScore Plus software. In order to estimate the lattice distortions and the crystallite size in the SPD samples, the Williamson–Hall (WH) and Halder–Wagner (HW) methods were used.

The lattice parameter of the material before and after SPD was obtained from a linear regression analysis of the measured lattice parameter, a_{app}, obtained from each peak, plotted against the Nilson–Reley (NR) function given by the following equation [18]:

$$NR = COS^2\frac{\theta}{2} \times (\frac{1}{\sin\theta} + \frac{1}{\theta}) \quad (4)$$

and extrapolated to NR = 0, that is 2θ = 180°.

For severe plastic deformed samples, dislocations are the main defects of which density, ρ, can be expressed in terms of crystallite size D and the mean lattice microstrains $< \varepsilon^2 >^{1/2}$ by [19-21]:

$$\rho = \frac{2\sqrt{3} < \varepsilon^2 >^{1/2}}{Db} \quad (5)$$

where b is the magnitude of the Burgers vector of dislocation equal to $a/\sqrt{2}$ for a fcc aluminum alloy; a is the lattice parameter of the material.

3. Experimental results and discussion

Microstructure examination by SEM

The micrograph in figure 2 shows typical SEM image of the microstructure of the undeformed aluminum alloy and in the Y plane of the samples processed by ECAP after two and four passes using route A. The microstructure of the undeformed alloy consists of equiaxed coarse grains with an average size of ~85 μm. The aluminum matrix contains precipitates located essentially at grain boundaries, even though some precipitates were also observed in the grain interior. The microstructure of the ECAP'ed (N=2) sample consists mainly of elongated grains and few equiaxed grains are also observed. After four passes, besides the elongated grains, the formation of equiaxed grains of about 0.2 μm is depicted in the ECAP'ed samples. Therefore, ECAP leads to intensive grain refinement of the aluminum alloy. Similar trend has been reported in microstructure investigation by TEM on high purity aluminum processed by ECAP [22,23]. In previous works [24,25], it was shown that the presence of the α-AlFeSi precipitates essentially at grain boundaries allowed to reveal the initial shape of the grains and to follow their evolution during ECA pressing, i.e. the grains appeared to be elongated in the side face of the sample (Y plane) [24]. In addition, the fraction of submicrometer grains (<1 μm) increased with the number of ECAP passes. However, the grain boundaries misorientation depended weakly on the number of passes and most subgrains were separated by low angle grain boundaries (< 10°) [25].

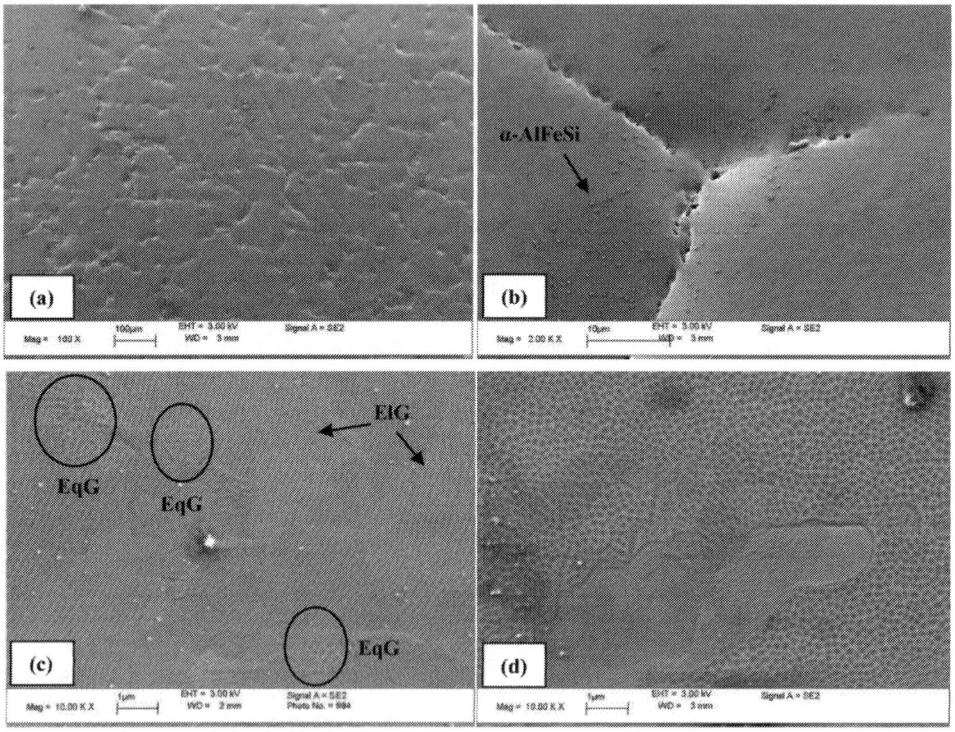

Figure 2. SEM micrographs of the (a-b) undeformed aluminum alloy and in the Y plane of the ECAP'ed samples for (c) N=2 and (d) N = 4. EqG: equiaxed grains, ElG: elongated grains.

Figure 3 shows SEM micrographs of the filed and 20 h milled powders. After filing, the powder particles are coarse and non-uniform in shape and size (figure 3a). After 20 h of milling, the particles became more regular and one can easily observe a significant change in the particle size and morphology (figure 3b). The particle size distribution became at least bimodal since very fine particles and coarser ones are examined. Some of the very fine particles welded to each other or to the surface of coarser ones.

IOP Conf. Series: Materials Science and Engineering **62** (2014) 012003 doi:10.1088/1757-899X/62/1/012003

Figure 3. SEM micrographs of the (a) filed and (b) 20 h milled samples.

Microstructure investigation by X-ray diffraction

The XRD patterns of SPD samples compared to that of the undeformed sample are given in figure 4. The intensities of the undeformed and ECAP'ed samples reflections are quite different from tabulated ones [26]. For example, for both samples, the 311 reflection is more intense than the 200 reflection and the 222 and 400 reflections are not observed in the undeformed material because of their very low intensities. Moreover, a significant decrease of the 200 reflection intensity is seen after ECAP. The change of the reflections intensities manifests a strong shear texture in the material. The SPD through filing and ball milling removes this texture since all the fundamental aluminum reflections in the range 2 = 20-120° are detected and their intensities are comparable to tabulated data [26].

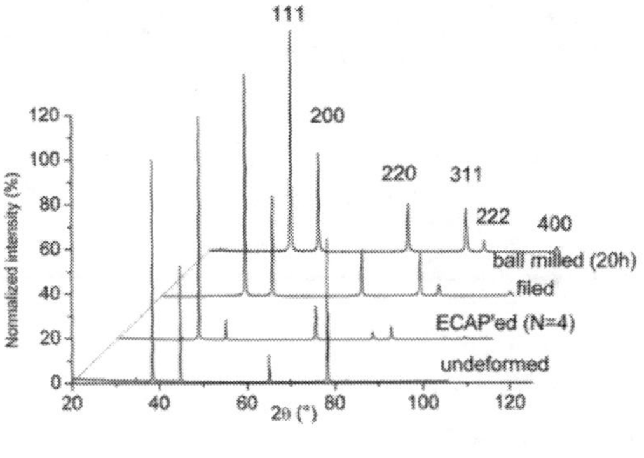

Figure 4. XRD patterns of SPD samples compared to that of the undeformed sample.

Figure 5 gives the enlarged view of the 111 and 200 reflections of SPD samples compared to that of the undeformed sample. One can easily observe a significant enlargement of these reflections accompanied with a slight shift. The $K_{\alpha 2}$ reflection is well resolved for the undeformed sample but not after severe plastic deformation because of the reflection broadening. The 111 reflection of the ECAP'ed sample shifts to higher 2 angle (Figure 5a), however the 200 reflection of the same sample shifts to lower 2θ angle (Figure 5b). This indicates a strong anisotropy of the deformation behavior of the ECAP'ed sample, i.e, the sample is subjected to compression stress along the [111] direction which, considering a constant volume of the pressed sample, provokes tensile behavior along different crystallographic directions, in particular along the [200] direction. On the other hand, all reflections of filed and ball milled samples shift towards higher 2θ angle, as depicted for the 111 (Figure 5c) and 200 (Figure 5d) reflections, indicating that these two deformation processes lead to isotropic compression of the aluminum powders. The reflection shift implies the variation of the lattice parameter. The lattice parameters of the undeformed and SPD samples calculated by the Nilson-Reley function presented in equation 4 are given in table 2. The lattice parameter of the undeformed aluminum is equal to 4.05097 Å. The SPD leads to a small decrease of the material lattice parameter. The lowest lattice parameter value (4.04418 Å), corresponding to a relative variation of about 0.16 %, was obtained after filing. The reflection shift could be attributed to planar faults [27], internal stresses, residual compressive or tensile stresses [28] and dissolution of a second phase containing elements with negative effect on the aluminum matrix (i.e. provoking the decrease of the lattice parameter) such as Fe and Si [28]. The enlargement of XRD reflections depicted in Fig. 5 is due mainly to small crystallite sizes and lattice microstrains.

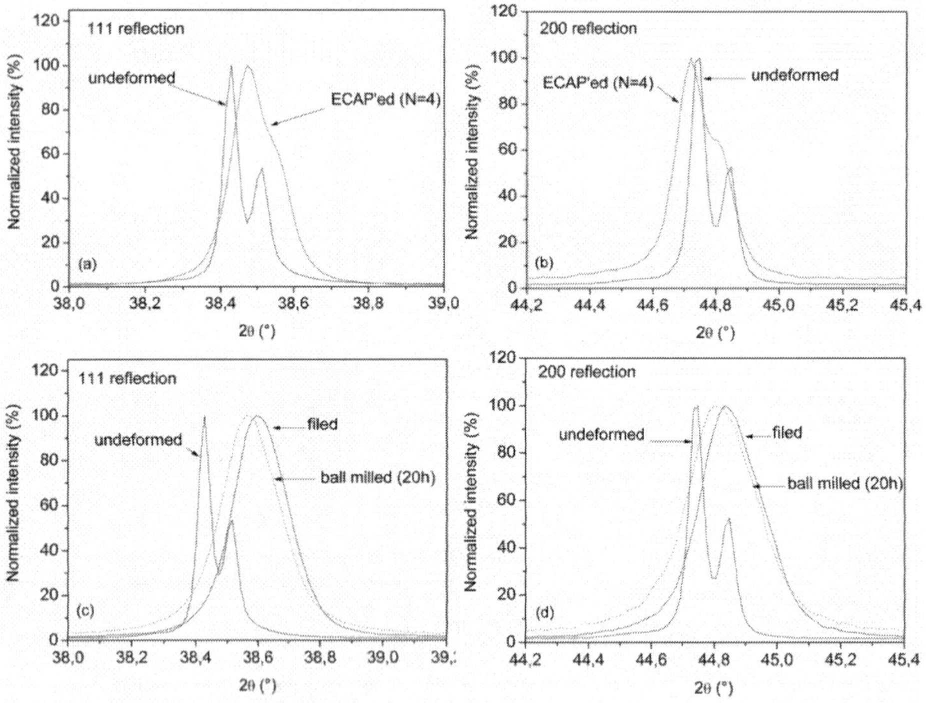

Figure 5. Enlarged view of the (a,c) 111 and (b,d) 200 reflections of SPD samples compared to those of the undeformed sample.

Figures 6a and b give the WH plot of the ECAP'ed sample and filed and 20 h milled samples, respectively. The WH plot shows the degree of isotropy of lattice strains. It is easy to note that the WH plot of the ECAP'ed sample manifests slightly higher scatter when compared to the filed and milled samples. The enhancement of scatter is also revealed by the increase of the adjust R^2 values of fitted linear plots from 0.7525 for ECAP'ed sample to 0.9206 and 0.9248, respectively, for filed and 20 h milled samples. Therefore, the strain is more isotropic after filing and milling than when using ECAP process. The HW method, using pseudo-Voigt approximation, was used in the present work to determine lattice distortions and crystallite sizes of the SPD alloy. The corresponding plots are shown in Figures 6c and d. The values obtained, depicted in table 2, show that SPD leads to lattice distortions (between 0.2 and 0.4%) and an intensive refinement of the microstructure (< 200 nm). Also, both filing and milling lead to nanometer crystallite sizes (<100 nm). Comparable results were obtained for Al-7075 milled for 20 h [21]. The crystallite size determined by HW method is lower than that deduced from direct observations by SEM analysis. The latter technique provides the size of grains that have well-defined, high-angle boundaries. Theses grains sometimes contain a substructure characterized by low-angle boundaries (crystallite). XRD allowed determining mean size values of these coherent domains or crystallites.

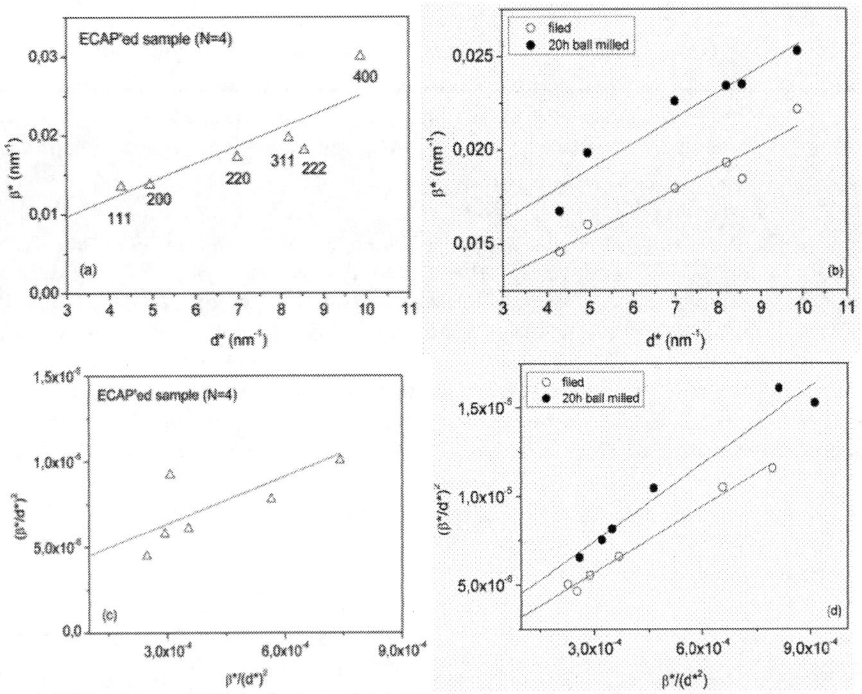

Figure 6. WH (a,b) and HW (c,d) plots of the ECAP'ed sample and filed and 20 h milled samples.

Table 2. Crystallite size and microstrains determined by the Halder-Wagner method and dislocation debnsiity of the SPD aluminium alloy.

	Crystallite size (nm)	Microstrains (%)	Lattice parameter (Å)	Dislocation density (10^{15} m^{-2})
ECAP'ed (N=4)	192	0.28	4.05057	0.17
Filed	80	0.28	4.04418	0.42
20h milled	68	0.34	4.04589	0.61

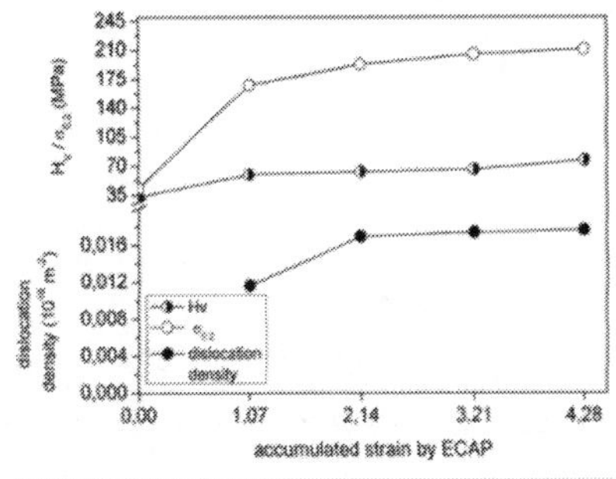

Figure 7. Evolutions of the Vickers microhardness (H_V), yield strength at 0.2 % of reduction ($\sigma_{0.2}$) and dislocation density during strain accumulation by ECAP.

3.1. Mechanical properties

The hardness could not be directly determined from filed and milled powders. It was shown that mounting powder in epoxy and forming discs of powder in a compaction die proved unsuccessful because the localized pressure of the hardness indenter displaced surrounding material, leading to inaccurate measurements [30]. Therefore, hardness and compression tests were only performed on the ECAP'ed samples. The evolutions of the Vickers microhardness (H_V) yield strength at 0.2 % of reduction ($\sigma_{0.2}$) and dislocation density during strain accumulation by ECAP are given in Figure 7. It is clearly seen that H_V and $\sigma_{0.2}$ increased significantly with the accumulated strain. For example, Hv increased from 33 for the undeformed sample to ~77 after four passes through the die and $\sigma_{0.2}$ became approximately five times higher after four passes ($\sigma_{0.2}$ =211 MPa) compared to that of the undeformed material ($\sigma_{0.2}$ =43 MPa). This increase is also associated with an increase in the dislocation density produced by ECAP. It should be noted that during SPD of polycrystalline materials, geometrically necessary dislocations (GNDs) and statistically stored dislocations (SSDs) are produced [29,31]. GNDs are related to strain gradient whereas SSDs are generated by the interaction of dislocations with lattice imperfections such as grain boundaries, twin boundaries, defects and tangled dislocations. The improvement of the mechanical properties is due to grain refinement and lattice dislocations accumulation during ECAP. We showed in a previous work [25] that, according to the Hall-Petch relationship, which predicts that the strength or hardness should increase when the grain size decreases, the dependence of the microhardness on the grain size, D, is consistent with a linear function: $H_V = 413 + 118D^{-1/2}$ and $\sigma_{0.2} = 35 + 76D^{-1/2}$. Transmission electron microscopy observations carried out on the same aluminum alloy indicated a high dislocation density generated after ECAP and revealed the formation of dislocation loops and dislocation walls [17]. The calculation of the dislocation density of the ECAP'ed alloy using equation 5 showed that a high dislocation density was obtained after two passes ($\rho = \sim 0.17 \times 10^{15}$ m^{-2}). Thereafter, only a slight increase was observed. The low variation of the dislocation density over N =2 could be explained by too many microstructure features, such a dynamic recovery which balances the creation of dislocations during ECA pressing [28], cell and grain boundary evolution, low- and high-angle misorientation evolution with strain [25] and tangled dislocation pinning by the existing fine particles.

4. Conclusion

In the present work, the experiments were carried out on an aluminum alloy that had been subject to SPD through ECAP, filing and ball milling. The microstructure of the alloy was examined by XRD and SEM. After annealing of the as-cast aluminum at 500 °C, the microstructure consisted of equiaxed coarse grains with an average size of ~85 μm. Independently of the deformation method used in the present work, lattice distortions, increased dislocation density and an intensive refinement of the microstructure were obtained after SPD. SEM analysis showed that the microstructure of the ECAP'ed sample consists of elongated grains and equiaxed grains of about 0.2 μm. The ECAP'ed material manifests a strong shear texture and anisotropy of the deformation behavior. Strain anisotropy was less pronounced in filed and ball milled powder particles. The highest dislocation density (0.61 10^{15} m^{-2}) and the lowest crystallite size (68 nm) calculated from X-ray reflections profiles were obtained after 20 h of ball milling. This was due to increased dislocation annihilation at higher strains. The SPD led to a small decrease of the material lattice parameter. The lowest value was obtained after filing. The decrease of the lattice parameter was attributed to lattice imperfections provoked by SPD.

Acknowledgment

The authors would like to acknowledge Professor N. Njah from the Laboratory of Applied Metallurgy -FSS (Tunisia) for useful helps.

References

[1] V.V. Stolyarov, Y. T. Zhu, T. C. Lowe, R.Z.Valiev, Microstructure and properties of pure Ti processed by ECAP and cold extrusion, Mater. Sci. Eng. A vol. 303 pp. 82-89, 2001.

[2] A. Korchef, N. Njah, J. Masmoudi, A.W. Kolsi, Y. Champion, S. Guerin, C. Leroux, Adv. Evolution of the mechanical properties of aluminum containing Al_8Fe_2Si precipitates during equal channel angular pressing, Adv. Eng. Mater. vol.6, no. 4, pp. 639-643, 2004.

[3] J. R. Weertman, Hall–Petch strengthening in nanocrystalline metals. Mater. Sci. Eng. A, vol. 166, pp.161-167, 1993.

[4] N.A. Smirnova, V.I. Levit, V.P. Pilyugin, Evolution of structure of f. c. c. single crystals during strong plastic deformation. Phys. Met. Metall. vol.61, pp.127-134, 1986.

[5] Y. Saito, N. Tsuji, H. Utsunomiya, T. Sakai, R.G. Hong, Ultra-fine grained bulk aluminum produced by accumulative roll-bonding (ARB) process, Scr. Mater. vol. 39, 1221-1227, 1998.

[6] Kohlar, M, Fritzsche, W, Nano Technology - An introduction to Nano structuring Techniques, Wiley - VCH- Verlag, Germany. p.96, 2004.

[7] V.M. Segal, V.I. Reznikov, A.E. Drobyshevskiy, V.I. Kopylov, Plastic metal working by simple shear. Izvestia Akademii nauk SSSR Metally vol.1, pp.115-123,1981.

[8] R.Z. Valiev, N.A. Krasilnikov, N.K. Tsenev, Plastic Deformation of alloys with submicro-grained structure. Mater. Sci. Eng. A vol.137, 35-40, 1991.

[9] Carl C Koch, Nano structured materials processing, properties and applications, William Andrew Inc. NY, USA. pp.122-130, 2006.

[10] S.C. Tjong, and H. Chen, Nanocrystalline materials and coating, Mater. Sci. Eng. R vol.45 pp.1-88, 2004.

[11] T. P. Yadav, R. M. Yadav, D. P. Singh, Mechanical milling: a top down approach for the synthesis of nanomaterials and nanocomposites, Nanoscience and Nanotechnology vol.2 pp. 22-48, 2012.

[12] R. Janot, and D.Guerard, Ball-milling in liquid media - Applications to the preparation of anodic materials for lithium-ion batteries, Progr. Mater. Sci. vol.50, pp. 1-92, 2005.

[13] G. Akshata Patil and S. Anandhan, Ball milling of Class-F Indian fly ash obtained from a thermal power station, International Journal of Energy Engineering IJEE, vol.2, pp. 57-62, 2012.

[14] Y. Iwahashi, J. Wang, Z. Horita, M. Nemoto, and T.G. Langdon, Principe of equal-channel angular pressing for the processing of ultra-fine grained materials. Scr. Materialia vol.35 pp. 143-146, 1996.

[15] V.M. Segal, Materials processing by simple shear, Mater. Sci. Eng. A vol.197, pp.157-164, 1995.

[16] J.I. Langford, National Institute of Standards and Technology Special Publication 846, in: Proceedings of the International Conference Accuracy in Powder Diffraction II, NIST, Gaithersburg, MD, pp. 241-244, 1992.

[17] A. Rebhi, T. Maklouf, N. Njah, Y. Champion, J.P. Couzinié, Characterization of aluminium processed by equal angular extrusion: effect of processing route. Mater. Charact. vol. 60, pp.1489-1495, 2009.

[18] L.V. Az`aroff, M. Buerger, The powder method in X-ray crystallography, McGraw Hill Book Co., McGrow Hill, New York, p.238, 1958

[19] Y.H. Zhao, H.W. Sheng, K. Lu, Microstructure evolution and thermal properties in nanocrystalline Fe during mechanical attrition, Acta mater. vol.49, pp. 365-375, 2001

[20] C. Aguilar, S. Ordóñez, J. Marín, F. Castro, V. Martínez, Study and methods of analysis of mechanically alloyed Cu–Mo powders, Mater. Sci. Eng. A vol.464, pp. 288–294, 2007.

[21] S. Bera, S. G. Chowdhury, Y. Estrin, I. Manna, Mechanical properties of Al7075 alloy with nano-ceramic oxide dispersion synthesized by mechanical milling and consolidated by equal channel angular pressing, Journal of Alloys and Compounds, vol.548 pp. 257–265, 2013

[22] Y. Iwahashi, Z. Horita, M. Nemoto, T. G. Langdon, The process of grain refinement in equal-channel angular pressing. Acta mater. vol. 9, pp. 3317-333, 1998.

[23]] M. Furukawa, Z. Horita, M. Nemoto, T. G. Langdon, Modeling the mechanical response of structural materials, Ed. E. M. Taleff, R. K. Mahidhara, The Minerals, Metals, Materials Society, 1997.

[24] A. Korchef, N. Njah, Y. Champion, S. Guerin and C. Leroux, J. Masmoudi, A.W. Kolsi, Material Flow During Equal Channel Angular Pressing of Aluminum Containing Al8Fe2Si Precipitate. Adv. Eng. Mater. Vol.6 pp. 222-228, 2004.

[25] A. Korchef, T. Baudin, Characterization of commercial aluminium processed by equal channel angular pressing, Metalurgia International 17 (2012) 46-53.

[26] Merck index, 8th ed. p.42; Am. Mineral. vol.65, p. 205, 1980.

[27] T. Ungar, Microstructural parameters from X-ray diffraction peak broadening. Scr. mater. vol.51, pp.777-781, 2004.

[28] A. Korchef, Y. Champion, N. Njah, X-ray diffraction analysis of aluminium containing Al8Fe2 processed by equal channel angular pressing. Journal of alloys and compounds vol.427, pp. 176-182, 2007.
[29] S.S. Satheesh Kumar, T. Raghu, Mechanical behaviour and microstructural evolution of constrained groove pressed nickel sheets, J. Mater. Processing Technol. Vol.213 pp.214-220, 2013.
[30] Adam J. Maisano, Master of Science, Cryomilling of aluminum-based and magnesium-based metal powders, Virginia Polytechnic Institute and State University p.71, 2006.
[31] G.E. Dieter, Mechanical Metallurgy. McGraw Hill Book Company, New York, 1986.

IOP Conf. Series: Materials Science and Engineering **62** (2014) 012004 doi:10.1088/1757-899X/62/1/012004

3-D loaded scaffolds obtained by supercritical CO_2 assisted process

S Cardea[1] and E Reverchon[1,2]

[1] Department of Industrial Engineering, University of Salerno, Via Giovanni Paolo II, 84084, Fisciano, Italy
[2] NANO_MATES, Research Centre for Nanomaterials and Nanotechnology at the University of Salerno, Via Giovanni Paolo II, 84084, Fisciano, Italy

E-mail: scardea@unisa.it

Abstract. In this work, a supercritical CO_2 (SC-CO_2) drying process for the formation of 3-D PVDF-HFP loaded scaffolds was tested. Experiments at pressures ranging between 150 and 250 bar and at temperatures ranging between 35 and 55°C were performed. The PVDF-HFP-acetone-ethanol solution at 15% w/w polymer was selected as the base case. The drug (amoxicillin) concentration was varied from 20 to 30% w/w with respect to PVDF-HFP. SC-CO_2 drying process was confirmed to be a valid alternative to generate loaded structures; indeed, scaffolds characterized by nanometric networks (with mean pore diameter of about 300 nm) with a homogeneous drug distribution were obtained. Drug controlled release experiments were also performed and a quasi-zero order release kinetic was observed.

1. Introduction

Poly(vinylidene fluoride) PVDF is frequently copolymerized with hexafluoropropylene (HFP), in order to obtain the PVDF-HFP, which is a material with a good resistance to acid environments, semi-crystalline and, furthermore, biocompatible. For these particular characteristics, various applications have been tested for PVDF-HFP such as electrical [1], catalytic [2] and biomedical [3-4] applications. Among biomedical applications, one of the most interesting is the use of loaded scaffolds, with one or more active principles, as controlled release devices; the aim is to control the release of a drug in a way that its administration to the patient is delayed or prolonged in time. Indeed, during the administration of a drug, haematic concentration peaks usually have to be avoided; however, with traditional pharmaceutical devices, such as tablets, this problem is very serious and worrying, and difficult to be solved [5]. Different mass transfer mechanisms are involved in the release of an active principle from a controlled release system, depending on the characteristics of the materials used and on the scaffolds morphology. In particular, the mechanisms of swelling, erosion and/or diffusion have to be considered in order to deeply study the phenomena. Regarding the characteristic of PVDF-HFP (non-swelling and non-degradable material), the use of scaffolds of this polymer in drug release processes implies the consideration of diffusivity mechanisms.

In the last 20 years, various processes have been tested for the generation of loaded scaffolds for controlled release applications such as phase separation [6-7], foaming combined with particulate leaching [8], solvent casting [9], gel drying [10-11], etc. This last process is probably the most interesting because by means of it, it is possible to assure the formation of homogenous and skinless nanostructured porous scaffolds. In any case, all the processes indicated above present several

IOP Conf. Series: Materials Science and Engineering **62** (2014) 012004 doi:10.1088/1757-899X/62/1/012004

problems; for example, long processing time (12-48 h) that can imply the stratification of the drug inside the scaffolds due to the separation of the loaded materials from the polymeric solutions. In this way, the efficiency of the generated devices sensibly decreases due to the inhomogeneous distribution of the drug. Moreover, tradition processes tested until now do not allow the complete removal of the organic solvents involved in the starting solutions, and the presence of solvent residue can compromise the efficiency of the controlled release device, too.

Considering the gel drying process, it is important to eliminate the solvent contained in the gel. Indeed, during gel drying, the polymeric structure can collapse due to the liquid surface tension; this effect is related to the action of cohesive forces that act between the liquid and the nanosized network of the polymer. In a previous work, our research group used a supercritical carbon dioxide (SC-CO_2) drying technique to generate PVDF-HFP aerogels with significant porosities [12]. This process can allow some benefits with respect to the traditional processes, thanks to the very interesting peculiarities of the CO_2 at the supercritical conditions, such as high diffusivity, low viscosity and high affinity to several organic solvents [13-14]. In particular, PVDF-HFP nanoporous aerogel was obtained without residual solvents and in very short processing times.

The target presented in this article is to test, for the first time, the SC-CO_2 drying technique to generate PVDF-HFP drug loaded scaffolds for controlled release applications. The morphology of the scaffolds and the behavior of the drug (i.e., amoxicillin, an antibiotic) in the system generated will be studied.

2. Materials and Methods

2.1 Materials

PVDF-HFP (density 1.78 g/cm3) was kindly provided by Solvay S.A. (Ixelles, Belgium); amoxicillin (purity 99.6%), acetone (purity 99.8%) and ethanol (purity 99.7%) were purchased from Sigma-Aldrich; carbon dioxide (purity 99.5%) was bought from Alfa Ossigeno (Fisciano, Italy).

2.2 Generation of loaded scaffolds

To prepare the scaffolds, PVDF-HFP was dissolved in acetone at a concentration of 15% w/w. The solution was stirred and warmed at 55°C, in order to obtain a homogeneous solution, to which different amounts of amoxicillin were added; a drug suspension (amoxicillin is insoluble in acetone) was then obtained. Then, the non-solvent (ethanol) was added. The obtained suspensions were placed in proper formation cells and put at -30°C in a deep-freezer. The residence time was properly chosen in order to obtain the gel formation.

The bench apparatus used for the SC-drying process was mainly constituted by a 316 stainless steel high-pressure vessel (I.V. = 80 mL). Once the gel was charged, the vessel was closed. Then, a high-pressure pump (Lewa, Germany) was used to produce the desired pressure, by pumping CO2 from the bottom of the vessel. The CO2 flux (at a constant flow rate of 1.5 kg/h) was assured in continuous mode for about 1.5 h. Then, the vessel was rapidly depressurized (at 25 bar/ min) and, once atmospheric pressure was reached, the scaffold was removed from the support. The experiments were performed at pressures ranging between 150 and 250 bar and temperatures ranging between 35 and 55 °C.

2.3 Scanning electron microscopy (SEM) and Energy Dispersive X-Ray analysis (EDX).

The generated scaffolds were cut with a microtome (Bio-optica S.p.A, Italy, Mod. Microm HM 550 OMVP); a sample was sputter coated with gold and observed with scanning electron microscope (SEM - mod. LEO 420, Assing, Italy), in order to study the morphology and drug distribution. An image analysis software program (Sigma Scan Pro 5.0, Jandel scientific, San Rafael, Canada)) and Origin 7 (Microcal, Northampton, USA) were used to measure the mean pore diameter.

To determine the amoxicillin dispersion in the scaffold, the samples were coated with chromium (layer thickness 150 A°). Then, an Energy Dispersive X-Ray (EDX) analyzer was used to detect the signal of Sulphur atoms (present in amoxicillin crystals, but not in the polymer).

2.4 Drug release analysis

Drug release tests were performed to study the kinetics of amoxicillin release from PVDF-HFP scaffolds. PVDF-HFP scaffolds were immersed in a glass flask with 1 liter of a physiological saline solution, at pH 7.2, used as drug release medium. The flask was sealed and put in a heater set at 37°C and with a stir of 250 rpm. Every 10 minutes, the amoxicillin concentration was measured using an ultraviolet spectrophotometer.

3. Results and Discussion

3.1 Effect of supercritical fluid

The use of supercritical fluids is very important in the generation of aerogels; indeed, one of the main problems of gel drying is related to the surface tension of the liquid solvent that has to be removed, which can determine solid structure collapse. Considering that a fluid at supercritical conditions has no surface tension, the use of SC-CO2 can allow the complete elimination of the organic solvents from the starting gel. However, it could also be insufficient to adequately process the gel; indeed, when SC-CO2 diffuses into gel, it forms a supercritical mixture with the solvents used during the preparation of the gel, in this case acetone and ethanol. When the mixture SC-CO2-acetone-ethanol is at supercritical conditions too, very low surface tension conditions are reached. Analyzing the temperatures used during the experimentation (ranging between 35 and 55°C), the mixture critical point (MCP) of both systems (i.e. CO2-ethanol and CO2-acetone) is at pressures lower than 100 bar [15-16]; considering that pressures higher than 150 bar (i.e. from 150 to 250 bar) were used, supercritical conditions were obtained in each operative condition.

Then, the gel structure is not destroyed and the scaffold generated by SC-CO2 drying process conserves sizes and tridimensional (3-D) form of the starting gel. Moreover, the high affinity of the SC-CO2 with the solvents used allowed complete elimination of them. The result was the formation of 3-D stable and dry devices that can work as scaffold for tissue engineering applications. Once the feasibility of the SC-CO2 drying technique to obtain 3-D scaffolds was verified, we focused our attention on the generation of loaded scaffolds.

3.2 Effect of drug concentration

A solution containing PVDF-HFP at 15% w/w, acetone at 60% w/w and ethanol at 25% w/w was processed, and added different quantities of drug (i.e. 20 and 30% w/w of polymer). SEM and EDX images, presented in Figure 1, report PVDF-HPF loaded scaffold generated at 250 bar and 35°C with 20% w/w of amoxicillin loaded inside. As seen in Figures 1a and 1b, the formation of an homogenous nanoporous structure is evident, very similar to those obtained for unloaded PVDF-HFP structures [12]; this result is very important because it confirms that the loading of an active principle in the starting solution did not condition the formation of the gel and the supercritical process efficiency. The porosity of the structure generated was very high (> 85%) and no residual solvents were present (lower than 5 ppm). Moreover, porous skins and pores with average diameter of about 300 nm were obtained.

Analyzing Figure 1c, in which EDX is reported, it is possible to observe the location of the drug crystals inside the scaffold; indeed, the sulfur atoms, which are the characteristic element of amoxicillin, are indicated with the green color. As a consequence, a very good distribution of the drug along the scaffold section can be put in evidence. This result is very important and interesting because, as reported in the introduction section, one of the main problems of the traditional loading techniques concerns the stratification of the drug along the structure, and the consequent low efficiency of the generated devices (possible burst effect phenomena).

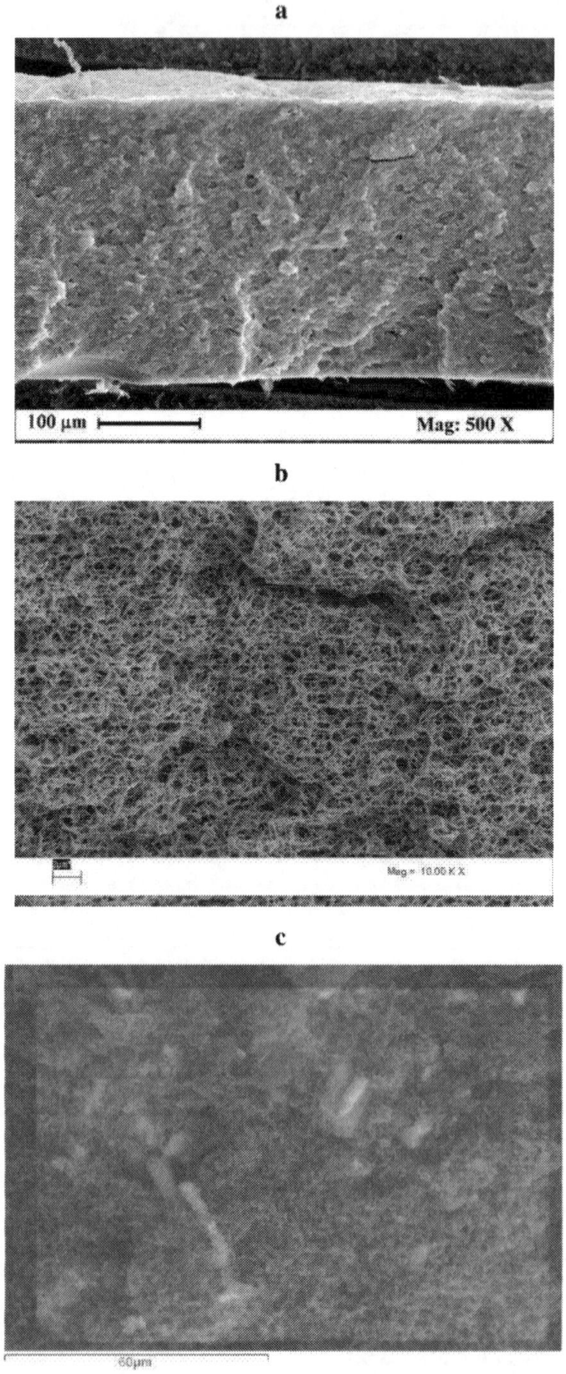

Figure 1 SEM and EDX analyses of scaffolds generated at 250 bar and 35°C with 20% w/w of amoxicillin: (a) scaffold section, (b) higher enlargement of scaffold section, (c) EDX.

3.3 Effect of SC-CO2 solvent power

Experiments were performed at different pressures (150, 200 and 250 bar) and temperatures (35, 45 and 55°C) maintaining the polymer concentration at 15% w/w and drug concentration at 20% w/w. It was observed that the modification of the process parameters did not affect the scaffolds' final morphology and characteristics. It is known that changing the operative conditions lead to the changes of the characteristics of the SC-CO_2, such as solvent power, viscosity and diffusivity. As indicated before, these characteristics are very important for the success of the supercritical drying procedure, but they do not interfere with the gel formation (that occurs in the freezer, before the drying step). As a consequence, the supercritical drying procedure can be considered as an on/off technique that can (or cannot) give efficient results in terms of gel drying, complete elimination of the solvents and preserving of the nanoporous network of the gel, but is not able to modify the aerogel final morphology.

3.4 Drug release analysis

In Figure 2, release curves of pure amoxicillin (. . .), PVDF-HFP scaffolds with 20% w/w of amoxicillin (——), PVDF-HFP scaffolds with 30% w/w of amoxicillin (- - -), generated at 250 bar and 35°C, are presented.

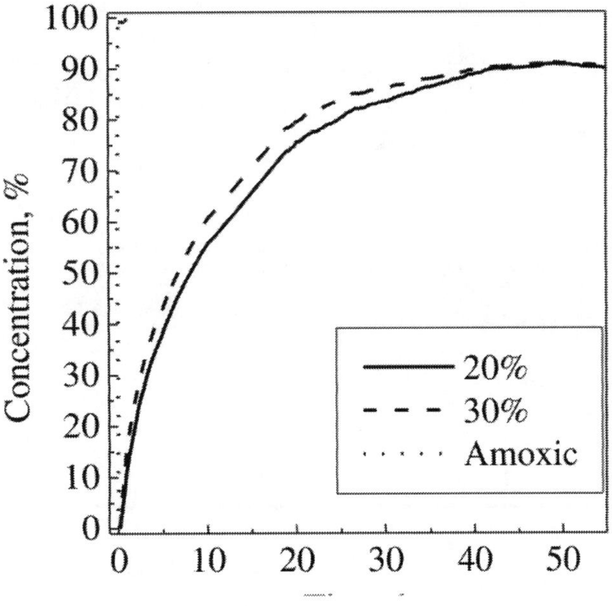

Figure 2 Release curves of pure amoxicillin (. . .), PVDF-HFP scaffolds with 20% w/w of amoxicillin (——), PVDF-HFP scaffolds with 30% w/w of amoxicillin (- - -), generated at 250 bar and 35°C

From the diagram presented, it can be observed that the drug was completely solubilized in 10 minutes, validating the idea that, in this case, a control of the release is necessary. From the release curves of the PVDF-HFP scaffolds, it is evident that the time of drug release increased to 50 h for both scaffolds (i.e. with 20% and 30% of drug). The concentration of amoxicillin in the starting solutions had no effect on the qualitative and quantitative trends of the release curves.

Therefore, the scaffolds morphology allowed control of the drug release. This result is related to the nanoporous structure of the PVDF-HFP scaffolds, which involves slow diffusions of the physiological solution into the scaffolds and of the physiological solution + drug mixture outside the scaffolds.

Moreover, it is important to put in evidence that in the first day of release (i.e. 24 h), a release kinetic of about zero order had been obtained: it is a well achieved target for a controlled drug release device.

References

[1] Gugliuzza A and Drioli E 2007 J. of Membr. Sci. 300 51.
[2] Buonomenna MG, Drioli E, Nugent W, Prins L, Scrimin P and Licini G 2004 Tetr. Lett. 45 7515.
[3] Guenard V, Valentini RF, Aebischer P 1991 Biomaterials 12 259.
[4] Chen H, Soldani G, Galletti PM, Goddard M, 1992 ASAIO J. 38 201.
[5] Simo C, Cifuentes A and Gallardo A, 2003 J. Chromatogr. B 797 37-49.
[6] Park YJ, Nam KH, Ha SJ, Pai CM, Chung CP and Lee S 1997 Journal of Controlled Release 43 151.
[7] Thombre AG, Cardinal JR, DeNoto AR, Herbig SM and Smith KL 1999 Journal of Controlled Release 57 55.
[8] Jang JH and Shea LD 2003 Journal of Controlled Release 86 157.
[9] Jackson JK, Smith J, Letchford K, Babiuk KA, Machan L, Signore P, Hunter WL, Wang K and Burt HM 2004 International Journal of Pharmaceutics 283 97.
[10] Cho JW and Lee GW 1996 Journal of polymer science: Part B: Polymer Physics 34 1605.
[11] Dasgupta D, Manna S, Malik S, Rochas C, Guenet JM, Nandi AK 2005 Macromol. Symp. 222 175.
[12] Cardea S, Gugliuzza A, Sessa M, Aceto MC, Drioli E and Reverchon E 2009 Applied Materials and Interfaces 1 171.
[13] Reverchon E, Cardea S, Rapuano C 2007 Journal of Applied Polymer Science 104 3151.
[14] Reverchon E, Cardea S, Schiavo Rappo E, 2008 The Journal of Supercritical Fluids 45 356.
[15] Polhler H and Kiran E 1997 J. Chem. Eng. Data 42 379.
[16] Polhler H and Kiran E 1997 J. Chem. Eng. Data 42 384.

Carboxy terminated rubber based on natural rubber grafted with acid anhydrides and its adhesion properties

P Klinpituksa[1], P Kongkalai[2] and A Kaesaman[3]

[1,2] Department of Science, [3] Department of Rubber Technology and Polymer Science, Faculty of Science and Technology, Prince of Songkla University, Pattani, 94000, Thailand

E-mail: kpairote31@gmail.com

Abstract. The chemical modification of natural rubber by grafting of various polar functional molecules is an essential method, improving the versatility of rubber in applications. This research investigated the preparation of natural rubber-graft-citraconic anhydride (NR-g-CCA), natural rubber-graft-itaconic anhydride (NR-g-ICA), and natural rubber-graft-maleic anhydride (NR-g-MA), with the anhydrides grafted to natural rubber in toluene using benzoyl peroxide as an initiator. Variations of monomer content, initiator content, temperature and reaction time of the grafting copolymerization were investigated. The maximum degrees of grafting were 1.06% for NR-g-CCA, 4.66% for NR-g-ICA, and 5.03% for NR-g-MA, reached using 10 phr citraconic anhydride, 10 phr of itaconic anhydride, or 8 phr of maleic anhydride, 3 phr benzoyl peroxide, at 85, 80 and 80°C for 2, 2 and 3 hrs, respectively. Solvent-based wood adhesives were formulated from these copolymers with various contents of wood resin in the range 10-40 phr. The maximal 289 N/in cleavage peel and 245.7 KPa shear strength for NR-g-MA (5.03% grafting) were obtained at 40 phr wood resin.

1. Introduction

Natural rubber is a well-known renewable resource obtained from the tree popularly known as *Hevea brasiliensis* in the form of a milky white fluid, and is a versatile raw material used mainly in the manufacture of tyres. Because of being highly unsaturated, it is less resistant to oxidation, ozone, weathering, various chemicals and solvents, in comparison to synthetic rubbers. Hence, chemical modification of natural rubber is essential to overcome some of its drawbacks. Graft copolymerization is one such modification technique used with natural rubber [1]. In particular, chemical modification with diene polymers of natural rubber has improved or produced useful materials. Among the many possibilities of chemical modifiers, the monomers used include maleic anhydride [2-3], styrene and methyl methacrylate [4], styrene and acrylonitrile [5], *n*-butyl and methyl methacrylate [6], ethyl acrylate and methyl methacrylate [7]. However, the grafting of acid anhydrides, particularly citraconic and itaconic anhydrides, onto natural rubber, and their effects on adhesion properties have not been thoroughly studied.

The present work focused on the grafting of citraconic anhydride, itaconic anhydride and maleic anhydride in toluene, using benzoyl peroxide as the radical initiator. The effects of these polar grafted copolymers on the adhesion properties were experimentally tested with plywood.

2.Experimental

2.1 Preparation of natural rubber grafted with acid anhydrides

A 250-mL three-necked round bottomed flask equipped with a mechanical stirrer, dropping funnel and N_2 inlet, immersed in a water bath at desired temperature (70-85°C), was charged with the solution of 5% dried natural rubber (air dried sheet rubber) in toluene, 1-4 phr benzoyl peroxide initiator, and stirred for 15 mins. An amount of 4-10 phr maleic anhydride, citraconic anhydride or itaconic anhydride (dissolved in 10 mL toluene) was introduced slowly through the dropping funnel into the reaction mixture. The resulting mixture was allowed to react for 180 mins. The grafted product was obtained by precipitating the reaction mixture in hot methanol, separating, and drying the precipitate in an oven at 50°C. The grafted rubber was purified before further investigation by dissolution in toluene and re-precipitating in methanol.

2.2 Characterization of grafted natural rubber

The natural rubber grafted maleic anhydride (NR-g-MA), natural rubber grafted citraconic anhydride (NR-g-CCA), and natural rubber grafted itaconic anhydride (NR-g-ICA) were characterized by FTIR spectrophotometry (Bruker Tensor 27 spectrophotometer). The grafting tendency of anhydrides onto the natural rubber molecules was monitored by using the peak height ratio of absorbance peaks at 1785 cm-1 and 838 cm-1:

$$Peak\ height\ ratio = \frac{A_{1785}}{A_{838}}$$

Here, A1785 represents the absorption peak of C=O (symmetric vibration) of cyclic anhydride, and A838 represents =CH (out-of-plane bending) of cis-1,4-polyisoprene (natural rubber)

The precise grafting of anhydrides onto natural rubber chain can be also determined using a titration method as follows.

0.25 g NR-g-MA, NR-g-CA, or NR-g-IA, was dissolved into 25 mL p-xylene in a 250-mL Erlenmeyer flask, with stirring using magnetic bar until a homogeneous solution was obtained. The grafted rubber solution was supplemented with 1 mL distilled water and refluxed for 1 hour. The hot resulting solution was then titrated with 0.005M KOH in ethanol solution using 1% thymol blue in dimethylformamide (DMF) as an indicator. The grafting percentage of anhydrides is determined by:

$$Anhydride\ content\,(\%w) = \frac{(v_1 - v_0) \times M \times MW \times 100}{2W \times 1000}$$

Here, M denotes the molar concentration of KOH/ethanol solution, v0 and v denote the volumes of KOH/ethanol solution for blank and sample titrations, and W and MW denote the sample weight and molecular weight of monomer, respectively.

2.3 Adhesion properties of natural rubber grafted with acid anhydrides

In order to study the adhesion properties of these graft copolymers, solvent-based contact adhesives were formulated by varying the amount of wood resin in the range 10-40 phr (Table 1). The adhesion properties were characterized in terms of cleavage peel strength and shear strength (ASTM D3807) [8] .

Table 1. Solution-based adhesive formulations based on NR-*g*-MA, NR-*g*-CCA or NR-*g*-ICA

	Formulation (phr)				
Ingredients	**A**	**B**	**C**	**D**	**E**
NR-*g*-CCA (%grafting=1.06) [a]	100	100	100	100	100
NR-*g*-ICA (%grafting=4.66) [a]					
NR-*g*-MA (%grafting=5.03) [a]					
20% K-oleate	5	5	5	5	5
Diphenylamine	1	1	1	1	1
Toluene	1000	1000	1000	1000	1000
Wood resin	0	10	20	30	40

[a] Titration method determination

3.Results and Discussion

3.1 Synthesis and characterization of natural rubber grafted acid anhydrides

The ATR-FTIR spectra of NR-*g*-MA (Figure 1), NR-*g*-CCA (Figure 2) and NR-*g*-ICA (Figure 3) present two principal absorption bands, as verified in previous work [2-3] : at 838 cm^{-1}, assigned to out of plane bending of =CH of *cis*-1,4-polyisoprene (natural rubber); and at 1785 cm^{-1} (strong peak) and 1859 cm^{-1} (weak peak) attributed to symmetric and asymmetric vibrations of C=O in a five-membered cyclic anhydride.

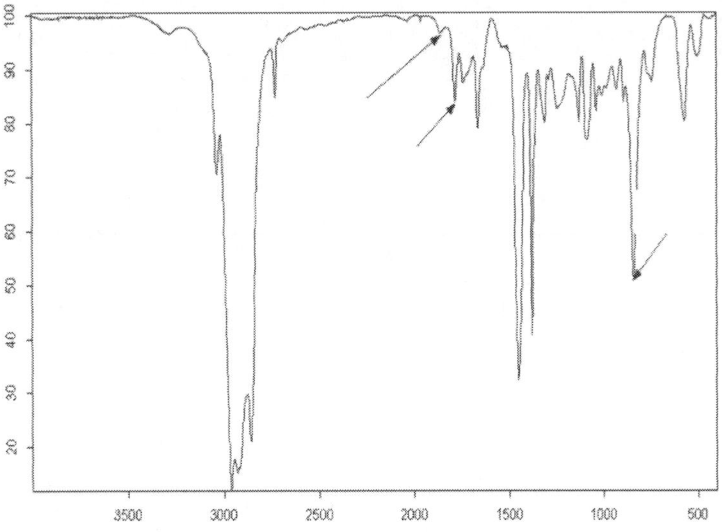

Figure 1. ATR-FTIR spectrum of NR-g-MA

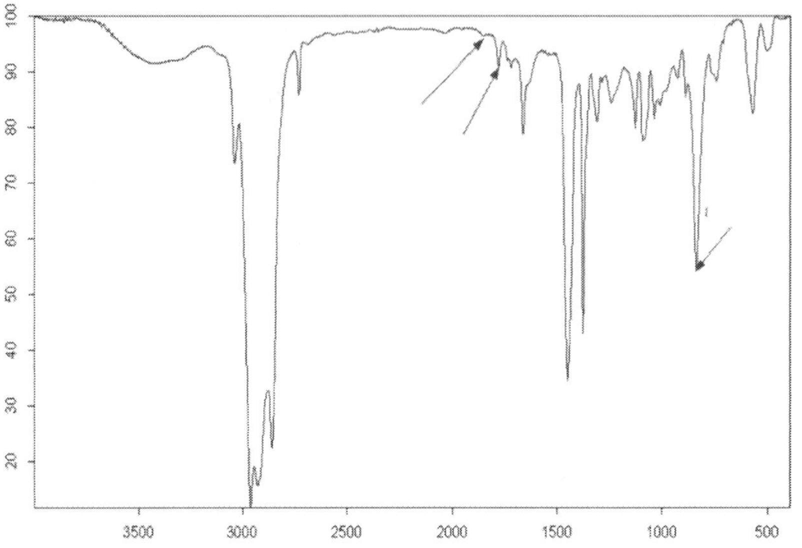

Figure 2 ATR-FTIR spectrum of NR-g-CCA

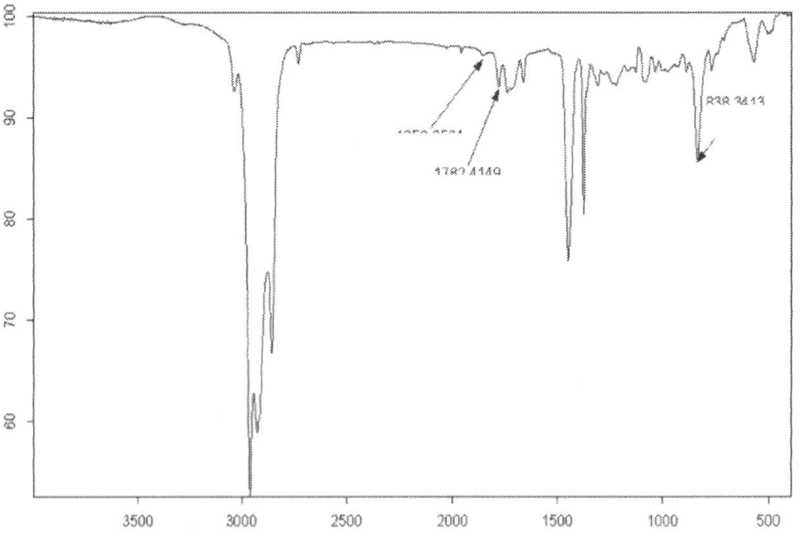

Figure 3 ATR-FTIR spectrum of NR-g-ICA

Free-radical graft copolymerization reactions of anhydrides with natural rubber involve the allylic hydrogen atoms in natural rubber chains, and these are susceptible to abstraction by interactions with free-radicals, generating reactive sites from which copolymerization can propagate [9] . The mechanism by which natural rubber grafted acid anhydrides are formed is given in Scheme 1.

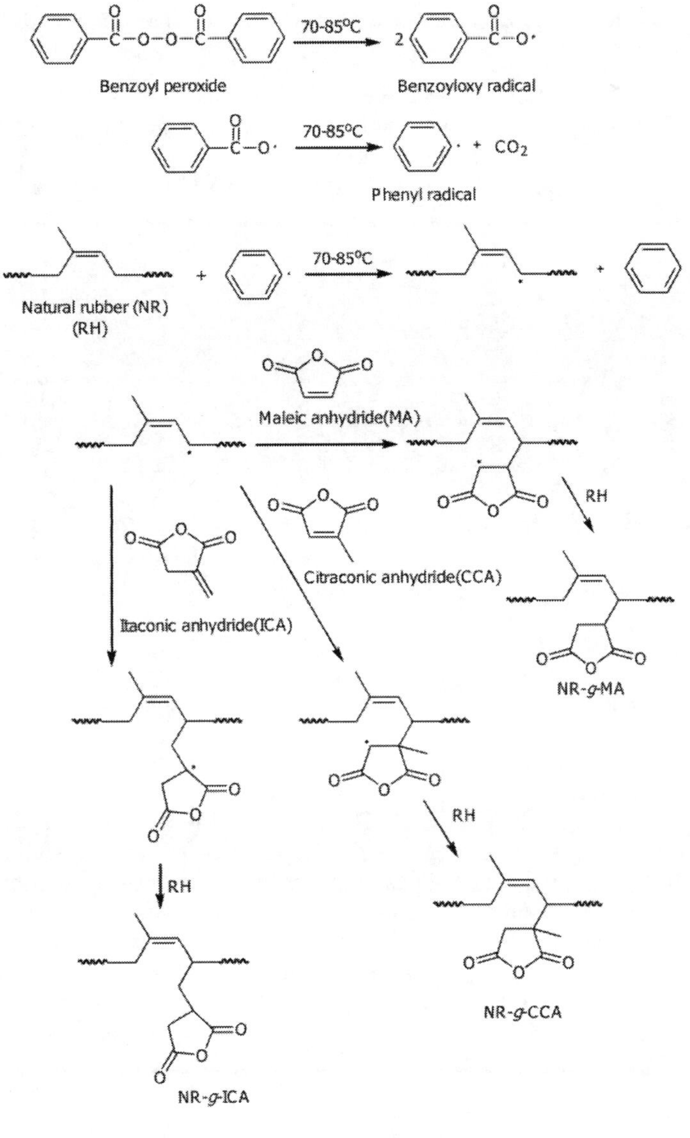

Scheme 1. A possible mechanism for preparation of NR-*g*-MA, NR-*g*-CCA and NR-*g*-ICA in the presence of benzoyl peroxide as an initiator.

3.2 Adhesion properties of natural rubber grafted anhydrides

Figure 4 illustrates that the NR-g-MA adhesives have better cleavage peel strength than either NR-g-ICA or NR-g-CCA. The cleavage peel strength increased with wood resin content. It is clear that the maximal 298 N/in2 peel strength for NR-g-MA was obtained at 40 phr of wood resin, because NR-g-MA increased the interfacial adhesion between carboxyl groups and cellulosic plywood. This may be due to higher grafting percentage of maleic anhydride in natural rubber, relative to the other anhydrides. Therefore, natural rubber grafted by maleic anhydride had enhanced interfacial adhesion to plywood. However, the peel strength of NR-g-CCA formulation was lower than that of ungrafted natural rubber. This may due to insufficient wetting of substrate by the rubber polymer chains. Similar results have been reported earlier in the case of n-butyl acrylate and methyl methacrylate [6], and grafting natural rubber with ethyl acrylate and methyl methacrylate [7]. In addition, the grafting percentage of citraconic anhydride on natural rubber was comparatively low.

Figure 5 shows that the shear strengths of these graft copolymer adhesives tended to increase with wood resin content. The NR-g-MA adhesives had better shear strengths than NR-g-ICA or NR-g-CCA. The maximal 245.7 KPa shear strength for NR-g-MA (at 5.03% grafting) was obtained with 40 phr wood resin. The lowest shear strength was found for NR-g-CCA (1.06% grafting) adhesive, and might be due to low grafting.

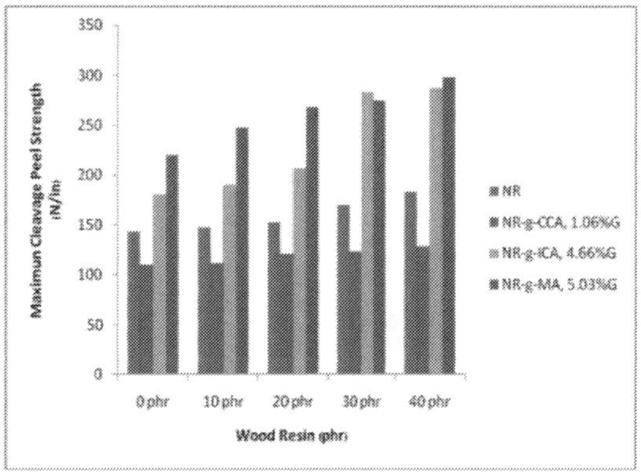

Figure 4. The maximum cleavage peel strength of NR-*g*-CCA, NR-*g*-ICA, NR-*g*-MA and NR (control) with various wood resin contents.

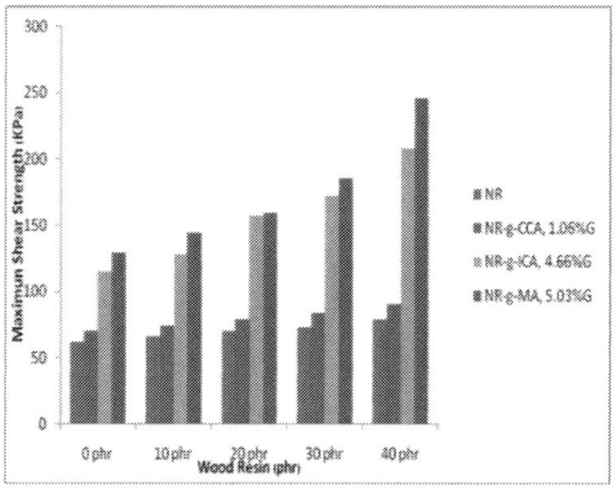

Figure 5. The maximum shear strength of NR-*g*-CCA, NR-*g*-ICA, NR-*g*-MA and NR (control) with various wood resin contents.

4.Conclusion

Natural rubber-graft-citraconic anhydride (NR-g-CCA), natural rubber-graft-itaconic anhydride (NR-g-ICA), and natural rubber-graft-maleic anhydride (NR-g-MA) were successfully prepared by graft copolymerization of natural rubber and the anhydrides in toluene, using benzoyl peroxide as the initiator. The maximum degrees of grafting were 1.06% for NR-g-CCA, 4.66% for NR-g-ICA, and 5.03% for NR-g-MA, reached using 10 phr citraconic anhydride, 10 phr of itaconic anhydride, or 8 phr of maleic anhydride, 3 phr benzoyl peroxide, at 85, 80 and 80oC for 2, 2 and 3 hrs, respectively. Solution-based wood adhesives were formulated from these copolymers, with various contents of wood resin in the range 10-40 phr. The maximal 289 N/in cleavage peel and 245.7 KPa shear strength for NR-g-MA (5.03% grafting) were obtained at 40 phr wood resin.

Acknowledgement

The authors gratefully acknowledge Prince of Songkla University for financial support.

References

[1] Mohapatra S and Nando G B 2013 *Ind. Eng. Chem. Res.* **52** 5951
[2] Nakason C, Kaesaman A and Supasanthitikul P 2004 *Polym. Test.* **23** 35
[3] Saelao L and Phinyocheep P 2005 *J. Appl. Polym. Sci.* **95** 28
[4] Arayapranee W, Prasassarakich P and Rampel G L 2003 *J. Appl. Polym. Sci.* **89** 63
[5] Prasassarakich P, Sintoorahat P and Wongwisetsirikul N 2001 *J. Chem. Eng.* **34** 249
[6] Chumsamrong P and Monprasit O 2007 *Suranaree J. Sci. Technol.* **14** 269
[7] Onyeagoro G N 2012 *Acad. Res. Inter.* **3** 387

[8] ASTM D3807 2012 *Standard Test Method for Strength Properties of Adhesives in Cleavage Peel by Tension Loading* (Engineering Plastics-to-Engineering Plastics)

[9] Blackley D C 1997 *Polymer Latices:Science and Technology* (England : Springer) p 523

IOP Conf. Series: Materials Science and Engineering **62** (2014) 012006 doi:10.1088/1757-899X/62/1/012006

Analysis of the usage of polymer building materials in China

Hai-yang REN [*]**, Su-fang LIU**

Nanyang Institute of Technology, Nanyang City, Henan Province, China

[*] Corresponding author email: Hycy625@126.com

Abstract: According to search and working practice, this article analyses the usage of polymer building materials in China, and then points out some problems that need to be solved, for which this article indicates the specific solution.

1. Introduction

With the fast development of new technologies and new materials, polymer building materials, as a new-style building material, have become an important developing trend. They have a bright prospect in the future building material market. However, there are still some shortcomings in the use of polymer building materials in China. The research and usage should be enhanced.

2. Application situation of polymer building materials in China

2.1 Low awareness of research and weak self-innovation

Innovation is the soul for national development. Innovation for an industry is also the endless power of the industry's development. However, there are many problems and issues in innovation of the material firms and construction units in China at present. Meanwhile, most designers only carry out the designing activities following the construction standards, rather than promoting the development of innovation in building materials.

2.2 Low awareness of application

The entities that have low awareness in using polymer building materials are building designers and building firms.

2.2.1. Most building designers do not have enough knowledge and information about polymer building materials. They do not know the good performances of polymer building materials, let alone specific application. Meanwhile, the designers do not realize the importance of polymer building materials to ecological design. The result is that most designers enjoy adopting old technology and materials that have been proved mature and perfect. Thus, they do not promote the development of innovation in polymer building materials. In the designs with special requirements such as natural lighting and waterproofing, polymer building materials is seldom applied because of outdated ideas. The advantage of polymer building materials cannot be further explored.

2.2.2. Most managers in building firms would not spend money and manpower on research about polymer building materials because what they think about is to get more profit first. The result is that the research about polymer building materials is stagnant among the building firms [2]. The major reason of this result is that related workers in building firms have not realized the advantages of polymer building materials in improving project quality and solving technical problems.

In the meantime, the conservative ideas of construction departments and firms are also a main reason that leads to polymer building materials being less applied. The management would rather apply conservative but qualified strategy that can ensure the project progress and smoother complement. They are reluctant to apply the improvement on technology to avoid problems on constructive activities and acceptance check. Immature technology of material firms is also why they are reluctant to apply new polymer building materials, because materials are key factor for construction quality and responsibility.

2.2.3 Relevant government departments have not provided proper encouragement and guidance in the use of polymer building materials; they still have conservative management thinking, which leads directly to construction companies hesitating to use active polymer building materials

2.3 Limited application

Certainly, we cannot deny that there is certain polymer building materials development in China, but the specific use is still relatively limited. Firstly, polymer building materials are less targeted; the technology is mainly from abroad, and such materials are even directly imported from overseas and very expensive. Thus, the concrete application has been restricted. Newer polymer materials for use in actual projects still experience some technical problems, so we should actively explore the application of polymer materials and the continuous improvement of polymer materials application technology.

Lack of talent is also one of the reasons restricting specific applications of such materials, and this can be seen from the relevant professional training at the school. Because the country has less sense of innovation, although the school culture is relatively mature in technology capabilities, it requires a lot of study and research of the industry, such as on polymer materials, while only a few individual scholars are doing this.

Construction enterprises in the application of polymer materials often just passively select polymer materials that have been developed. Even so, due to geographical constraints, information and other factors, the scope of application of new polymer building materials is significantly compromised. Some managers in the construction industry lack of new ideas in the application of polymer materials; they do not combine polymer materials with other materials to optimize architectural effect.

3. How to make a reasonable use of polymer materials in construction

3.1 Enhanced application awareness

Proper measures should be made by relevant government to promote polymer materials, Construction sector and relevant business managers should be encouraged to spend manpower and funds to improve building standard. Meanwhile construction sector and workers in relevant firms must realize the significance of polymer materials on improving the quality of construction and solving technical problems. Proper measures may take out to enhance the use of polymer materials, such as bonus and reward [1].

Research departments and construction firms should often communicate with each other on the use of polymer materials. Then construction firms can know and master polymer materials better [3]. As a result, they can use the materials correctly. With the wide use of polymer materials, conservative strategy on polymer materials will be changed. Constructors and building designers must face the use of polymer materials and spend more time and energy to master them. All of the opinions above are benefit to promote polymer materials.

3.2 Improve the application of technology

Construction companies can not only passively select the polymeric materials that have been developed; they should also actively study higher-performance materials according to their needs. Construction sector cannot always be bound by the traditional thoughts of architecture, local effects of polymer materials in the application should be also considered in addition to the overall effect of architectural design.

Managers in the construction industry should continue broadening the application of polymer materials new ideas, try to optimize the polymer material applied in combination with other materials to achieve architectural effect, and continue improving application methods in order to meet high expectations of molecular polymer materials applications.

Less developed applications of the technology must also be addressed. Polymer materials involve multiple disciplines such as chemistry and materials science fields, including many aspects of fire, water, radiation, pressure, heat and wear, etc. When conducting research on existing polymer materials, researchers must base on their characteristics and performance for the targeted research. Meanwhile research achievement on polymer materials should be checked amended in practical engineering, and then the research achievement will become mature technology. As a result, polymer materials will be used widely.

3.3 Training professionals

Construction-related colleges and universities, building departments and enterprises should establish a specialized research institution on polymer materials, which can gather together technical personnel to promote the rational use of polymer materials [4]. Construction companies also need to actively foster technical personnel with experience in the application of polymer materials, and relevant technical staff to fully grasp the possible polymer building materials-related technologies, particularly the use of polymer building materials, to enhance the research and promotion.

4. Conclusion

Polymer materials in the construction industry will surely have a bright future, we should start with government departments, higher education and construction industry to improve R & D capabilities of polymer building materials, and keep expanding applications of polymer materials in construction engineering, in order to make the polymer building materials better serve the country's construction industry.

References

[1] Hu Z,Chen L,Zhao B,Luo Y,Wang DY,Wang YZ.A novel efficient halogen-free flame retardant system for polycarbonate[J].POlym.Degrad.Stab,2011,96:320-327

[2] G. P. Trubitskaya,A. N. Bokov, A. I. Polyak, R. F. Komarova.Allergenic characteristics of chemical compounds migrating from polymer building materials.Bulletin of Experimental Biology and Medicine.1997.02

[3] HUNTINGTON C G.Permanent architectural fabric structure-performance of the new materials technology [J].Construction & Building Materials,1987.01

[4] SALLIS R.High perormance coated fabric structural materials [J].Journal of Industrial Textiles, 1984(14)

Research on Bamboo Charcoal Bonded Grinding Wheel and Its Mechanical Properties

WeiLi[1], **Minjie Xu**[1], **Fangyong Zhan**[1] **and Mingsheng Jin**[1]

[1]Key Laboratory of E&M Zhejiang University of Technology, Ministry of Education,No.6 Zhaohui District, Hangzhou 310014, P.R. China

Corresponding Author / E-mail: weilihz@163.com

Abstract. In this paper, a new type of grinding wheel and its manufacturing production process are introduced. The new BCB (Bamboo Charcoal Bond) grinding wheel was made of bamboo charcoal, phenolic resin and abrasive powder with higher press and temperature. To investigate its mechanical features, such as Rockwell hardness, resistance to abrasion, and resistance to pressure, some experiments on three BCB samples with different Resin weight ratios 20%, 25%, 30%, were carried out. The results showed that the BCB sample with proper moulding process and Resin weight ratio had better performance.

Keywords: BCB Grinding Wheel; Surface Clean Grinding; Mechanical Properties
NOMENCLATURE: BCB = Bamboo Charcoal Bond

1. Introduction

The development trend for new machining methods it that they should be more convenient, more precise and pollution-free for workpiece surface. For these new machining process demands, BCB (Bamboo Charcoal Bond) grinding wheel was developed. Bamboo charcoal has the characteristics of high density, porous structure, good wear resistance, higher conductivity, etc. [1]. For the ELID grinding process, the grinding wheel should have good conductivity property; the BCB diamond grinding wheel is made by bamboo charcoal mainly, therefore it has the good conductivity and higher strength for achieving the ultra-precision machining of new advanced materials such as silicon and sapphire wafers, even the non-hard metal materials.

2. Experimental set-up

The production process of BCB sample is shown in Fig. 1. The bamboo charcoal powder, phenolic resin and abrasive powder were mixed and put in the mould controlled by a temperature regulator. Then, the mould was set on a pressing machine. Fig. 2 shows the pressing system with mould surrounded by a heating tape.

[1] Key Laboratory of E&M Zhejiang University of Technology, Ministry of Education, No.6 Zhaohui District, Hangzhou 310014, P.R. China

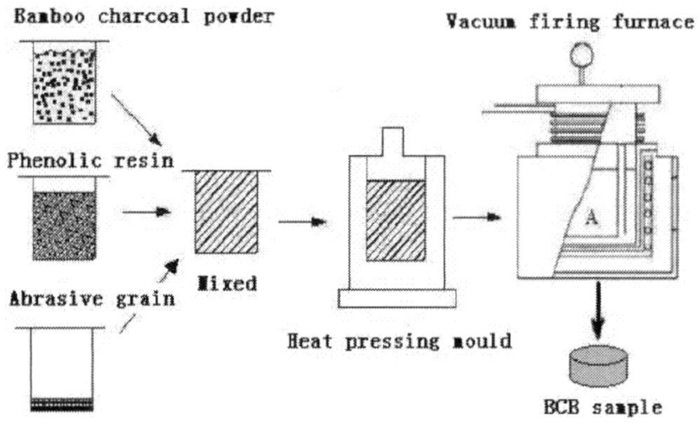

Fig. 1 The process of BCB sample preparation

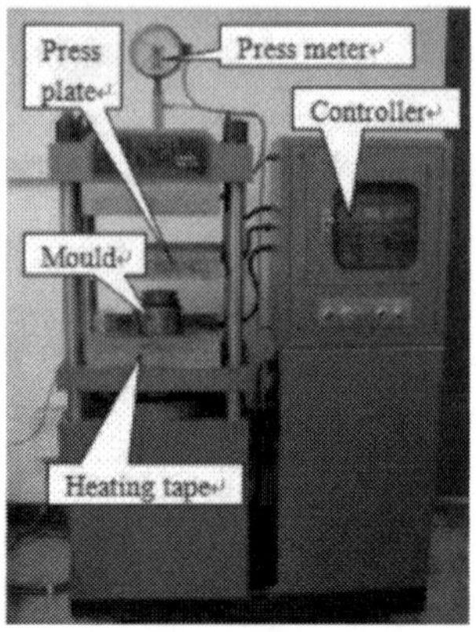

Fig. 2 Pressing system

The process at molding temperature is shown in Fig. 3. Every sample was made with different ratios among bamboo charcoal powder, phenolic resin and abrasive powder. The weight ratios of phenolic resin were 20%, 25% and 30%. The heat pressed BCB sample is shown in Fig. 4.

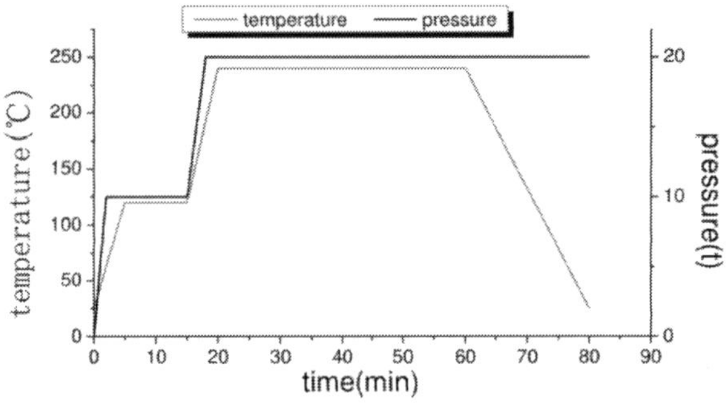

Fig. 3 Temperature and pressure process in heat pressing

Fig. 4 ample of BCB grinding wheel

3. Mechanical performance test

3.1 Rockwell hardness

A material's hardness is a main mechanical property for keeping its higher machining efficiency. Six points on the workpiece surface were measured with XHR150 Rockwell tester to get the average Rockwell hardness. The average Rockwell hardness is presented in Fig. 5. The sample's Rockwell hardness of the two sides were improved when the phenolic resin's content was increased. Anyway, it was shown that the Rockwell hardness of the samples were higher when the weight ratios of phenolic resin was higher, because the higher content of phenolic can make a sample of higher strength.

3.2 Resistance to abrasion

Resistance to abrasion is very crucial for grinding wheel to have long service life. The performance of three BCB samples were tested and compared, as shown in Fig. 6. It is shown that the BCB sample c (30%) has better frictional wear resistance performance than sample b (25%), and sample a (20%); sample a (20%) has the highest lost weight, which means it has worse frictional wear resistance performance. The BCB sample with higher composition of phenolic resin (30%) has higher Rockwell hardness based on the results above in section 3.1; therefore, it has better performance in resistance to abrasion.

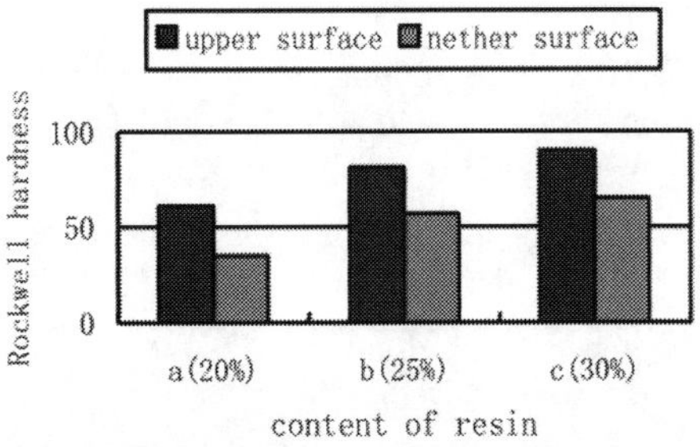

Fig. 5 The Rockwell hardness

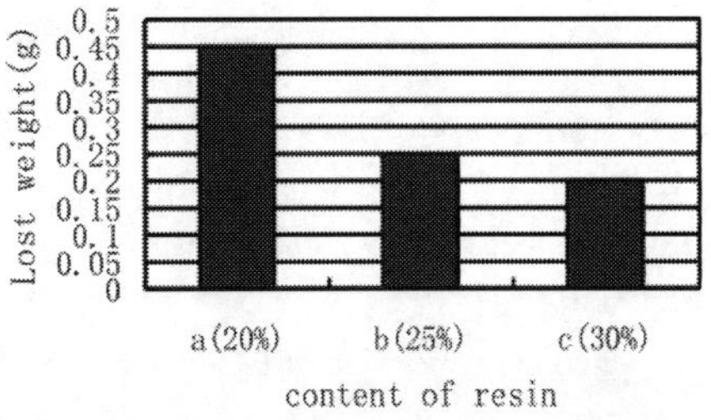

Fig. 6 The weight lost at different contents of phenolic resin

3.3 Resistance to pressure

The resistance to pressure is also one of grinding wheel's necessary mechanical properties. The experimental data and results are shown in Fig. 7. In Fig.7 (a), the non-carbonized sample with resin content of 30% can endure the pressure of almost 100 MPa, while the other one with that of 20% can only endure pressure of about 50 MPa. This is because the higher content of resin can make BCB sample harder than others when resin solidifies. In Fig.7 (b), (c) and (d), it is shown that the carbonized samples' resistance to pressure is much lower than those of the non-carbonized samples, which is because the structure of the resin in the sample was destroyed significantly during the carbonization process.

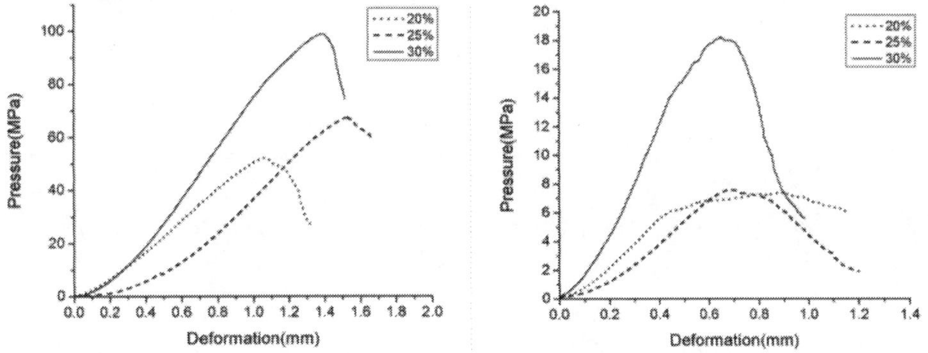

Fig. 7(a) Resistance to pressure of non-carbonized samples (b) Resistance to pressure of samples carbonized at 500℃

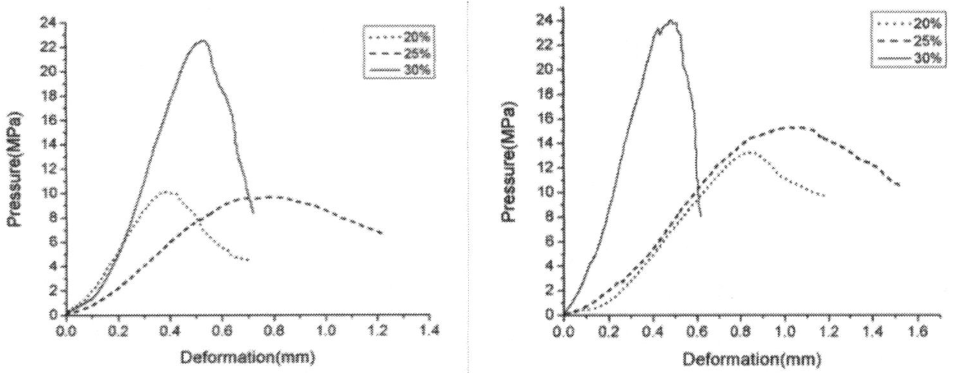

Fig. 7(c) Resistance to pressure of samples carbonized at 750℃ (d) Resistance to pressure of samples carbonized at 900℃

4. Conclusions

Aiming to develop surface clean grinding technique, a new kind of BCB grinding wheel was proposed, and the manufacturing process and some of its basic mechanical properties were introduced and studied in this paper. Based on the results discussed above, the following conclusions were drawn.

(1) With different weight ratios of resin, the BCB sample with the higher ratio of resin has better resistance to abrasion than the one with low ratio of resin.

(2) Also, the BCB sample with the higher ratio of resin has higher Rockwell hardness, but the increasing trend become slow, and the upper side of BCB sample has better performance than the nether side.

(3) Resin content affects the BCB sample's performance in resistance to pressure. The carbonized sample's resistance to pressure is much lower than that of the non-carbonized sample. Moreover, higher carbonization temperature can make the BCB sample have higher bond strength.

Acknowledgement

This project is supported by the Natural Science Foundation of China (Grant No.51205358 and 50775207). This work was granted financial support from the Key Laboratory of Mechanical Manufacture and Automation and the Key Subject Open Foundation of the Zhejiang University of Technology (2009EP017).

References

[1] T Hirose, T Fujino, T Fan, H Endo, T Okabe and M Yoshimura, "*Effect of Carbonization Temperature On The Structural Changes of Woodceramics Impregnated With Liquefied Wood*," Carbon , Vol. 40, No. 5, pp.**761-765**, 2002.

IOP Conf. Series: Materials Science and Engineering **62** (2014) 012008 doi:10.1088/1757-899X/62/1/012008

Single Electron Transfer Living Radical Polymerization *via* a New Initiator

Xiongxiong Bai, Ying Hu, Xu Zhang, Lingling Ai, Chuanjie Cheng*

School of Chemistry and Chemical Engineering, Jiangxi Science Technology Normal University, Nanchang, Jiangxi, PR China, 330013

Abstract. Research and development of novel initiating system such as single electron transfer living radical polymerization (SET-LRP) is of high importance in polymer chemistry. A new SET-LRP initiator was synthesized and applied to prepare end-functionalized poly(methyl methacrylate) (PMMA) in this study. α-Trichloromethyl benzyl alcohol was firstly synthesized, followed by preparation of PMMA under SET-LRP conditions. Conversion of MMA was 81.9%, and the molecular weight of PMMA was about 2.5 kDa at 60 °C for 1 h. Consistency of the number-average molecular weight of PMMA from NMR, GPC and theoretical calculation indicated that the polymerization featured controllable property. Broad molecular weight distribution (MWD) may be ascribed to branched polymers formed by initiation and chain transfer.

Keyword: Single electron transfer living radical polymerization (SET-LRP); α-Trichloromethyl benzyl alcohol; initiator; poly(methyl methacrylate) (PMMA)

Introduction

Controlled/living radical polymerization methods opened a new door for precise syntheses of polymers or macromolecules with different topological architectures. Among those methods, single electron transfer living radical polymerization (SET-LRP) approach has attracted more and more attention in recent years since its report by Percec and coworkers [1-4]. Various polymers with different topological architectures can be prepared by SET-LRP method. For example, Ding et al. has studied the polymerization of styrene[5], and Chen and his colleagues reported preparation of poly(vinyl acetate) *via* SET-LRP[6]. SET-LRP has opened a new area of living polymerization, which has advantages such as low catalyst loading, rapid reaction rate, and higher molecular weight of the prepared polymers [7-9].

α-Trichloromethyl benzyl alcohol is an important intermediate of some fragrance and flavor compounds, which can be readily synthesized by cheap benzaldehyde and chloroform in the presence of a strong base. To the best of our knowledge, there is no report on the use of this functional molecule as an initiator of controlled/living radical polymerizations.

1. Experimental

1.1 Instruments and reagents

BrukerAV 400MHz nuclear magnetic resonance meter (CDCl3 as solvent, TMS as internal standard) was used for the experiments.

* Corresponding author. Tel: 86-791-83815321, Fax: 86-791-83823357, Email: chengcj530@gmail.com

Benzenaldehyde, chloroform, tetrabutylammonium bromide (TBAB) and dimethyl sulfoxide (DMSO) were all AR grade and used as received. Methyl methacrylate (MMA) was treated with NaOH to remove possible inhibitor.

1.2 Synthesis of α-trichloromethyl benzyl alcohol

A solution of benzaldehyde (10.6 g, 100 mmol), chloroform (16 mL), acetone (2 mL) and TBAB (2.0 g, 6 mmol) was cooled to 0 oC, followed by adding 50 wt% of aqueous KOH (9 mL). The mixture was stirred at 10 oC for 3 h. Water was added to the mixture, and the organic layer was separated, washed with 5 wt% aqueous HCl, water, dried with anhydrous Na2SO4, and concentrated in vacuo to afford the crude product. Water vapour distillation gave pure product as an oil (17.6 g, 78.1%). 1H NMR (400 MHz, CDCl3): δ 7.58-7.51 (m, 2H), 7.34-7.30 (m, 3H), 5.15 (s, 1H); 13C NMR (100 MHz, CDCl3): δ 134.9, 129.5, 129.2, 127.8, 103.1, 84.5.

1.3 Preparation of PMMA via SET-LRP

A mixture of MMA (5.0 g, 50 mmol), copper power (0.05 g, 0.2 mmol), α-trichloromethyl benzyl alcohol (0.05 g, 0.2 mmol), and 2,2'-bipyridyl (0.156 g, 1 mmol) in DMSO (10 mL) was stirred at 60 oC for 1 h under N2 atmosphere. Excess cold methanol was added, and the precipitate was filtered to give crude polymer. The crude PMMA was purified by Al2O3 column chromatography to give white product (4.15 g, 81.9%).

2. Results and Discussion

2.1 Characterization of PMMA

Molecular weight and molecular weight distribution (MWD or PDI) are important parameters of polymers. Firstly, PMMA was characterized by 1H NMR spectrum as shown in Figure 1. The peak at lower field corresponded to aryl protons, indicating that the PMMA polymer had a phenyl ring as the end-group, which proved that α-trichloromethyl benzyl alcohol as an initiator had successfully initiated the polymerization of MMA under SET-LRP conditions. The integral area of the peak at 7.4 ppm was designated as 5, as there were 5 protons in the only phenyl ring of the polymer, and integration values of other peaks were based on this peak. At the chemical shift of 3.5 ppm, the integration value was 731, indicating that there were about 731/3 = 244 of methoxy groups in PMMA, i.e. about 244 of MMA units. Thus, the number-average molecular weight of PMMA can be calculated as follows: Mn,NMR = 244 × 100 + 225 = 24625 ≈ 24.6 kDa, where 225 was the molecular weight of the initiator, while 100 was the molecular weight of MMA monomer. The theoretical molecular weight of PMMA was Mn,th = 20.7 kDa when the conversion of MMA monomer was 81.9%. Thus, the molecular weight of PMMA from 1H NMR was consistent with the theoretical one.

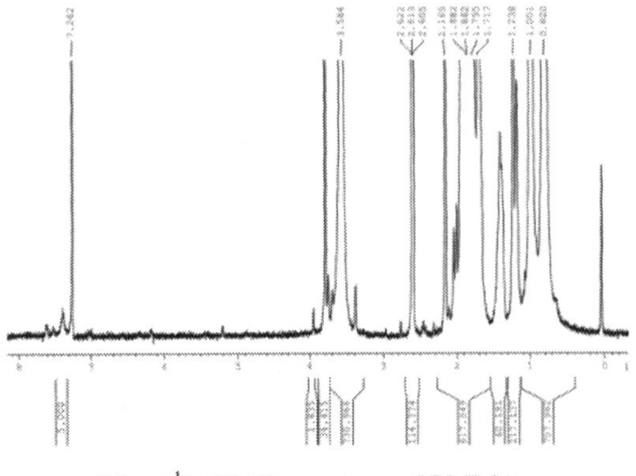

Fig. 1 ^{1}H NMR spectrum of PMMA

To further characterize the PMMA polymer, GPC was conducted and the diagram is demonstrated in Figure 2. The number-average molecular weight of PMMA measured by GPC was Mn,GPC = 23.4 kDa. Thus, we have obtained the following data: Mn,th = 20.7 kDa, Mn,NMR = 24.6 kDa, Mn,GPC = 23.4 kDa, and primary consistency of the data indicated that the polymerization featured controlled property. However, molecular weight distribution (MWD) of the PMMA was not ideal, i.e., with a broad value of 1.8. This may be explained as follows: there were three chlorine atoms in the initiator α-trichloromethyl benzyl alcohol, and branched polymers might be formed during polymerization process.

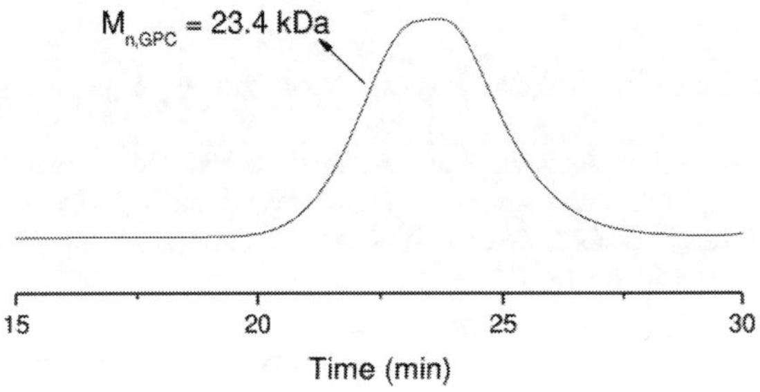

Fig. 2 GPC diagram of the prepared PMMA

2.2 Effect of temperature on conversion of MMA

As shown in Figure 3, polymerization can undergo at 20 oC with low reaction rate. The conversion was 81.9% when the temperature was increased to 60 oC, and further increasing temperature to 80 oC did not improve the conversion. Therefore, the optimized temperature of the reaction was 60 oC.

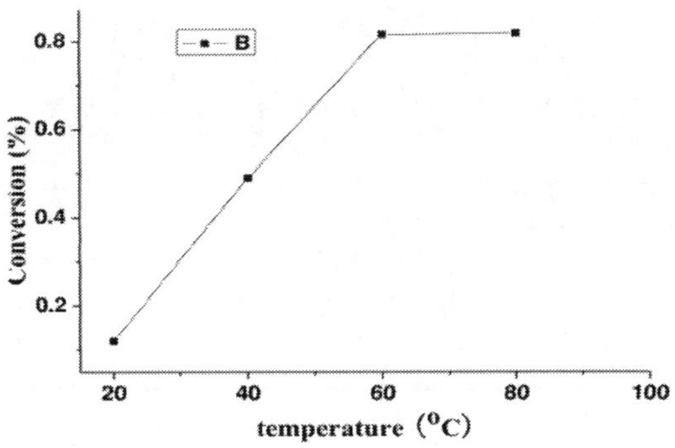

Fig. 3 Effect of temperature on conversion of MAA (reaction time: 1 h)

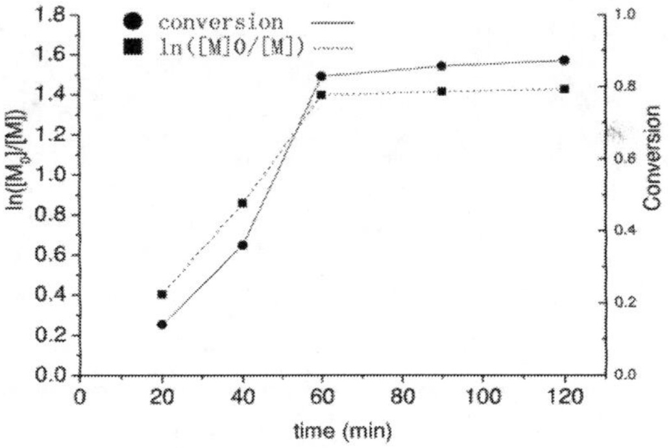

Fig. 4 Relationship of conversion, $ln([M]_0/[M])$ and reaction time

2.3 Effect of time on conversion of MMA

At 60 oC, the influence of time on the conversion of MMA was determined as demonstrated in Figure 4. The conversion was gradually improved with the increase of reaction time, and when the time was 60 min, the conversion was 81.9%. Longer reaction time has no evident positive effect on the conversion, so 60 min was the optimal reaction time. Kinetic curve of ln([M]0/[M])-t showed that linear relationship between ln([M]0/[M]) and reaction time conforming to first-order reaction kinetics.

3. Conclusions

The functional molecule α-trichloromethyl benzyl alcohol was used as a SET-LRP initiator for the first time to polymerize MMA. The conversion of MMA was 81.9% under optimized conditions: time of 1 h, and temperature of 60 oC. The benzyl alcohol-ended PMMA product is expected to further undergo ring opening polymerization to prepare biocompatible block copolymers [10-11], and the related work is still underway.

Acknowledgement

The work was financially supported by the Natural Science Foundation of China (No. 21264008) and the 7th Foundation of Creativity and Study for College Students in Jiangxi Science Technology Normal University.

References

[1] V. Percec, T. Guliashvili, A. V. Popov, E. Ramirez-Castillo. Catalytic effect of dimethyl sulfoxide in the Cu(0)/tris(2-dimethylaminoethyl)amine-catalyzed living radical polymerization of methyl methacrylate at 0–90 °C initiated with CH_3CHCll as a model compound for α,ω-di(iodo)poly(vinyl chloride) chain ends. *J. Polym. Sci. A: Polym. Chem.*, **2005**, 43, 1935-1947.

[2] A. D. Asaadei, V. Percec. From metal-catalyzed radical telomerization to metal-catalyzed radical polymerization of vinyl chloride: toward living radical polymerization of vinyl chloride. *J. Polym. Sci. A: Polym. Chem.*, **2001**, 39, 3392-3418.

[3] M. E. Levere, N. H. Nguyen, X. F. Leng, V. Percec. Visualization of the crucial step in SET-LRP. *Polym. Chem.*, **2013**, 4, 1635-1647.

[4] Q. Zhang, Z. B. Zhang, W. X. Wang, Z. P. Cheng, J. Zhu, N. C. Zhou, W. Zhang, Z. Q. Wu, X. L. Zhu. In Situ Cu(0) Catalyzed SET-LRP: the First Attempt. *J. Polym. Sci. Part A, Polym. Chem.*, **2011**, 49, 4694–4700.

[5] W. Ding, C. F. Lv, Y. Sun, H. Jing, J. J. Wei, H. X. Luan, T. Yu, G. M. Qu. Single electron transfer living radical polymerization of styrene. *Chin J Appl Chem*, **2011**, 28(9), 1000-1005. (in Chinese)

[6] C. Chen, W. Y. Ren, L. Jiang, Y. Dan. Single electron transfer living radical polymerization of vinyl acetate. *Polymer Materials Science & Engineering*, **2013**, 29(1), 44-47. (in Chinese)

[7] W. Ding, Y. Sun, C. F. Lv, J. J. Wei, T. Yu. Current situation and perspectives of single electron transfer living radical polymerization. *Chin J Appl Chem*, **2011**, 28(3), 85-91. (in Chinese)

[8] X. Y. Tang, G. Q. Zhai, L. Z. Kong. A novel controlled radical polymerization technique: single electron transfer living radical polymerization (SET-LRP). *Chin Polym Bull*, **2010**, 8, 85-90. (in Chinese)

[9] G. L. Lu, Y. J. Li, H. Guo, W. Y. Du, X. Y. Huang. SET-LRP synthesis of novel polyallene-based well-defined amphiphilic graft copolymers in acetone. *Polym. Chem.*, **2013**, 4, 3132-3139.

[10] M. A. Kryuchkov, C. Detrembleur, C. G. Bazuin. Synthesis and thermal properties of linear amphiphilic diblock copolymers of L-lactide and 2-dimethylaminoethyl methacrylate. *Macromolecule*, **2011**, 44(13), 5209-5217.

[11] A. H. Soeriyadi, C. Boyer, F. Nyström, P. B. Zetterlund, M. R. Whittaker. High-order multiblock copolymers via iterative Cu(0)-mediated radical polymerizations (SET-LRP): toward biological precision. *J. Am. Chem. Soc.*, **2011**, 133(29), 11128-11131.

The stress intensity factor of the crack on the interface between homogeneous material and FGMs

Li Shimin[1], Zhao Xinjun[1], Li Minjie[2]

[1]Beijing Special Vehicle Institute, Beijing 100072, People's Republic of China
[2]The Automobile Proving Ground of the General Armament Department, Nanjing 210028, People's Republic of China

E-mail: lishimin0625@163.com

Abstract. Based on higher-order asymptotic stress fields for the interfacial crack between homogeneous material and functionally graded materials (FGMs), stress intensity factors were calculated by finite element and node collocation methods. The method was verified by comparing the results with numerical solutions obtained in other studies, with varying degrees of homogeneity. The comparison showed that the obtained results agree well with the numerical solutions.

Keywords: Stress intensity factor; Interface crack; Functionally graded materials

1. Introduction

Stress intensity factor (SIF) is the most important parameter in the fracture analysis of linear elasticity. Analytical method, numerical solution and experimental calibration [1] are the common approaches to calculate SIFs. Analytical methods, including Westergaard stress function method, general complex variable function method, integral transform method, series expansion, Green function method, can be used to solve simple problems such as crack in an infinite plate. For complicated problems, analytical solutions are difficult to obtain, and numerical methods must be applied to obtain SIFs. Numerical methods contain include finite difference method (FDM), boundary collocation method (BCM), finite element method (FEM), boundary element method (BEM), etc. A drawback of numerical methods is that they cannot show the influence of structure size and non-homogeneity on the SIFs explicitly. Experimental calibrations include flexibility method, method of lattice, photo-elastic method, laser holography and laser speckle method, moiré interferometry, etc. Experimental calibrations can only measure the crack tip SIFs in special structures and under external loads. Complex crack problems are usually solved by semi-analytical numerical methods, including integral transform method, Williams' stress function-boundary element method and conformal transformation-boundary element method.

Among all the mentioned methods for calculating SIFs, the integral transform method is the most popular among researchers; it has been used to solve a large number of crack problems by Erdogan [2], Wu [3], Dag [4] and Ozturk [5]. As of the other methods, the conformal transformation-boundary element method has been used by Parameswaran [6] to calculate the former six items in the crack tip stress field of mode Ⅰ, and discuss the non-homogeneous effect on stress constant value line. Using the Westerguard stress function method, Chalivendra [7] determined the oblique crack tip stress field with the elastic modular varying along the gradient in exponent form. The same method has also been used to determine the crack tip field of anisotropy materials [8]. The Williams' stress function-boundary element method, however, has rarely been reported for obtaining SIFs, possibly because (1) it is difficult to obtain the Williams' solution of stress crack tip field for non-homogeneous and composite structures, and (2) the collocation region is affected by the number of items, although the relation in between has not been proved.

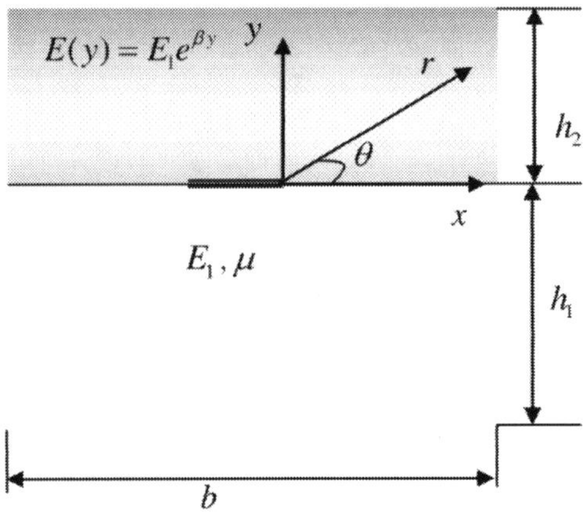

Figure 1. Interfacial crack of FGMs-coating

Figure 1 presents a structure where the interface parallels to the x-axis, and the gradient direction parallels to the y-axis. Below the interface is a matrix of homogeneous material, while above the interface is functionally graded materials (FGMs). The elastic modulus function of the FGMs is assumed to be $E(y) = E_1 e^{\beta y}$. Measures of the structure are shown in the diagram. To simplify the analysis, the corresponding polar coordinate has been established. The purpose of this study is to calculate the stress intensity factor of this structure.

2. Higher-order asymptotic stress fields of FGMs with homogenous matrix

According to Reference [9], the first six items of the interface crack tip higher-order asymptotic stress fields are as follows, where $-\pi \le \theta \le 0$.

$$
\begin{aligned}
\sigma_y = {} & r^{-\frac{1}{2}}[\frac{3}{4}B_{11}(5\cos\frac{\theta}{2} - \cos\frac{5\theta}{2}) - \frac{1}{4}B_{12}(\sin\frac{\theta}{2} - \sin\frac{5\theta}{2})] + r^{\frac{1}{2}}[(-\frac{15}{4}(B_{31} + \frac{1}{4}\beta B_{12})(3\cos\frac{\theta}{2}\theta + \cos\frac{3\theta}{2}) \\
& - \frac{3}{4}(B_{32} + \frac{3}{4}\beta B_{11})(\sin\frac{\theta}{2} + \sin\frac{3\theta}{2})] + r^{\frac{3}{2}}[-\frac{35}{12}(B_{51} + \frac{9}{35}\beta B_{32} - \frac{3}{14}\mu\beta^2 B_{11} - \frac{27}{140}\beta^2 B_{11})(3\cos\frac{\theta}{2} \\
& + \cos\frac{3\theta}{2}) + \frac{15}{4}(-B_{52} - \beta B_{31} + \frac{1}{30}\mu\beta^2 B_{12} - \frac{13}{60}\beta^2 B_{12})(\sin\frac{\theta}{2} + \sin\frac{3\theta}{2})] + r^2(6A_{61}\cos 2\theta - 6A_{62}) \\
& + \cdots
\end{aligned}
$$

In the FGMs region ($0 \le \theta \le \pi$),

IOP Conf. Series: Materials Science and Engineering **62** (2014) 012009 doi:10.1088/1757-899X/62/1/012009

$$\begin{aligned}\sigma_{fy} =\; & r^{-\frac{1}{2}}\Big[\frac{3}{4}B_{11}\Big(5\cos\frac{\theta}{2}-\cos\frac{5\theta}{2}\Big)+\frac{1}{4}B_{12}\Big(\sin\frac{\theta}{2}-\sin\frac{5\theta}{2}\Big)\Big]+r^{\frac{1}{2}}\Big[-\frac{15}{4}B_{31}\Big(3\cos\frac{\theta}{2}+\cos\frac{3\theta}{2}\Big)+\frac{3}{4}B_{32} \\ & \Big(\sin\frac{\theta}{2}+\sin\frac{3\theta}{2}\Big)-\frac{3}{16}\beta B_{11}\Big(\sin\frac{7\theta}{2}-5\sin\frac{3\theta}{2}-4\sin\frac{\theta}{2}\Big)+\frac{1}{16}\beta B_{12}\Big(\cos\frac{7\theta}{2}-17\cos\frac{3\theta}{2}-11\cos\frac{\theta}{2}\Big)\Big]+ \\ & r^{\frac{3}{2}}\Big[-\frac{35}{12}B_{51}\Big(3\cos\frac{\theta}{2}+\cos\frac{3\theta}{2}\Big)-\frac{15}{4}B_{52}\Big(\sin\frac{\theta}{2}+\sin\frac{3\theta}{2}\Big)-\frac{3}{16}\beta B_{32}\Big(10\cos\frac{\theta}{2}+5\cos\frac{3\theta}{2}+\cos\frac{5\theta}{2}\Big). \\ & -\frac{15}{16}\beta B_{31}\Big(2\sin\frac{\theta}{2}+3\sin\frac{3\theta}{2}+\sin\frac{5\theta}{2}\Big)+\frac{1}{128}\beta^2 B_{11}\Big(3(63\mu+71)\cos\frac{\theta}{2}+15(7\mu+5)\cos\frac{3\theta}{2}+ \\ & 3(9\mu-1)\cos\frac{5\theta}{2}-(\mu-3)\cos\frac{9\theta}{2}\Big)+\frac{1}{128}\beta^2 B_{12}\Big(7(\mu-7)\sin\frac{\theta}{2}+7(5\mu-11)\sin\frac{3\theta}{2}+(5\mu-29)\sin\frac{5\theta}{2} \\ & +\Big(1-\frac{\mu}{3}\Big)\sin\frac{9\theta}{2}\Big)\Big]+6r^2 B_{61}(\cos 2\theta-1)+\cdots\end{aligned} \tag{1}$$

In Eq. (1), A_{ij} and B_{ij} （$i=1\cdots 6, j=1,2$） are unknown constants obtained by the far field boundary conditions. Then, the SIFs of mode Ⅰ and Ⅱ can be obtained as

$$K_I=\lim_{r\to 0}\sqrt{2\pi r}\bullet\sigma_{yy}(r,0)=3B_{11}\sqrt{2\pi}\,,\;K_{II}=\lim_{r\to 0}\sqrt{2\pi r}\bullet\sigma_{xy}(r,0)=-B_{12}\sqrt{2\pi} \tag{2}$$

3. Finite element simulation

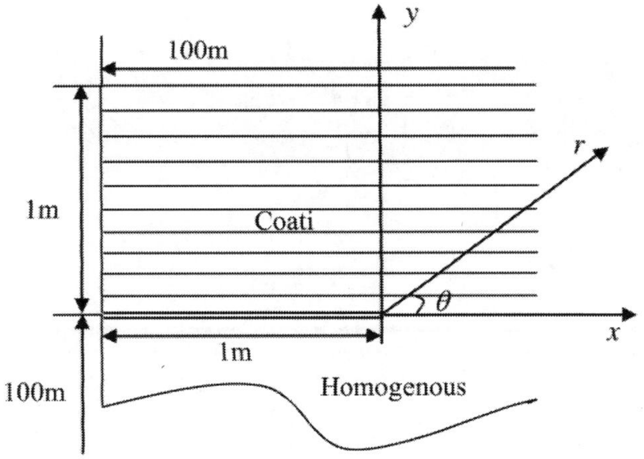

Figure 2. Simplified FGM model of the interfacial

The composites model was used to simulate the change of the FGMs, and then the model of the FGMs was simplified, as shown in Figure 2. A quadrilateral element was adopted in the lattice division and refined in the field of $r/a \le 1$. In the plane stress condition, a uniform load $\sigma_0 = 20$ Mpa was applied to the upper face, while the lower face remained fixed. Because of the symmetrical loads, the x-direction displacement on the left boundary was 0. The field $r/a \le 1$ near the crack tip field was divided into two fields, with the quadrilateral elements of $a/50$ and $a/100$, respectively. The SIFs were calculated accordingly by finite element method and node collocation method, and the difference between the calculated SIFs was only 1.2%. Therefore, the quadrilateral element of $a/50$ was adopted to meet the precision requirement and save calculation time. The stress results were obtained from the element nodes in the field where $0 \le \theta \le \pi$ and $0.1 < r \le 1.0$, and the node collocation region was $R_{i-1} < r \le R_i$ ($R_i = 0.1a \times i, i = 2 \cdots 10$).

Based on Eq. (1), the unknown coefficients of the asymptotic stress field were calculated by least square method. According to the definition of SIFs, $K_I = 3\sqrt{2\pi}B_{11}$, K_I was acquired. The normalized SIFs in different node collocation regions, as shown in Figure 3, were compared with the results of Reference [2]. The black triangles in Figure 3 denote the normalized SIFs. The node collocation calculation stops when the difference of solution between the study and the references is greater than 10%. The field adopted in this study was basically within $R/a \le 0.6$.

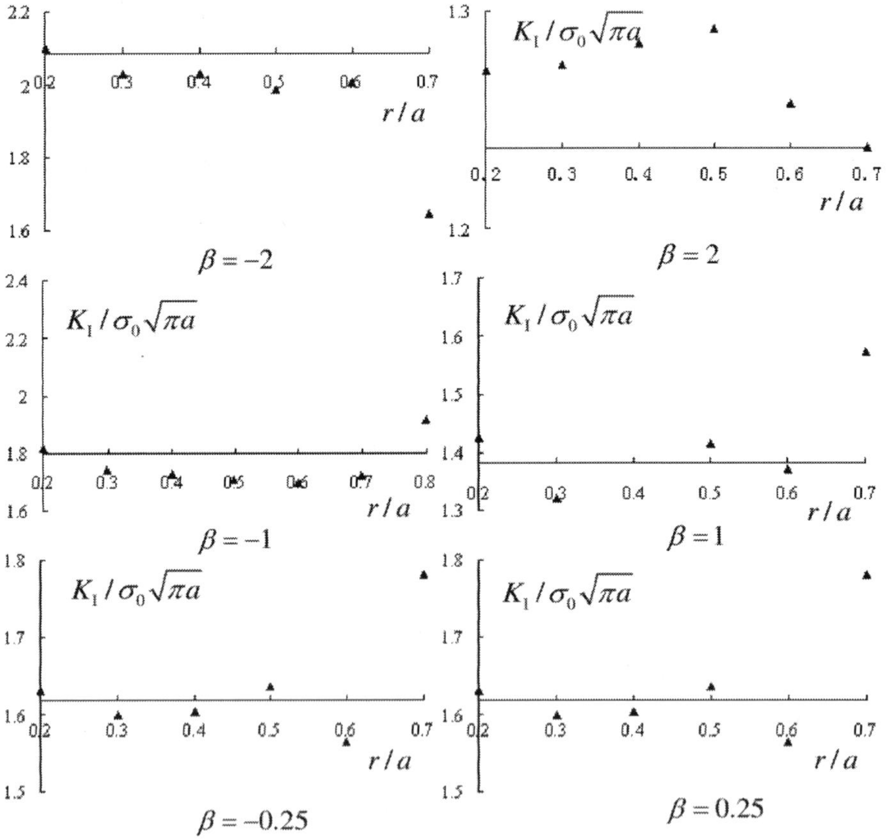

Figure 3. Normalized SIFs in different node collocation

Figure 4 shows the comparison between the numerical solutions obtained in this study and those of Reference [2]. In the diagram, $E_1 = 4800$ MPa, $\mu = 0.3$ and the measures of the structure are $h_1 = 100$ mm, $h_2 = 1$ mm, $a = 1$ mm, $b = 100$ mm. It can be seen that as the degree of non-homogeneity of the exponent FGMs or transition layer increases, the SIFs on the physical discontinuous line will minish.

The relative errors of the obtained results are listed in table 1. It can be seen that the errors are all less than 4% , indicating good accuracy of the results. The reasonability and validity of the crack tip field on the physical discontinuous line are thus proved.

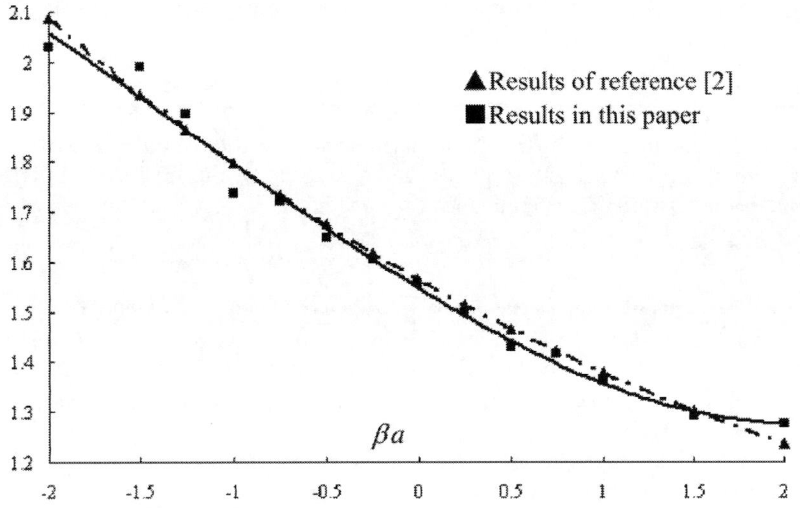

Figure 4. Variation of normalized SIFs with the degree of non-

Table 1. Relative error of the results obtained in this study based on those of Reference [2]

βa	Results of Reference [2]	Results in this study	Relative error
-2	2.087	2.03	-2.7311931
-1.5	1.936	1.992	2.874175153
-1.25	1.866	1.895	1.548264206
-1	1.799	1.738	-3.390772651
-0.75	1.735	1.722	-0.756442486
-0.5	1.675	1.649	-1.545406861
-0.25	1.618	1.607	-0.679851669
-0.01	1.566	1.556	-0.662334129
0.25	1.514	1.5	-0.924702774
0.5	1.466	1.428	-2.581329598
0.75	1.422	1.417	-0.321812436
1	1.38	1.362	-1.304347826
1.5	1.304	1.292	-0.930761041
2	1.237	1.276	3.152789006

4. Conclusions

By comparing the results obtained in this study with numerical results, the reasonability and validity of the crack tip field on the physical discontinuous line were proved. Meanwhile, this field can effectively describe the whole field stress and the effect of the degree of non-homogeneity on the crack tip stress field. The following conclusions were made.

(1) The first six items of the crack tip stress field on the physical discontinuous line of exponent FGMs/homogeneous materials can exactly depict the crack tip stress in the range of $r/a \leq 0.6$. With this range, the three-dimensional effects of the crack tip can be avoided.

(2) As the degree of non-homogeneity of the exponent FGMs or transition layer increases, the SIFs on the physical discontinuous line will minish.

(3) The results obtained in this study were a full-field asymptotic description. The physical discontinuous line possesses the characters of the general solutions. It can explicitly depict the crack tip stress field in the condition of different structures, external loads and effect of elastic modulus gradient. This method in the study can improve the assessment of the fracture coefficient in fracture experiments, and provides valuable reference for future studies.

References

[1] Jin Cheng, Shushan Zhao. *Fracture mechanics*. Beijing: Science Publishing Company, 2006

[2] Chen Y F, Erdogan F. The Interface Crack Problem for an Inhomogeneous Coating Bonded to a Homogeneous Substrate [J]. 1996 *Journal of Mechanics and Physics of Solids*, vol. 44(5): pp771~787

[3] Erdogan F, Wu B H. The Surface Crack Problem for a Plate with Functionally Graded Properties [J]. 1997 *Journal of Applied Mechanics*, vol. 64(3): pp449~456

[4] Dag S, Erdogan F. A Surface Crack in a Graded Medium under General Loading Conditions [J]. 2002 *Journal of Applied Mechanics*, vol. 69(3): pp580~588

[5] Erdogan F, Ozturk M. Periodic Cracking of Functionally Graded Coating [J]. 1995 *International Journal of Engineering Science*, vol. 33(15): pp2179~2195

[6] Parameswaran V, Shukla A. Asymptotic Stress Fields for Stationary Cracks Along the Gradient in Functionally Graded Materials. 2002 *Journal of Applied Mechanics*, vol. 69: pp240~243

[7] Chalivendra V B, Shukla A, Parameswaran V, Kingston, Island R. Quasi-static stress fields for a crack inclined to the property gradation in functionally graded materials [J]. 2003 *Acta Mechanica*, vol. 162: pp167~184

[8] Chalivendra V B. Mode-I crack-tip stress fields for inhomogeneous orthotropic medium [J]. 2008 *Mechanics of Materials*, vol. 40: pp239~301

[9] Yao Dai, Shimin Li. The Discontinuous Higher Order Fields for Anti-plane Crack at Interface [J].2008 *Advanced Materials Research*, vol.7-50: pp1027~1030

Brief Discussion on Green Building Materials

Jia-wei CAI[1],Jian SUN[2]

[1]Nanyang Institute of Technology, Nanyang City, Henan Province, China
[2]Fangshan Branch of BIT, Beijing, China

[1]Nycjw@163.com

Abstract. With more and more emphasizes on the environment and resources, the concept of green buildings has been widely accepted. Building materials are vectors of architectures, only if green building materials and related technical means are used, can we construct green buildings to achieve the purpose of energy conservation and environmental protection. This paper introduces the relationship between green building materials and green buildings, the current situation of green building materials in China, as well as the measures to accelerate the development of green building materials.

1. Introduction

In recent years, every kind of industrial production has increased efforts in environmental protection; environmental pollution of cities, the nature of industrial production caused by the traditional extensive mode, and the negative impact of ecological deterioration are gradually curbed. The building materials industry has no exception; more green building materials are applied in constructions, and materials enterprises pay more attention to the development of green building materials in production. When constructors or users choose building products, they also have specific requirements related to environmental issues. Productions and applications of green building materials are now beginning to promote each other.

Moreover, since the reform and opening up of China, with the rapid development of economy, society and living standards, people have much higher requirements of quality of housing and environment, thus the research, development and use of green building materials are more in-depth and wide-ranging. The "green" of construction and decoration materials is not only a new requirement of human to the field of construction materials since ancient times, but also a necessary way to sustainable development of building materials.

2. Current situation of application of green building materials in China

Green building materials, known as ecological building materials, environmental building materials or healthy building materials, refers to a kind of building materials that use clean production technology, less natural resources and energy, and a large number of industrial or municipal solid waste, and are non-toxic, non-polluting, and non-radioactive, as well as beneficial to environment and human health. At present, the following characteristics are shown by China's usage of green building materials.

2.1 Attention to comprehensive operation of various new building materials

For example, a new century green house built in Beijing is a combination of dozens of green building materials. This concept is better because only green building materials are not enough for the standard, people should pursue to live environmentally-friendly. This green house applies the latest scientific and technological achievements, fully considers the control and adjustment of energy, and the central ventilation systems, vacuum systems, and heating systems are also applied. The choice of building materials, interior decoration and household goods all follow the principle of environmental protection and ecological balance in maximum. The environmental protection composite floor used in the green house is the first product obtaining environmental labelling certification in China, and other application and designs such as the whole kitchen, integrated decoration, furniture and bathroom also have distinctive environmental characteristics.

2.2 Emphasis on rapid application of advanced technology products

China is gradually accelerating the application of new technologies. For example, environmental paint produced by nano-technology has been considered as specific product of many important construction projects. The product utilizes a unique photocatalytic technology to decompose and eliminate toxic gases in the air, with good properties of gloss, color and temperature retention, and strong UV resistance. Its small diameter particles can penetrate into walls, and react with silicates in the wall so that they combine together firmly with no peeling, as well as the properties of anti-aging and permanent. Therefore, once the product was in the market, it has been rapidly and widely accepted and applied, and much welcomed by construction units. Accelerating application of leading technology products has promoted eliminating those outdated technology products.

3. **Application of new building materials create a lot of invisible economic profits**

The difference between green building materials and traditional building materials determines the economic profits brought by green building materials. As the raw materials used in the production of green building materials use natural resources as little as possible, it greatly use tailings, waste, garbage and other waste materials to reduce manufacturing costs and the waste of resources. Green building uses low-energy manufacturing processes and pollution-free production technology, reducing environmental hazards and controlling expenses. The design of green building products regards improving the production environment and the quality of life as the purpose. It not only does no harm to human health, but also is beneficial to human health; the product is multifunctional, such as antibacterial, sterilization, mildew proof, deodorization, heat insulation, flame retarding, regulating temperature and humidity, degaussing, anti-radiation, antistatic and so on, creating potential added value. Green building materials are easy to recycle, and the waste does not pollute the environment, which can meet the requirement of sustainable development, and achieve the harmony between the development and environmental protection, as well as the combination of the current and the future. At the same time, green building materials has led to the development of its related industries in the supply chain.

4. Increasing the research and development of green building materials and products, improving equipment level of producing technology in building materials industry

On the basis of introducing, digesting and absorbing foreign advanced technology and equipment, it is important to set new green material as target, targeting the promising new products and technology in the market, and investigate new materials, new technology and new equipment that are in line with China's national conditions. Today's development of science and technology has gone beyond borders, and in order to develop our green building materials at a higher starting point, we should introduce focused foreign advanced production technology and equipment selectively, while enhancing the ability of self-innovation, investigate green building materials production technology with intellectual property, study design procedures and construction techniques that suit the new green building materials and products, enhance the development of environmentally friendly energy, multi-functional and social beneficial products, and strengthen the research of engineering technology to accelerate the pace of the application and promotion of new materials and products. In order to promote the development of green building materials, foreign developed countries start with formulating, implementing "green" logo certification system of green building materials. We can learn from the advanced experience of foreign countries, propose related national authorities to strengthen cooperation, formulate and implement standards and specific management measures of new green building materials as soon as possible, and emphasize and solve scientific research, production, building design, construction and various other practical problems in the process of the development of new building materials, in order to let "green materials" with "healthy, beautiful, multifunctional" features become the leading products of building materials industry.

References

[1] Gu Z A, Shuai C X. 2002 J. Current Status and Prospect of Materials Industry Development in China. Journal of East China University of Science and Technology 441-448

[2] China Building Materials Academy. 2002 J. China Green Building Materials Development Forum in 2002. Beijing. 11, 28-29

[3] China Building Materials Academy. 2003 J. Retrospect and Prospect of Economic Situation of Materials Industry(2002—2003). China Building Materials, (2)

[4] Qiu B X. 2009 J. From Green Building to Low Carbon Eco-City. Urban Development Studies, 7.1-11

[5] Liu B Y. 2012 J The Development of Green Materials in Construction Industries under Sustainable Development. Economic Research Guide. 201-202

2014 Global Conference on Polymer and Composite Materials (PCM 2014) IOP Publishing
IOP Conf. Series: Materials Science and Engineering **62** (2014) 012011 doi:10.1088/1757-899X/62/1/012011

The Manufacture of Spherical Titanium Alloy Powder in Plasma

HUANG Wei[1], CHEN Ziming, LIU Yong, SHANG Fujun, HUANG Weiming

Ningbo Branch of China Academy of Ordnance science

E-mail:hw315@126.com

Abstract. In this study, the preparation technology of spherical titanium alloy powder（tc4）was studied by using continuous induction plasma powder synthesis system. The flow-ability, particle size and impurity content of the powder before and after plasma treatment were measured and compared. The result showed that, by applying induction plasma preparation technology, the powder's flow-ability was improved by 20% while oxygen and carbon content were decreased significantly. The powder's sphericization rate and recovery rate were above 95% and 90% respectively with plasma spherical treatment.

1. Introduction

The material of titanium alloy has many excellent properties such as high strength, low density, good corrosion resistance, and so on. Titanium is becoming a very important material in automobile manufacturing, because of the energy saving and emission reduction projects in China. However, there is a critical problem limiting the titanium alloy's usage—its high cost. Many researches are focused on reducing the price of titanium, and the results of them show that the near net shape technology through the powder metallurgy and laser rapid prototyping is the best research direction and technology. The spherical titanium alloy powder is the best raw material for the near shape technology, and the demand for this type of powder keeps increasing [1-4].

Currently, the spherical Titanium alloy powder can be obtained by spray method, but the oxygen content of the powder will increase to a level of 1000 ppm, because of the high affinity between oxygen and titanium. In this study, we manufactured spherical titanium alloy powder through induction plasma technology, and controlled the particle size distribution and the oxygen content of powder by adjusting parameters.

2. Experiments

Material: normal titanium alloy powder(tc4);

Plasma gases: argon and hydrogen;

Quench gas: recycle gas of plasma gas;

Power: 40~60 KW;

Feeder rate: 2~4 kg/h.

[1] No.199 Linyun Road, Jiangdong district, Ningbo, China.

We can collect the spherical Titanium powder from two positions. We added oxygen in this system to passivate the powder.

We can get the spherical Titanium alloy powder's properties by testing of laser particle size, SEM, oxygen content, hall flow-ability.

The number of parameters of induction plasma manufacture powder technology is more than 130. In this study, we adjusted six parameters, including feeder rate, the power of plasma, reactor pressure, particle size of raw powder, flow of quench gas, and position of probe.

Table 1. Processing parameters and value ranges

Parameter	Range
Flow of center gas	30 slpm
Flow of sheath gas	80～90 slpm(argon)＋40～50 slpm(hydrogen)
Reactor Pressure	85～120 KPa
Powder of the plasma	40～60 KW
Feeder rate	2-5 kg/h
Flow of quench gas	800～2000 slpm

3. Results And Discussion

3.1 Flow-ability of the powder

In this study, we selected two types of raw Titanium alloy powder to treat, and we collected the powder from two positions. The flow-ability is as follows.

Table 2. The flow-ability testing result

	No.1			No.2		
	Raw powder	Position 1	Position 2	Raw powder	Position 1	Position 2
Average size（μm）	3	0.7	0.2	2.6	0.42	0.17
Hall flow-ability	25s/50g	14 s/50g	20s/50g	26s/50g	17s/50g	20s/50g

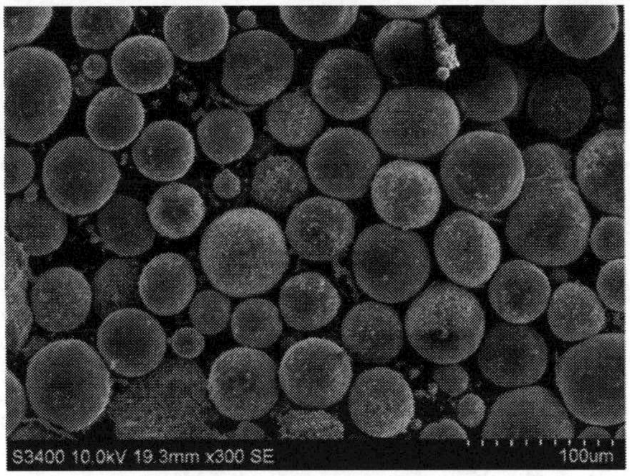

Figure 1. SEM of treated powder

It is well known that when the powder's size becomes smaller, its flow-ability becomes much worse. However, we found that the flow-ability of the powder manufactured through plasma technology becomes better than that of raw powder, even though the powder's particle size is much smaller than that of raw powder. The reasons for the better flow-ability are the powder's smoothness, spherical morphology and high sphericization rate.

IOP Conf. Series: Materials Science and Engineering **62** (2014) 012011 doi:10.1088/1757-899X/62/1/012011

3.2 Oxygen content of the powder

In this study, we selected two types of raw titanium alloy powder to treat. One was with oxygen content of 0.063%, the other with oxygen content of 0.12%. And the parameters are: plasma power: 40~60 KW; quench gas: 1500 slpm; plasma gas: Argon and Hydrogen; feeder rate: 0.5~1.0 kg/h

The results of oxygen content test are as follows.

Table 3. Oxygen content testing result

	No.1			No.2		
	Raw powder	Position 1	Raw powder	Position 1	Raw powder	Position 1
Average size (μm)	3	0.497	0.327	3	0.448	0.181
Oxygen content(%)	0.063	0.039	0.051	0.12	0.043	1.03

It can be found that the oxygen content of the last sample is very high, because its particle size is too small and there are oxidation happening on its surface. There is a very important factor leading to the increase of the powder's oxygen content, which is the oxygen adsorption. In fact, the oxygen content of the powder treated by induce plasma technology is lower than that of raw powder. There are two reasons behind the increase of the powder's oxygen content:

(1) The powder's particle size becomes small, and small particle size will make the powder's surface increase. Its surface activity becomes very high, thus oxidation will happen easily.

(2) The process of passivation will make the oxygen content increase. We always use passivation treatment to prevent oxidation, and the passivation treatment will generate an oxide film about 2~3 nm in thickness.

3.3 Particle size of the powder

The parameters for manufacturing the powder and the particle sizes are listed in Table 4.

Table 4. Particle size testing result

The NO. of sample		Parameter	Particle Size(nm)
1	Position 2	Feeder rate 0.5kg/h	38.4
	Position 1		780.0
2	Position 2	Quenching gas 1200slpm, raw powder size<74um	191.0
	Position 1		448.0
3	Position 2	Quenching gas 800slpm, raw powder size<74um	209.7
	Position 1		467.1
4	Position 2	Power 40KW, Feeder rate 0.5kg/h	138.0
	Position 1		890.1
5	Position 2	H_2 50slpm, Quenching gas 1500slpm, power 40KW	152.2
	Position 1		527.2

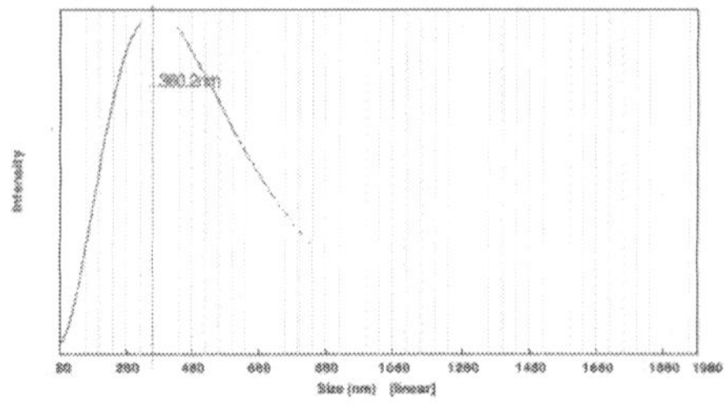

Figure 2. PSD of the powder collected in NO.1

IOP Conf. Series: Materials Science and Engineering **62** (2014) 012011 doi:10.1088/1757-899X/62/1/012011

We can find that this technology has the capacity of controlling the powder's particle size. There are four rules in the process of manufacturing the powder, including (1) The powder's particle size will significantly decrease by increasing the plasma power; (2) The powder's particle size will significantly decrease by adding Hydrogen, because the Hydrogen can increase the plasma's power; (3) The powder's particle size will significantly decrease by decreasing the feeder rate; and (4) The powder consists of three parts, big particle of unvaporization, nano-powder and residual particle after vaporization. We can conclude that the powder's pretreatment and post-treatment will help control the particle size more precisely.

4. Conclusions

First, the flow-ability of the spherical powder manufactured through induction plasma technology was better than that of raw powder. On the basis of particle size decreased to 1/10, the flow-ability of the spherical powder increased by 20% than that of raw powder.

Secondly, manufacturing the powder through induction plasma technology can decrease the powder's oxygen content.

Thirdly, the size distribution of the power manufactured through induction plasma technology is uniform. We can control the particle size and get power of desired size in an interval by adjustment of the parameters.

References

[1] Tankala, K. and DebRoy, T., Modeling of the role of atomic hydrogen in heat transfer during hot filamentassisted deposition of diamond. Journal of Applied Physics,1992, 72,7 12-7 18.

[2] Ye R, Proulx P and Boulos M I 1999 Int. J. Heat MassTransfer 42 1585–95

[3] Zhao, G. Y.. Mostaghimi, J. and Boulos, M. I., The induction plasma chemical reactor: Part II. Kineticmodel. Plasma Chemistry Plasma Processing, 1990, 10,151-166.

[4] Jiang xianliang,M.I.Boulos, Effect of precess parameters n induction plasma reactive deposition of tungsten carbide from tungsten metal powder,Trans.Nonferrous.Met.China,2001,10:639-6435.

IOP Conf. Series: Materials Science and Engineering **62** (2014) 012012 doi:10.1088/1757-899X/62/1/012012

Research on the characteristics of temperature field of asphalt pavement in seasonal frozen region

Jiangang Qiao[1], Weizheng Liu[1]

[1]College of Civil Engineering, Hebei University of Technology, Tianjin 300401, China

Email: qiaojg369@126.com

Abstract. The characteristics of climate in seasonal frozen area are low temperature and a large range of temperature variation between day and night in winter. These characteristics often lead to problems of asphalt pavement, especially transverse cracks. To reduce the problems of asphalt pavement, it is necessary to examine the distribution of the temperature range of asphalt pavement. A three-dimensional finite element model was used, taking the SMA asphalt pavement as an example with solid70 and plane55 unit features of ANSYS software. It can obtain the relationship between temperature gradient and time and the relationship between temperature gradient and depth. In addition, a function relation model of stress and time was also established. It can provide a theoretical basis for the prevention and treatment of problems of asphalt pavement in seasonal frozen area. Moreover, it has an important significance for improving asphalt pavement design.

1. Introduction

Pavement materials will produce temperature stress under the effect of the change of external temperature. If the stress exceeds the threshold strength of the material, pavement will produce cracks. The two main solutions used to study temperature fields are finite difference method and finite element method. Finite difference method requires a large number of iterative calculations, but the number of iterations will make a high impact on the accuracy [1]. Finite element method can get a relatively high accuracy by using ANSYS software [2]. Northern Hebei in China is a typical seasonal frozen area, and SMA pavement is a typical pavement structure in the seasonal frozen area. The analysis can provide a theoretical basis for the prevention and treatment of problems.

2. Temperature field theory

Because of the change of external temperature, the internal structure of the pavement will generate an unstable heat flow. The heat exchange model can be expressed as follows:

$$q=R+P. \tag{1}$$

According to the Newton equation, thermal convection can be expressed as follows:

$$P=B\,(T_{air} - T_0), \tag{2}$$

where B is the heat transfer coefficient (W/m2°C); Tair is the air temperature(°C); and T0 is the road surface temperature (°C).

Thermal radiation, including solar radiation and active radiation, can be expressed as follows:

$$R=\alpha_s*Q-F, \tag{3}$$

where αs is the radiation absorption coefficient; Q is the energy of solar radiation; and F is the efficient radiation.

Heat flow can be obtained by Fourier's law:

$$q=-\lambda\frac{\partial T_b}{\partial n}, \tag{4}$$

where Tb is the road surface boundary temperature (°C).

Air temperature including the efficient radiation can be drawn by expanding the temperature amplitude method [3] :

$$T_{ac}=T_1+\left(T_2+C_f\ a_s\right)\left(0.96\ \sin\left(\omega\ (t-t_0)\right)+0.146\ \sin(2\ \omega\ (t-t_0))\right), \tag{5}$$

where T1=(Tmax + Tmin)/2, T2=(Tmax - Tmin)/2; Cf is the effective emissivity coefficient; Tmax is the maximum temperature(°C); Tmin is the minimum temperature(°C); t0 is the initial moment, t0=3(h); and ω=2π/24 (rad).

Therefore, the unsteady heat flow above the upper boundary can be expressed as follows:

$$-\lambda\frac{\partial T_b}{\partial n}=B\left(T_{ac}+\frac{\alpha_s Q}{B}-T_0\right). \tag{6}$$

3. Finite element model of the pavement

To efficiently distribute load stress of the base layer and protect the permafrost roadbed, the gravel cushion whose thermal conductivity is small was added. The material parameters and geometry parameters of structural layers are shown in Table 1.

Table 1 Material parameters and geometry parameters

Structural layer	Elasticity modulus (Mpa)	Poisson's ratio	Density (kg/m^3)	Thermal conductivity (w/m*k)	Specific heat Capacity (j/kg*k)	Temperature shrinkage coefficient (10-5/°C)	Thickness (cm)
SMA	1400	0.25	2323	1.2	1050	2	3
AC	1200	0.25	2334	1.2	1050	2	7
base layer	1100	0.25	2275	1.3	1080	1.02	20
gravel cushion	200	0.35	2140	1.8	1020	50	18
soil base	50	0.35	1800	1.5	1040	50	-

A three-dimensional pavement model was established with the unit solid70 and plane55 of ANSYS software, which are suitable for the transient thermal analysis. It can calculate the temperature range, and then take it as a temperature load for the stress analysis. The pavement model's size is 6 m * 6 m * 6 m. In the division of grid generation of pavement, the surface layer and the base layer are refined. Contact conditions between layers are completely continuous. The surface boundary condition can be obtained by Equation (1-6), and the lower boundary is soil base temperature. Other boundaries are adiabatic state. A constraint is that the driving direction and the direction of road width have the constraints of respective management and the bottom of the pavement is fully constrained.

The most adverse weather condition in Chengde was chosen for the study of temperature field. The soil base temperature is 13.6°C, and the maximum and minimum air temperatures are -3°C and -20°C respectively. Surface boundary condition of temperature is shown in Figure 1. The time of initial temperature field is 20:00, and the distribution of the initial temperature field is shown in Figure 2.

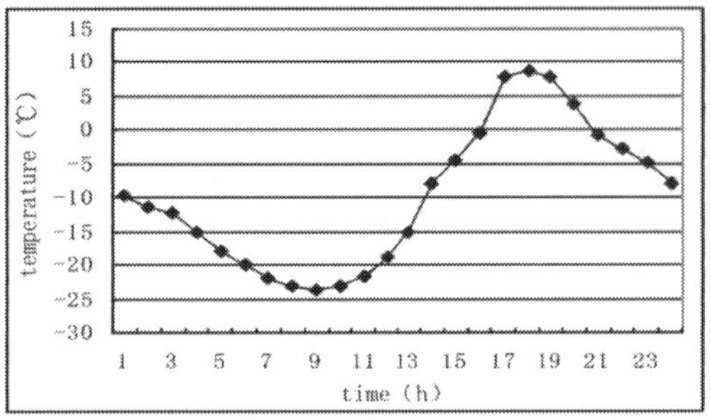

Figure 1. Road surface temperature.

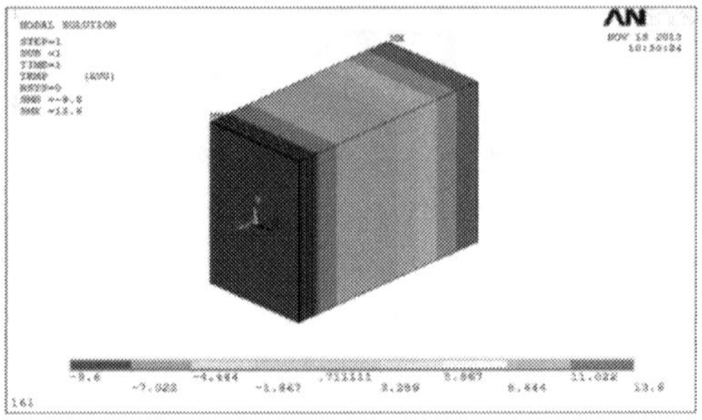

Figure 2. Initial temperature field.

4. Finite Element Analysis

4.1 Analysis of temperature field

The distribution of temperature field affects the stress of the pavement. By comparing the temperatures of road surface and the bottom of upper surface layer, it was found that the time of appearance of the minimum and maximum temperatures gradually goes backwards with the increase of depth.

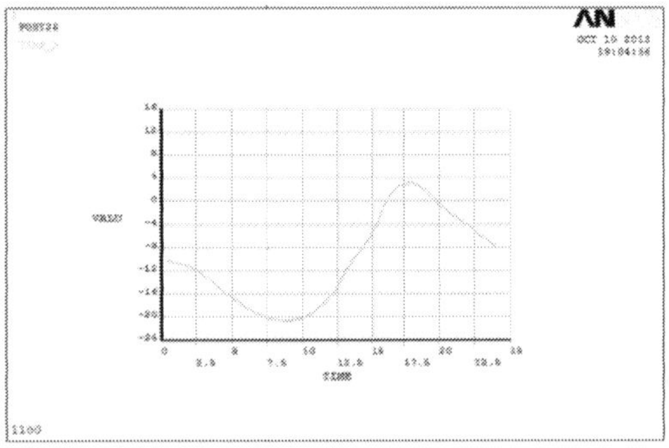

Figure 3. Temperature range of the bottom of upper layer within a day.

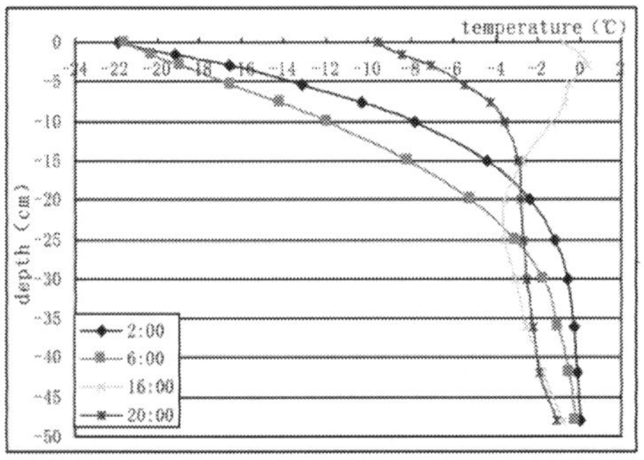

Figure 4. Temperature distribution in the depth direction at different time.

Figure 4 shows that the temperature fluctuation amplitude with the increase of depth gradually decreases, which indicates that the external environmental impact on pavement is reduced. Therefore, the thickness of the surface layer or the base layer is important to decrease the temperature fluctuation amplitude.

4.2 Analysis of temperature gradient

The value of the temperature gradient in the depth direction can roughly reflect tensile stress, which can lead to transverse cracks. Figure 5 shows that the change of the temperature gradient is substantially consistent with the temperature. As can be seen in Figure 6, with the depth increasing, the magnitude of temperature gradient change of the surface layer is reduced. Meanwhile, it is almost a constant in the base layer and the cushion layer.

IOP Conf. Series: Materials Science and Engineering **62** (2014) 012012 doi:10.1088/1757-899X/62/1/012012

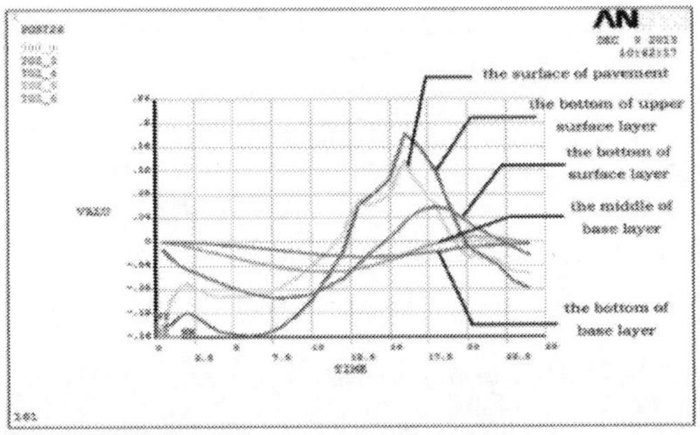

Figure 5. Temperature gradient - time diagram.

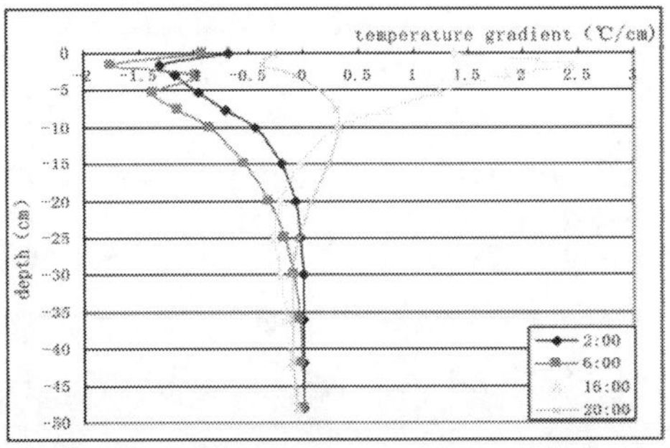

Figure 6. Temperature gradient distribution in the depth direction at different time.

4.3 Analysis of temperature stress

The distribution of pavement flexural stress is shown in Figure 7 and Figure 8. The surface layer was subjected to compressive stress under the positive temperature gradient and subjected to tensile stress under the negative temperature gradient, while the base layer has been in a state of tensile stress.

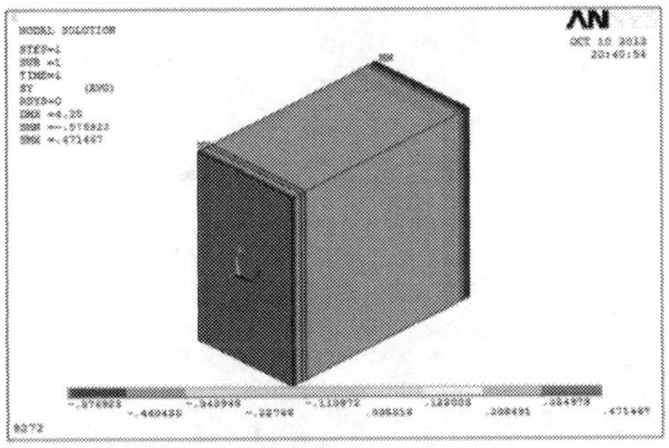

Figure 7. Stress cloud under negative temperature gradient.

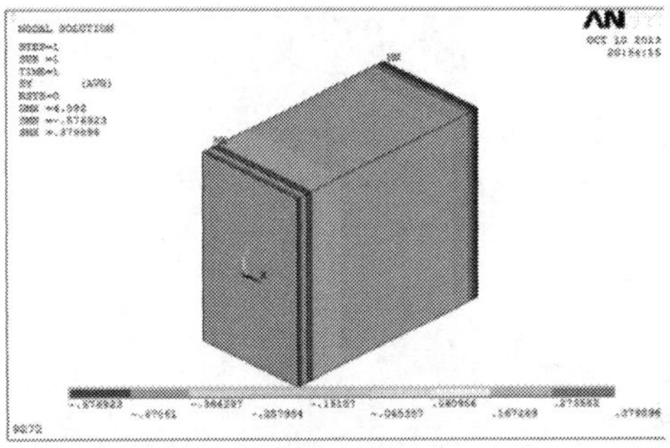

Figure 8. Stress cloud under positive temperature gradient.

Figure 9 shows that pavement stress will change with time and the state of stress of the surface layer is alternated. Figure 10 shows that the surface layer is in tension for half of the time, it is prone to fatigue damage and cracking. In addition, the base layer is in tension for most of the time so that its relaxation properties will drop at low temperature, which makes the base layer cracks occur easily.

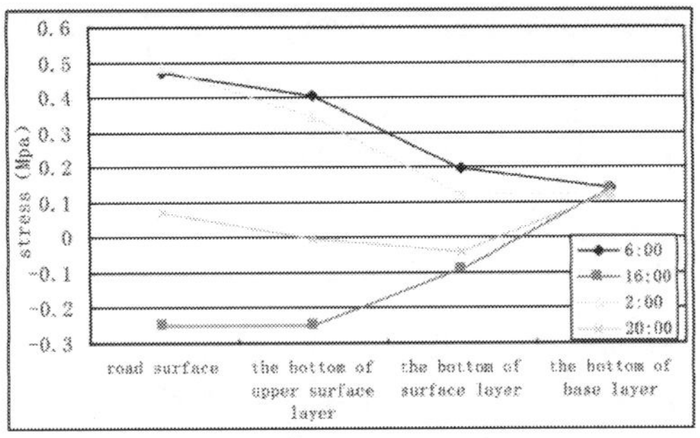

Figure 9. Stress of each structural layer.

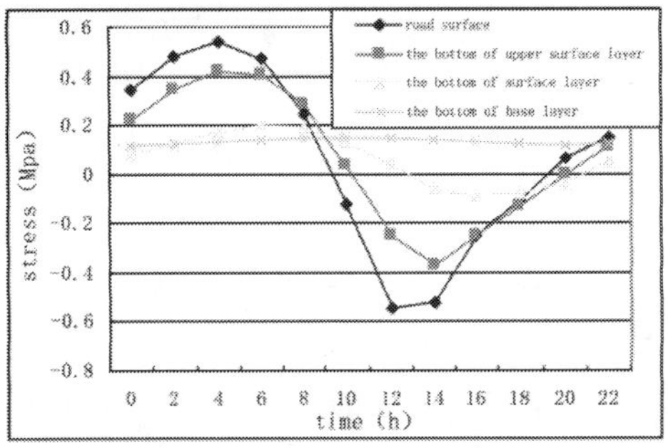

Figure 10. Time - stress diagram.

The change trend between temperature stress and time of the surface layer is a sinusoidal relationship. Then, the sinusoidal model about the temperature stress and time was established to describe or predict stress value and stress state of the pavement interior structure. The model is as follows:

$$\sigma = \sigma_1 + \sigma_2 \sin(\omega(t\text{-}t_0)) \quad (7)$$

where $\sigma 1 = (\sigma max + \sigma min)/2$, $\sigma 2 = (\sigma max - \sigma min)/2$; σmax is the maximum temperature stress (Mpa); σmin is the minimum temperature stress (Mpa); $\omega = 2\pi/24$; t0 means the average moment of the maximum tensile stress and the maximum compressive stress corresponding bit.

The following equations can be obtained by bringing the stress data into Equation (7).

(1) The relationship between stress and time of the bottom of the upper surface layer is as follows:

$$\sigma = 0.022 + 0.395 \sin\left(\frac{\pi}{12}(t-10)\right), \quad (8)$$

(2) The relationship between stress and time of the bottom of the surface layer is as follows:

$$\sigma = 0.052 + 0.138 \sin\left(\frac{\pi}{12}(t-12)\right). \quad (9)$$

5. Summary

Through the analysis of the characteristics of pavement temperature range in seasonal frozen region, it was obtained that stress of the surface layer alternately change with time, and the sinusoidal function design about the stress and time was obtained at the same time. The surface layer is in tension for half the time so that the surface layer is prone to fatigue damage and cracking. The base layer is in tension for most of the time in winter, thus its relaxation properties will drop at low temperature, which makes the base layer cracks occur easily. Therefore, the external temperature affects the stress state of the pavement greatly, and pavement design should take full account of the temperature stress.

Acknowledgements

This work was financially supported by Research Department of transportation standard traffic Project.

References

[1] Lu Jia, Lijun Sun,Likui Huang and Jian qin 2007 A numerical temperature predictionmodel for asphalt concrete pavement. Tongji University (Natural Science) 35 1040-43

[2] Jinsheng Zhou 2005 Fatigue properties of asphalt pavement under the coupling effect of load and temperature. Ms D Thesis. Xi'an: Chang'an University

[3] Zuoren Yan 1982 Temperature field analysis of layered pavement Tongji University

[4] Manuel JC Minhoto, Jorge C Pais and Paulo AA Pereira 2005 Prediction of asphalt pavement temperature using 3D finite element method. TRB 2005 Annual Meeting

[5] Setyawan A, Zoorob S E and Hasan K E 2013 Investigating and comparing traffic induced and restrained temperature stresses in a conventional rigid pavement and semi-rigid Layers The 2nd International Conference on Rehabilitation and Maintenance in Civil Engineering 53 875-884

IOP Conf. Series: Materials Science and Engineering **62** (2014) 012013 doi:10.1088/1757-899X/62/1/012013

A Quantitative Study on Packing Density and Pozzolanic Activity of Cementitious Materials Based on the Compaction Packing Model

Jianqing Gong*, Kai Chou, ZhengYu Huang, Minghua Zhao

* Associate Professor, Civil Engineering College of Hunan University, Changsha 410082, P. R. China.

Email: 130171860@qq.com

Abstract. A brief introduction to the theoretical basis of compaction packing model (CPM) and an over-view of the principle of the specific strength method provided the starting point of this study. Then, research on quantitative relations was carried out to find the correlation between the contribution rate of the pozzolanic activity and the contribution value of packing density when CPM was applied to fine powder mixture systems. The concept of the contribution value of the packing density being in direct correspondence with the contribution rate was proved by the compressive strength results and SEM images. The results indicated that the variation rule of the contribution rate of the pozzolanic activity is similar to that of the contribution value of packing density as calculated by CPM. This means the contribution value of the packing density could approximately simulate the change tendency of the contribution rate of the pozzolanic activity, which is of significant value for the future of mix designs for high and ultra-high performance concrete.

Key words: compaction packing model; specific strength method; packing density; pozzolanic activity; mix design

1. Introduction

In recent years, new components such as fibers, and various gel and chemical admixtures have been added to composite materials like high performance and ultra-high performance concrete, resulting in complicated concrete mix designs [1]. There have been high-performance concrete mix designs and optimization methods proposed [2-6], but the results are not yet perfect. More research is needed to fully understand the role of mineral admixtures and chemical additives, and the ideal mix proportions [7]. Due to the wide variety of high performance concrete admixtures and their computational complexity, the accuracy and economy of mix designs could be greatly improved with the use of computer technology [8]. It has become increasingly urgent to make full use of modern computer technology the relatively mature theory to establish a coordinated, reasonable, science-based method to develop high-performance concrete mix designs [9, 10, 11]. In this investigation, the quantitative relations between the packing density and the pozzolanic activity of high performance concrete were investigated based on the CPM, which offers a new choice for the mix design of high-performance concrete.

IOP Conf. Series: Materials Science and Engineering **62** (2014) 012013 doi:10.1088/1757-899X/62/1/012013

2. Cpm And Specific Strength Method

To date, there has been extensive research conducted to predict the packing density of a mixture of solid particles by using mathematical models in theory. For examples, the Aim-Goff model [12], and Furnas, Toufar, and Dewar models [13, 14] use mathematics for their theories, but most are based on the erroneous assumption that solid particles are of a single size. Compressible Packing Model, referring to CPM models, first proposed by De Larrard of France [15], obtained the actual packing density by calculating the virtual packing density according to the stacking structures of solid particles of different particle size distributions and by solving a nonlinear equation based on compacting Index K [15]. The CPM model considers the particle size distribution and the impact of different stacking methods on the particle packing density. It has been used worldwide, generally for making high performance concrete. Another valuable tool is the MATLAB program, which is prepared on the basis of the CPM model, and can calculate the particle packing density of multivariate powder mixtures [16, 17].

The specific strength method was initially used to describe the pozzolanic activity of mineral powders in high strength (HSC) and high performance concrete (HPC) [18], in which the strengths are governed by both the strength contribution of the hydration effect and of the pozzolanic activity. To explore the quantitative strength contribution of the two parts, indicators including the specific strength (Rspecific), the contribution rate of the pozzolanic activity (Ppozzolanic,%), the contribution rate of hydration (Phydration,%), and activity indicators (A) were put forward. Inactive admixtures were required to prepare a cement concrete to obtain a specific strength, which was used as the reference to calculate the strength contribution value and its contribution rate of the pozzolanic activity and hydration, respectively. This method has proved practical to quantitatively analyze the strength contribution of the pozzolanic effect of mineral powder in HPC. Less attention, however, was paid to the quantitative influence on the packing density [18].

In this investigation, the specific strength method was employed to quantitatively analyze the contribution value of packing density and the strength contribution rate of the pozzolanic activity. The difference between the packing density of the concrete with mineral powder or without mineral powder can be defined as the contribution value of packing density (Dpacking), which can be calculated in the equation:

$$D_{packing} = D_{doped} - D_{reference} \quad (1)$$

D_{doped} refers to the packing density of concrete mixed with mineral powder;

$D_{reference}$ refers to the packing density of the reference concrete;

Based on the CPM model, the packing density contribution was analyzed and the quantitative relationship between the packing density and the pozzolanic reaction was established accordingly in this study.

3. Experimental Program

3.1 Materials

Table 1 Physical properties of the four cementitious materials

Performance	PC	SS	FAI	SF
Density (g/cm³)	3.14	2.94	2.4	2.21
Average particle diameter (μm)	13.9	12.96	5.14	0.12

A type of Portland cement (PC), 52.5 grade, ordinary granulated blast furnace slag (SS), class I fly ash (FAI), silica fume (SF), and quartz sand were used in this study. The physical properties of PC, SS, FAI, and SF were tested and the results are shown in Table 1.

Quartz sand was 40-70 mesh with a density of 2.626g/cm3. Moreover, polycarboxylate superplasticizer was used in the study with its water reduction rate of approximately 30%.

3.2 Test Procedures

1) Mixtures

In the tests mentioned in this paper, the total mineral powder dosage was controlled unchanged at 30%, with 5% as a variable. (If one mineral powder dosage increased from 5% to 25%, another was reduced from 25% to 5%.) To facilitate this study, the water/glue ratio throughout the test was fixed at 0.25, the sand/cement ratio at 1.3; and the superplasticizer dosage amounted to 1.0% of the cementitious material; the specific mixing ratios are shown in Table 2.

Table 2 Mix Design details and compressive strength results

Item	PC (%)	SF(%)	SS (%)	FAI (%)	packing density	compressive strength 3days (MPa)	compressive strength 28days (MPa)
W0	100	--	--	--	0.7489	47.3	67.4
W1	70	--	--	30	0.7722	40.1	72.0
W2	70	5	--	25	0.7818	45.6	80.3
W3	70	10	--	20	0.7895	47.6	84.5
W4	70	15	--	15	0.7975	54.3	87.9
W5	70	20	--	10	0.8034	54.9	91.2
W6	70	25	--	5	0.8010	58.4	89.1
W7	70	--	30	--	0.7518	45.3	69.5
W8	70	5	25	--	0.7567	51.5	73.4
W9	70	10	20	--	0.7708	53.5	77.4
W10	70	15	15	--	0.7789	56.7	81.2
W11	70	20	10	--	0.7846	59.8	84.5
W12	70	25	5	--	0.7932	61.1	86.9
W13	70	30	--	--	0.7985	49.4	87.9

2) Test Procedures

(1) Compressive strength test methods:

The specimen size was 40mm × 40mm × 160mm, and the compressive strength was measured and calculated according to ISO 679:2009 Cement - Test methods - Determination of strength. All samples were formed in the curing room for one day with mold conservation. Conservation was maintained after the mold was taken apart in the curing room for three days and 28 days, and then its compressive strength was tested.

(2) Packing density calculation:

A laser particle size analyzer was first used to measure each powder particle size distribution, and then the calculation of the packing density of the mineral powder mixtures was achieved through the MATLAB CPM model. For the convenience of the calculation of the packing density of the entire concrete, it was assumed that the remaining packing density of the mix of each size was equal.

(3) SEM analysis:

The sample was observed and analysed by electron microscopy, which integrated micro and macro perspectives and provided a more detailed description of the test results.

4. Test Results And Analysis

4.1 Test Results

Table 2 lists the test results of the respective mix ratios.

4.2 Results and Discussion

1) Analysis of Packing Density and Compressive Strength

The changing trend of packing density and compressive strength of the PC-FAI-SF and the PC-SS-SF ternary systems at various mineral powder dosages are shown in Figure 1 and Figure 2, respectively, as calculated by the CPM model.

As can be seen in Figure 1, in the ternary system of the PC-FAI-SF, as the SF content increases and FAI dosage decreases, the packing density gradually increases. When the silicon ash content is kept at 20%, and the FAI content at 10%, the bulk density of the system reaches its maximum, which is due to the fact that the average particle diameters of the three materials (silicon ash, fly ash and cement), were in three different magnitudes, and thus the micro aggregate gradation was optimized for close packing and filling, with the mineral powder compacting effect at its full potential. The trend of the packing density of the ternary system of PC-FAI-SF that is revealed in Figure 2 presented a roughly similar change to the compressive strength trend at 28 days, which indicated that the compressive strength growth of the concrete at 28 days was closely related to the mineral powder compacting effect. On the contrary, the compressive strength growth of the concrete at three days was slightly different than the change of packing density. It was, in essence, a linear upward trend, which might be that the age of the concrete was too short and its early strength was primarily determined by the contribution of the cement hydration and the pozzolanic reaction of the silica fume.

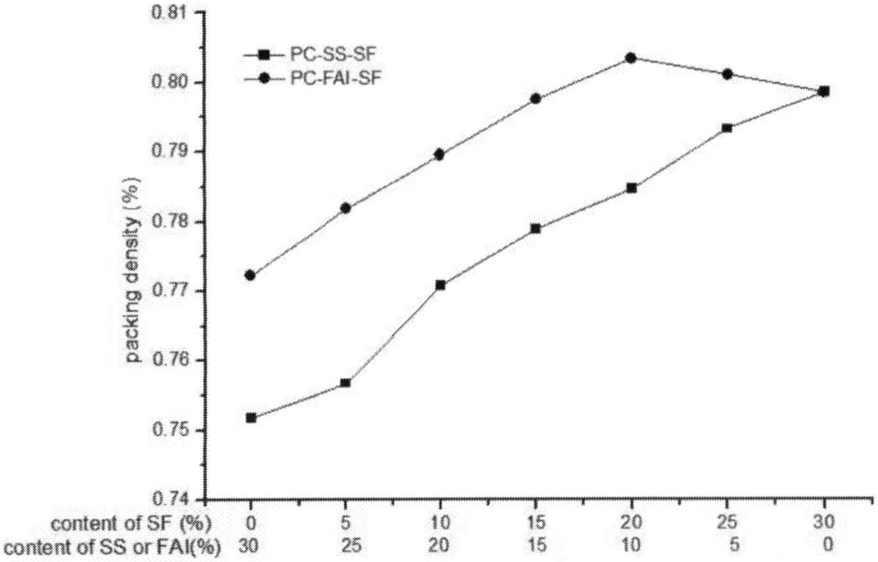

Fig. 1 Effect of SF dosage on packing density

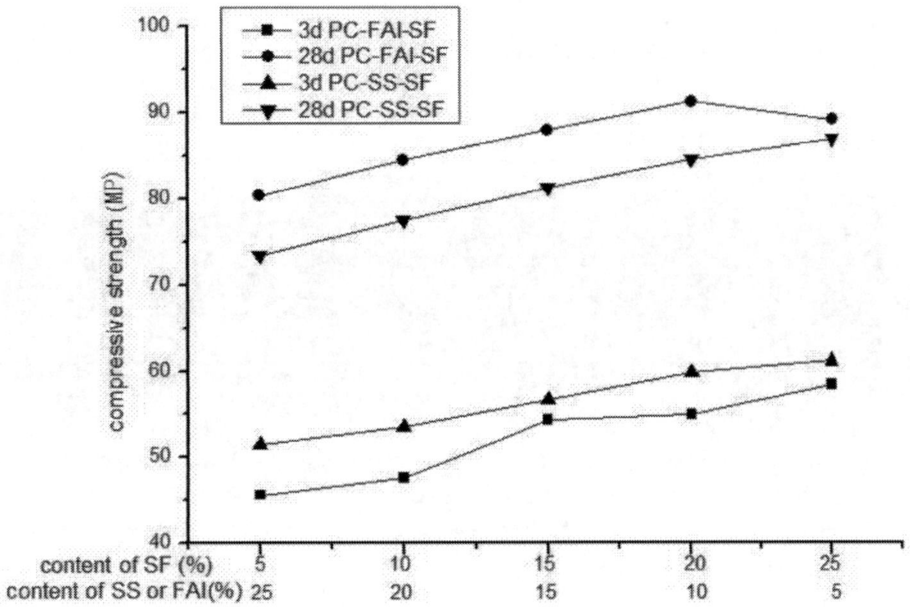

Fig. 2 Effect of SF dosage on compressive strength

For the PC-SS-SF system, as can be seen in Figure 1, the silica fume increases, and the packing density shows a slightly different trend from that of PC-FAI-SF. It is largely straight up, with a significantly lower value than that of the PC-FAI-SF system. This is due to the fact that the particle size of the slag was larger than that of fly ash, and roughly the same size as the cement particle, resulting in a slightly lower grading of the PC-SS-SF than that of the PC-FAI-SF system. However, the composite incorporation of the slag and silica fume is still better than concrete with slag only. For the ternary system of PC-SS-SF, the growth trend of the compressive strength at three days and 28 days was consistent. In particular, its three-day compressive strength, as well as its compressive strength after hardening for 28 days, coincided with the bulk density trends, indicating that the growth of the concrete compressive strength was closely related to the mineral powder compacting effects.

Through a comparative study of the PC-FAI-SF and PC-SS-SF systems, it was found that the compacting effect of the mineral powder in the two systems was directly related to the compressive strength of the concrete, which shows that the increase of compressive strength of high-performance concrete is a result of both the mineral powder compacting effect and the pozzolanic reaction of secondary hydration. In the early stages, the hydration of the mineral powder plays a major role in the strength of the concrete, but in the latter stages, the compressive strength of the concrete depends primarily on the mineral powder compacting effect. As shown in Figure 2, when the same amount of silica fume was added, the compressive strength of the early PC-FAI-SF system was slightly lower than that of the PC-SS-SF system. However, as it aged, the compressive strength of the PC-FAI-SF system gradually became higher than that of the PC-SS-SF system. This is because in the early stage, the slag was more active than the fly ash, and the C-H generated by the cement hydration reaction became an alkaline activator, which accelerated the secondary hydration of the slag. The participation of the fly ash in the secondary hydration was much slower, contributing little to the compressive strength of the concrete in its early stages. Even though the PC-SS-SF system did not show good particle size distribution, the link between bulk density and compressive strength was masked by the chemical activity of the silica fume and slag in the early stages, thus the compressive strength of the early PC-SS-SF was greater than that of the PC-FAI-SF system. As it aged, the fly ash became more active and increased its contribution to the compressive strength, and complemented the particle gradation of the PC-FAI-SF system. In the late stages, the compressive strength of the PC-FAI-SF system was ultimately higher than that of the PC-SS-SF system. In addition, the silica fume itself largely improved the early strength of the concrete. As the silica fume increased, the development of the early strength gradually increased. For example, the PC-SS-SF system, when mixed with 5% silicon ash, showed the strength at three days was increased by 13.7%, and when the dosage was increased to 15%, the strength increased to 25.2% at three days.

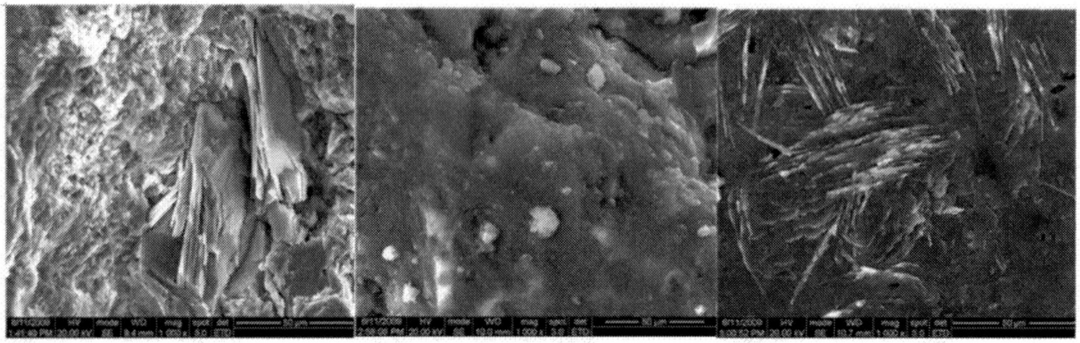

(a)SEM image of concrete without mixing any powder | (b) SEM image of the PC-FAI-SF system | (c)SEM image of the PC-SS-SF system

Fig. 3 SEM image of hardened concrete

Figure 3 shows the reference concrete and electron microscopy images of the ternary systems at 28 days. Graph (a) is the electron micrography of the reference concrete without mixing any powder. More coarse pores can be found between the hydration products of the reference concrete, and many flocculent CSH gel and slender rod-shaped crystals of ettringite can also be seen; apart from those, the layered $Ca(OH)_2$ crystal structure is relatively free and features overall looseness. Graph (b) shows the concrete of the PC-FAI-SF system under the electron microscope; the structure is clearly denser, the amount of fibrous calcium silicate hydrate gel is significantly reduced, and the gel is filled with a large number of mixed particles of various dimensions. Graph (c) shows the concrete of the PC-SS-SF system under the electron microscope; fibrous calcium silicate hydrate can be seen, but the fibers are smaller and more uniformly distributed; in addition, the fibers are interweaved and form a dense mesh, and in places where there is no fiber distribution, the gel is very dense, reflecting a high compressive strength when considered from a macro perspective.

In summary, when a certain dosage of mineral powder is added to the PC-FAI-SF and PC-SS-SF, the greater the bulk density of the mineral powder, and the greater the compressive strength of the system, proving a close correlation. And whether it was from a macro perspective or from the microscopic point of analysis, the packing density and compressive strength of the PC-FAI-SF system were superior to that of the PC-SS-SF system.

2) Pozzolanic Effect Analysis

Figures 4 and 5 indicate the changes of the contribution rate of the pozzolanic activity on the mineral powder under conditions of various aging times corresponding to SF dosage changes.

Judging from Figures 4 and 5, in the test dosage range, although doped species and aging times varied within the two systems, in general, the contribution rate of the pozzolanic activity on various systems increased with the increase of the dosage of silica fume.

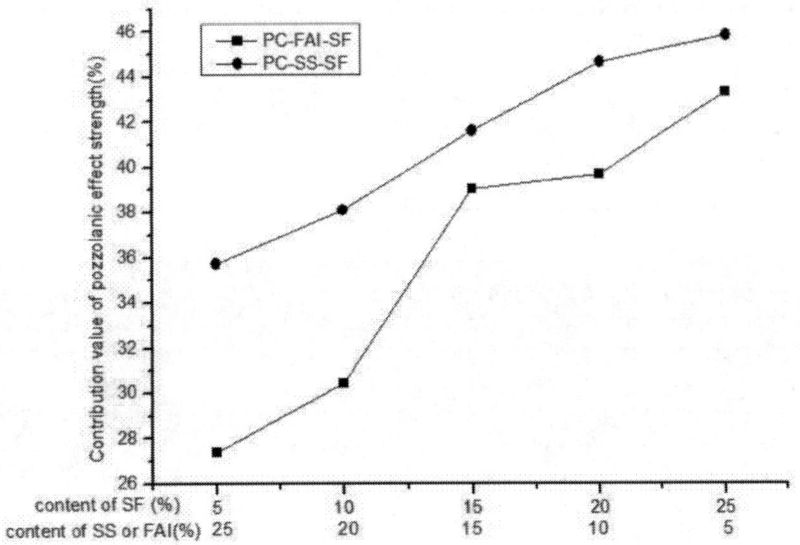

Fig. 4 Effect of SF dosage on pozzolanic effect (at 3d)

For the ternary system of PC-FAI-SF, when the silica fume dosage exceeded 20%, the contribution rate of the pozzolanic activity at 28 days decreased slightly, which might be the reason that when the compacting effect of the mineral powder increased, the contribution rate of the pozzolanic activity decreased with the decline in the compacting effect. It is worth noting that the fly ash itself became relatively more active with age, and the particle size was between that of cement and silica fume, which was conducive to being denser and more closely stacked. Meanwhile, as it had a greater density than the silica fume, the fly ash was more beneficial to the strengthening of the cement. Thus, the pozzolanic activity of the mineral powder after 28 days presented a much more significant increase than that of the mineral powder after three days.

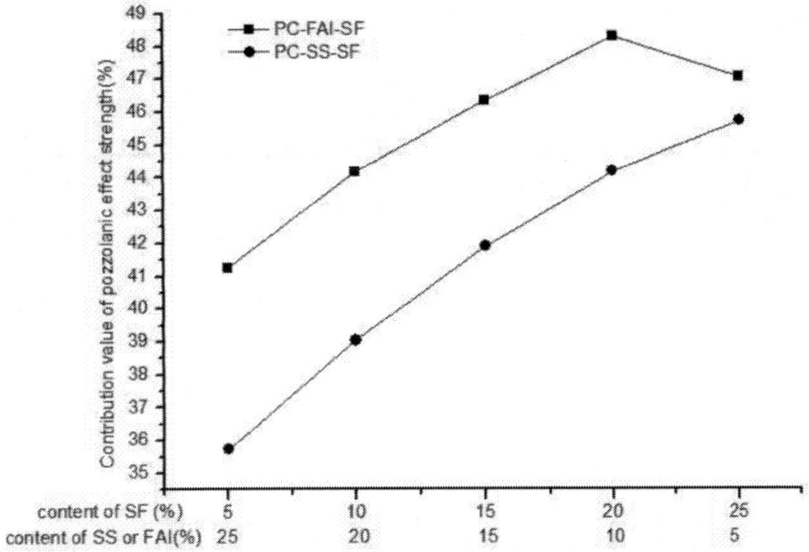

Fig. 5 Effect of SF dosage on pozzolanic effect (at 28d)

In the ternary system of PC-SS-SF, as the slag particle size was on the same order of magnitude as the cement, it did not work for dense packing. On the contrary, the silica fume particles could only be filled in a passive way between the slag particles, which helped fill the pores and condense the mix, thus contributing to the improvement of the cement strength. Therefore, in the ternary system, as the silica fume increased, the packing density was greater, and the contribution rate of the pozzolanic activity was greater.

A comparison of the two systems shows that in the instance of the same dosage of silica fume, the contribution rate of the pozzolanic activity at 28 days within the PC-SS-SF system was less than that of the PC-FAI-SF. However, at three days, its contribution rate was greater than that of the PC-FAI-SF system and the contribution rate of the pozzolanic activity tended to rise with the increase of silica fume. This is because fly ash is slightly less active than slag, and the early involvement of the secondary hydration reaction is slow. For the two systems, the chemical activity of the mineral powder plays a dominant role in the contribution rate of the pozzolanic reaction strength in the early stages, while the role of dense particles is not obvious. Although the particle size distribution of the PC-FAI-SF system is better than that of the PC-SS-SF system, the contribution rate of the pozzolanic effect strength in the early PC-SS-SF system is always greater than that of the PC-FAI-SF system. With the passage of time, advantages of the particle size distribution in the PC-FAI-SF system is reflected and the particle compacting effect plays a major role in the contribution rate of the pozzolanic activity. The changes of the contribution rate of the pozzolanic activity and the packing density gradually coincided in the two systems in the 28-day test, as is shown in Table 3. Thus, at 28 days, the contribution rate of the packing density can be modeled as the contribution rate of the pozzolanic activity.

Table 3 Specific strength, contribution rate of pozzolanic effect and of packing density

Item	Specific strength of concrete cement quantity（MPa）		Specific strength of pozzolanic effect（MPa）		Contribution value of pozzolanic effect strength（%）		contribution value of packing density（10^{-2}）
	3days	28days	3days	28days	3days	28days	
W0	0.473	0.674	--	--	--	--	0
W1	0.573	1.029	0.100	0.355	17.43	34.47	2.33
W2	0.651	1.147	0.178	0.473	27.39	41.25	3.29
W3	0.680	1.207	0.207	0.533	30.44	44.17	4.06
W4	0.776	1.256	0.303	0.582	39.02	46.33	4.86
W5	0.784	1.303	0.311	0.629	39.69	48.27	5.45
W6	0.834	1.273	0.361	0.599	43.30	47.05	5.21
W7	0.647	0.993	0.174	0.319	26.91	32.12	0.29
W8	0.736	1.049	0.263	0.375	35.71	35.72	0.78
W9	0.764	1.106	0.291	0.432	38.11	39.04	2.19
W10	0.810	1.160	0.337	0.486	41.60	41.90	3.00
W11	0.854	1.207	0.381	0.533	44.63	44.17	3.57
W12	0.873	1.241	0.400	0.567	45.81	45.71	4.43
W13	0.706	1.256	0.233	0.582	32.98	46.33	4.96

Through the above analysis, the contribution rate of the pozzolanic activity on the composite mixed mineral powder is not only related to the activity of the mineral powder itself, but also to the bulk density of the concrete, which shared a similar variation to the contribution rate of the pozzolanic activity. Therefore, it is reasonable to model the contribution rate of the pozzolanic activity by the packing density contribution value.

3) Bulk Density Modelling the Pozzolanic Effects

It was shown in the study that there was good correlation between the mineral powder bulk density and the contribution rate of the pozzolanic activity, and the latter varied according to the change of the former. Thus, a mathematical fitting tool was used to fit the relationship between the contribution rate of the pozzolanic activity and the contribution value of the mineral powder intensity of each system under the conditions of 28 days and acquired the corresponding fitting formula. Although there was a certain correlation between its bulk density contribution value and the contribution rate of the pozzolanic activity for the concrete at three days, the actual engineering significance was of little use due to its short life span. As such, this article's aim was to fit the relationship between the contribution rate of the mineral powder and its contribution value to bulk density at 28 days.

Figure 6 and Table 4 show the fitting diagram and the corresponding parameters and formulas between the contribution value of the packing density and the contribution rate of the pozzolanic activity under the standard curing conditions of 28 days.

As can be seen from Figure 6 and Table 4, the relationship between the contribution rate of the packing density and the contribution rate of the pozzolanic effect was a polynomial equation, with the correlation coefficient exceeding 98.5%, which shows a good correlation.

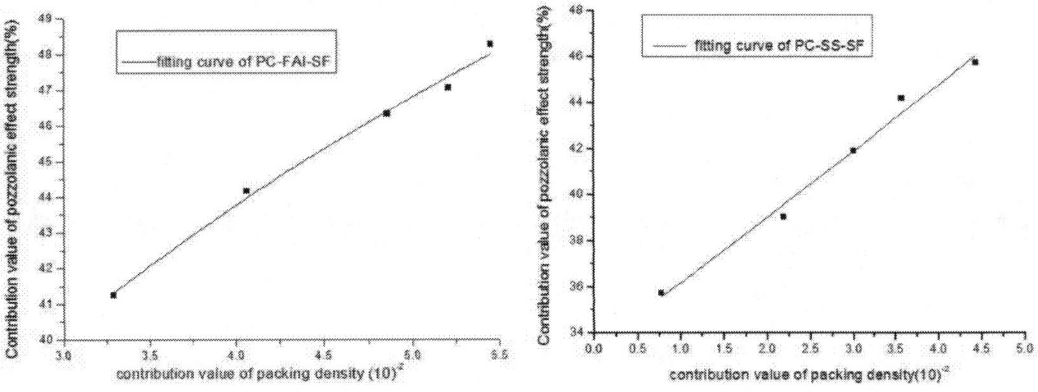

Fig. 6 Relation between contribution rate of pozzolanic effect and packing density (at 28d)

Table 4 Relation between contribution rate of pozzolanic effect and packing density (at 28d)

System Name	Fitting Formula	Correlation Coefficient	Standard differential
PC-FAI-SF	$y=26.77+5.23x-0.24x^2$	0.99252	0.33797
PC-SS-SF	$y=33.38+2.78x+0.015x^2$	0.98742	0.63601

A coefficient of a polynomial equation is the s Tab1lope of the curve, representing the rising trend of the packing density contribution value per unit to the contribution rate of the pozzolanic activity. Obviously, the better the particle size distribution, the more noticeable the contribution value of the packing density on the contribution value of the pozzolanic activity, with the equation coefficient of the PC-FAI-SF system equation coefficient greater than that of the PC-SS-SF system. However, the quadratic coefficient of the PC-FAI-SF system is negative, and this is because the increase in the bulk density reveals a downward trend on the increase of the contribution rate of the pozzolanic activity within the PC-FAI-SF system.

In summary, the packing density had a strong correlation with the contribution rate of the pozzolanic activity of the composite mixed mineral powder. When cured for 28 days under standard conservation conditions, the contribution value of the packing density can be used to calculate the contribution rate of the pozzolanic activity, which provides a foundation for the future design of high-performance concrete mix.

5. Conclusion

The two systems, PC-FAI-SF and PC-SS-SF, were both mixed with dosages of different materials. Under a certain dosage, a greater packing density of the mineral powder correlated with a greater compressive strength of the system. And whether it was from a macro perspective or micro perspective, the packing density and compressive strength of the PC-FAI-SF system were greater than those of the PC-SS-SF system.

The contribution rate of the pozzolanic activity of the composite mixed mineral powder was not only related to the activity of the mineral powder itself, but also to the bulk density of the concrete, which shared a similar variation to the contribution rate of the pozzolanic activity. Therefore, it was reasonable to model the contribution rate of the pozzolanic activity by the packing density contribution value.

The packing density had good correlation with the contribution rate of the pozzolanic activity of the composite mixed mineral powder. When cured for 28 days under standard conservation conditions, the contribution value of the packing density could be used to calculate the contribution rate of the pozzolanic activity, which provided a new idea for the pursuit of future designs of high-performance concrete mix and even the ultra-high-performance concrete mix.

References

[1] M.F Alves, R.A Cremonini, D.C.C Dal Molin (2004) A comparison of mix proportioning methods for high-strength concrete. Cement and Concrete Composites, 26 (6): 613-621.

[2] F. +de Larrard (1990) A method for proportioning high-strength concrete mixtures. Cement and Concrete Research,12 (2): 47-52.

[3] H.-G. Kessler (1994) Spheres model for gap grading of dense concretes. BFT, (11): 73-75.

[4] F.de Larrard, T. Sedran (1994) Optimization of ultra high-performance concrete by the use of a packing model. Cement and Concrete Research, 24 (6): 997-1009.

[5] Chul -Hyun Lim, Young-Soo Yoon, Joong-Hoon Kim (2004) Genetic algorithm in mix proportioning of high-performance concrete. Cement and Concrete Research, 34 (3): 409-420.

[6] M.I.Khan (2012) Mix proportions for HPC incorporating multi-cementitious composites using artificial neural networks. Construction and Building Materials, 28 (1): 14-20.

[7] Konstantin Sobolev (2004) The development of a new method for the proportioning of high-performance concrete mixtures. Cement and Concrete Composites, 26 (7): 901-907.

[8] Ma Baoguo, Wang Xingang, Li Xiangguo, Wang Kai (2005) Design of High-performance concrete and its existing problems. Concrete, (2): 12-15.

[9] F. de Larrard, T. Sedran (1996) Computer-aided mix-design: Predicting final results. Concr. Int, (10): 39-41.

[10] Dewar J D (1999) Computer Modelling of Concrete Mixtures. London: E & FN Spon.

[11] M.F.M. Zain, Md. Nazrul Islam, Ir. Hassan Basri (2005) An expert system for mix design of high performance concrete. Advances in Engineering Software, 36 (5): Pages 325-337.

[12] R. Ben Aïm, P.Le Gof (1968) Effet de paroi dans les empilements désordonnés de sphères et application à la porosité de mélanges binaires. Hand-book Powder Technology, 29 (4): 281-290.

[13] W. Toufar, E. Klose, M. Born (1977) Berechnung der Packungsdichte von Korngemischen Aufbereit.-Tech,

(11): 603-608.

[14] Jones M R, Zheng L, Newlands M. D (2002) Comparison of particle packing models for proportioning concrete constituents for minimum voids ratio. Materials and Structures, 35 (2): 301-309.

[15] De Larrard F, Sedran T (2002) Mixture-proportioning of High-performance Concrete. Cement and Concrete Research, 32 (10): 1699 -1704.

[16] J.Q.Gong. *Study on the gradation effect of ultra-high performance concrete*: (Ph.D. Hunan University, China 2007).

[17] D.Mao. *Study on the gradation effect of mineral powder in cement-based composite cementitious materials*: (MS. Hunan University, China 2004).

[18] Pu Xincheng (1998) A numerical analysis of pozzolanic effect of the high-strength and high-performance concrete. Concrete, (6): 13-23.

IOP Conf. Series: Materials Science and Engineering **62** (2014) 012014 doi:10.1088/1757-899X/62/1/012014

Facile Preparation of a Macromolecular Benzophenone Photoinitiator

Qinghua Huang, Lingling Gu, Xiongxiong Bai, Chuanjie Cheng*

School of Chemistry and Chemical Engineering, Jiangxi Science Technology Normal University, Nanchang, Jiangxi, PR China, 330013

E-mail: chengcj530@gmail.com

Abstract. Photoinitiators play important roles in the preparation of photo-cured resins. Macromolecular as well as reactive photoinitiators have attracted much attention both in industry and in academia due to the disadvantages of conventional small molecular photoinitiators such as volatility and mobility. A macromolecular benzophenone photoinitiator was designed and efficiently synthesized in this study. Hydroxyl-containing Michler's ketone was firstly synthesized in 82% yield, followed by reacting with toluene di-isocyanate (TDI) to prepare polyurethanetype macromolecular benzophenone photoinitiator.

1. Introduction

Photo-cured resins have found wide applications in many areas such as coatings, oil inks, and dental materials due to their unique advantages (e.g., rapid curing, energy-saving, environmental-friendliness, and adaptability to large-scale manufacturing)[1-4]. However, conventional photo-cured resins are usually produced by small molecular photoinitiators which are gradually restricted due to their negative effect on the environment and the polymer products. In recent years, researchers are seeking to design and synthesize macromolecular and polymeric photoinitiators to solve the problems [5-9]. Herein, we designed and synthesized a hydroxyl-containing Michler's ketone, followed by forming a polyurethane-type macromolecular photoinitiator.

2. Experimental

2.1 Instruments and reagents

Bruker AV 400 MHz nuclear magnetic resonance meter ($CDCl_3$ as solvent, TMS as internal standard). The FTIR spectra were recorded *via* the KBr pellet method by using a Bruker V70 FTIR spectrophotometer.

4-Methyl benzophenone, 40% aqueous HBr, 30% H_2O_2, urea, diethanolamine, toluene diisocyanate (TDI), dibutyltin dilaurate (DBTDL), ethylenediamine tetraacetic acid (EDTA) were all AR grade and used as received without treatment.

* Corresponding author. Tel: 86-791-83815321, Fax: 86-791-83823357, Email: chengcj530@gmail.com

2.2 Synthesis of urea hydrogen peroxide (UHP)

Into a 250 mL of three-necked flask were added 30% hydrogen peroxide (95 mL, 0.85 mol), urea (60.0 g, 1 mol) and EDTA (1.6 g, 5.5 mmol). The mixture was stirred at room temperature for 30 min, filtered, and the filtrate was cooled at 0 °C for 5 h. The resulting precipitate was filtered, and the filtered cake was collected and dried at 40-45 °C to obtain UHP product as white crystals (32.1 g, 40%).

2.3 Synthesis of 4-bromomethyl benzophenone (1)

4-Methyl benzophenone (1.25 g, 6.4 mmol) and UHP (1.2 g, 12.8 mmol) were dissolved in dichloromethane (30 mL) and were heated to reflux. Under irradiation of a 100-watt bulb, aqueous HBr (1.4 mL, 6.9 mmol) was added dropwise within about 20 min, followed by refluxing for additional 40 min. The reaction mixture was cooled, and water was added, separated, and the organic layer was dried with anhydrous Na_2SO_4, filtered, and finally concentrated in *vacuo* to afford a white solid, which was recrystallized in anhydrous ethanol to give pure 4-bromomethyl benzophenone (1.60 g, 91%). M.p. 109-110 °C. IR: ν (cm^{-1}) 3040, 1668, 1610, 1463, 1419, 852, 599. ^{1}H NMR ($CDCl_3$): δ 7.80-7.77 (m, 4H), 7.60-7.58 (m, 1H), 7.51-7.47 (m, 4H), 4.53 (s, 2H). ^{13}C-NMR ($CDCl_3$): δ 196.0, 142.1, 137.5, 137.4, 132.6, 130.6, 130.0, 129.0, 128.4, 32.3.

2.4 Synthesis of 4-[di(hydroxyethyl)aminomethyl] benzophenone (2)

4-bromomethyl benzophenone (0.28 g, 1 mmol) was dissolved in dichloromethane (20 mL), followed by adding sodium hydroxide (0.2 g, 5 mmol) and diethanol amine (0.21g, 2 mmol). The mixture was stirred at room temperature for 10 h, until which TLC results indicated that the reactant 4-bromomethyl benzophenone was almost completely consumed. To the resulting mixture was poured water (20 mL) and the mixture was stirred vigorously for 10 min. The organic layer was separated, dried with anhydrous sodium sulfate, filtered, and the filtrate was concentrated in vacuo to afford the crude product. Further purification on silica gel chromatography with petroleum ether/ethyl acetate (5 : 1, v : v) as eluant gave pure oily product. (0.25 g, 82%). ^{1}H NMR ($CDCl_3$): δ 7.74-7.72 (m, 4H), 7.57-7.53 (m, 1H), 7.46-7.41 (m, 4H), 3.78 (s, 2H), 3.64 (t, 4H, J = 5.2 Hz), 3.51 (s, br, 2H), 2.71 (t, 4H, J = 5.2 Hz). ^{13}C-NMR ($CDCl_3$): δ 196.5, 143.6, 137.6, 136.6, 132.4, 130.3, 130.0, 128.9, 128.3, 59.6, 58.9, 55.8.

2.5 Synthesis of polyurethane type macromolecular photoinitiator (3)

Toluene diisocyanate (TDI) (1.045 g, 6 mmol) was dissolved in DMF (10 mL) at room temperature. Then 4-[di(hydroxyethyl)aminomethyl] benzophenone (0.9 g, 3 mmol) in DMF (10 mL) was added dropwise to the above solution, followed by the addition of a drop of dibutyltin dilaurate (DBTDL). The resulting mixture was stirred at 65 °C for 4 h, cooled, poured slowly into diluted aqueous NH_3, and the precipitate was filtered. The filter cake was washed with distilled water, ethanol, and distilled water again. Finally, the solid product was dried in a vacuum oven at 40 °C for 24 h, and the product mass was 1.43 g.

3. Results and Discussion

3.1 Synthesis, characterization and properties of 4-[di(hydroxyethyl)aminomethyl] benzophenone (2)

First, 4-(bromomethy) benzophenone (**1**) was easily prepared from 4-methy benzophenone in excellent yield by using light-promoted oxidation and bromination reactions. Under the reaction conditions, Baeyer-Villiger reaction may be a competitive side reaction, but this ester-forming reaction usually requires harsh conditions or very long reaction times, and the presence of phenyl ring further retarded the reaction. For example, only 47% yield of ester was obtained when 4-methy benzophenone underwent Baeyer-Villiger reaction with the time of as long as 8 days [10]. On the other hand, HBr can be easily converted to bromine radical under UHP and light irradiation, and bromination readily occurred at benzyl site of 4-methy benzophenone to form the product. As a matter of fact, more than 90% yield of 4-(bromomethy) benzophenone (**1**) was obtained within about 1 h. Then, the synthesis of compound **2** was conducted by 4-bromomethyl benzophenone (**1**) and diethanol amine in the presence of a strong base (Fig. 1).

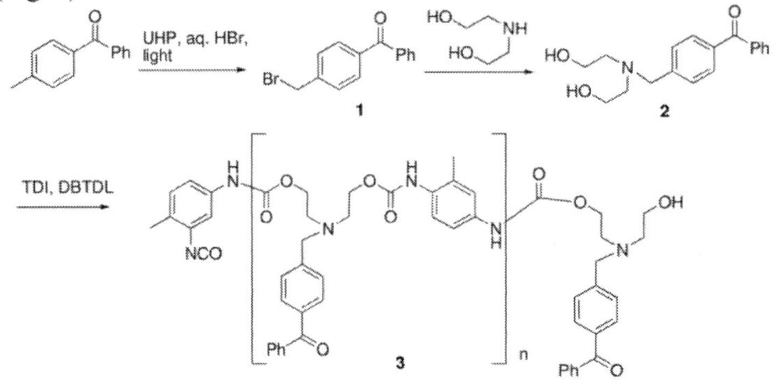

Fig. 1 Synthesis of polyurethane type macromolecular photoinitiator (3)

^{1}H NMR spectrum of compound **2** was shown in Fig. 2A. Nine protons at the chemical shifts around 7-8 ppm represented aryl hydrogens; a single peak at 3.78 ppm corresponded to the protons from the benzylic CH_2; two triplet peaks at 3.6 and 2.7 ppm corresponded to the two hydroxyl ethyl protons; a broad peak at 3.5 ppm represented the two hydroxyl active hydrogens. The structure of **2** was further confirmed by ^{13}C NMR spectrum in Fig. 2B.

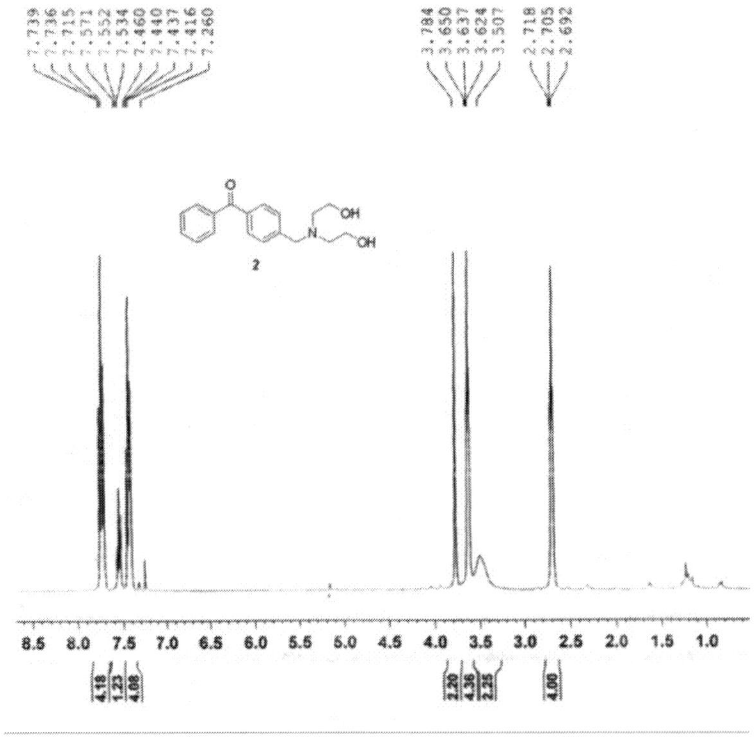

(A)

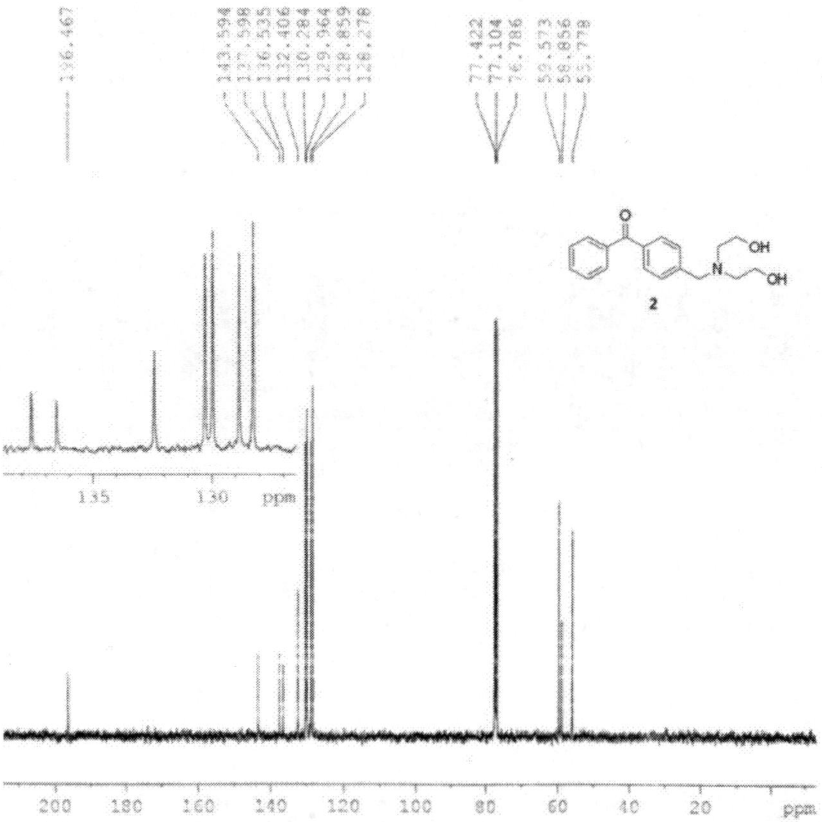

(B)

Fig. 2 ^{1}H NMR and ^{13}C NMR spectra of compound **2**

The intermediate **2** is a functional molecule that contains benzophenone moiety, tertiary amine and hydroxyl groups. The benzophenone part is a hydrogen-trapping photoinitiator, tertiary amine acts as co-photoinitiator, and dihydroxyl groups are hydrophilic and can react with other functional groups. Therefore, compound **2** is expected to be a water-based Michler's ketone type photoinitiator.

3.2 Preparation of polyurethane type macromolecular photoinitiator (3)

The compound **2** contains two hydroxyl groups that can react with diisocyanates to form polyurethanes under proper conditions. The polyurethane is a useful functional polymer as the macromolecular type II photoinitiator. Furthermore, tertiary amine groups can act as both co-photoinitiators and curing agents for polyurethanes.

Acknowledgement

The work was financially supported by the Natural Science Foundations of China (No. 21264008) and the 7th Foundation of Creativity and Study for College Students in Jiangxi Science&Technology Normal University.

References

[1] Z. L. Yang, D. A. Wicks, J. J. Yuan, et al. Newly UV-curable polyurethane coatings prepared by multifunctional thiol- and ene-terminated polyurethane aqueous dispersions: Photopolymerization properties. *Polymer*, 2010, 51, 1572-1577.

[2] V. Jaččovičová, M. Mikula, B. Havlínová, et al. Influence of UV-curing conditions on polymerization kinetics and gloss of urethane acrylate coatings. *Prog. Org. Coat.*, 2013, 76, 432-438.

[3] C. J. Cheng, X. X. Bai, W. Q. Fan, et al. Synthesis of a Photo-Active Gemini Surfactant and Its Use in AGET ATRP Miniemulsion Polymerization and UV Curing. *Chem. Pap.*, 2014, 68, 136-144.

[4] C. J. Cheng, X. X. Bai, S. J. Liu, et al. UV Cured Polymer Based on a Renewable Cardanol Derived RAFT Agent. *J. Polym. Res.*, 2013, DOI: 10.1007/s10965-013-0197-2.

[5] C. J. Cheng, W. Q. Fan, J. Sun, et al. Facile synthesis of water-soluble polymeric benzophenone photoinitiators. *Chemical Research and Applications*, 2013, 25(8), 1200-1203. (in Chinese)

[6] C. J. Cheng, Z. J. Wei, Z. B. Liu, et al. Synthesis and UV-curing properties of a photo-active cardanol derivative. *Chin. J. Appl. Chem.*, 2012, 29(4), 392-396. (in Chinese)

[7] Z. B. Liu, C. J. Cheng, Y. Zheng, et al. Synthesis of UV-Curing Acrylic Resin by the Combination of RAFT and Click Chemistry. *Polymer Materials Science and Engineering*, 2012, 28(4), 156-159. (in Chinese)

[8] A. F. Luo, X. S. Jiang, J. Yin. Thioxanthone-containing renewable vegetable oil as photoinitiators. *Polymer*, 2012, 53, 2183-2189.

[9] Y. Chen, J. Loccufier, L. Vanmaele, et al. Novel Multifunctional Polymeric Photoinitiators and Photo-Coinitiators Derived from Hyperbranched Polyglycerol. *Macromol. Chem. Phys.*, 2007, 208, 1694-1706.

[10] M. Renz, B. Meunier. 100 Years of Baeyer-Villiger Oxidations. *Eur. J. Org. Chem.*, 1999, 737-750.

IOP Conf. Series: Materials Science and Engineering **62** (2014) 012015 doi:10.1088/1757-899X/62/1/012015

Effect of tempering on microstructure and mechanical properties of cast iron rolls laser alloyed with C-B-W-Cr

R. Zhou[1, 3], G.F. Sun[2,*], K.K. Chen[1], Y.Q. Tong[3]

[1] School of Mechanical and Metallurgical Engineering, Jiangsu University of Science and Technology, Zhangjiagang 215600, China
[2] School of Mechanical Engineering, Southeast University, Nanjing 211189, China
[3] School of Mechanical Engineering, Jiangsu University, Zhenjiang 210013, China

Email: gfsun82@gmail.com

Abstract: Laser alloyed layer with C-B-W-Cr powders on high-Ni-Cr infinite chilled cast iron roll was tempered at 500℃ for 60 mins and cooled to room temperature in the furnace. Optical scanning microscopy and X-ray diffractometer were used to investigate the microstructure and phases of the layer. A microhardness tester and wear tester were used to test the microhardness distribution, friction and wear behavior. Ferrite was formed in the tempered layer. Metastable carbides changed to stoichiometric carbides after tempering. FWHM (full width at half maximum) of the first primary peak was decreased by tempering. Microhardness decreased by 44%. Friction coefficient of the tempered layer exhibited gentle increase and a stable value of 0.6 was achieved for both the tempered layer and substrate. Relative wear resistance of the tempered layer was 1.11 times that of the tempered roll substrate. Mechanism of the effect of tempering on microstructure, FWHM, microhardness and wear properties were discussed.

1. Introduction

Rolls are the main deformation tools in rolling mills. They withstand very high loads, thermal fatigue, and severe environmental attack. As a result, they often fail due to cracks, spalling, and wear [1, 2]. High-Ni-Cr infinite chilled cast iron rolls are widely used in manufacturing hot rolled plate, wire rod, bar, et al due to their good wear resistance, viscous resistance, thermal cracking resistance and spalling resistance. The requirement for high wear resistance is improved due to the increased rolling speeds and loads [3]. Strengthening technologies are effective ways to prolong service life of rolls.

Laser surface modification has some unique advantages over other surface modification techniques, such as low heat input to the material, strong metallurgical bond between the composite layer and substrate, the formation of a non-equilibrium or amorphous phase as well as homogenization and refinement of the microstructure, all without affecting the substrate properties [4-7]. Laser surface alloying (LSA) has improved the wear resistance of different rolls to 1.6-8.8 times [2]. Ceramic materials are often added to metal matrix to form composite coatings to improve the wear resistance of the matrix [8,9].

In view of the preceding information, the authors have investigated the microstructure and microhardness of the high-Ni-Cr infinite chilled cast iron rolls laser alloyed with C-B-W-Cr composite powders [10]. Results indicated that there were lots of cracks in the alloyed layers. Microhardness can be improved to an average value of $1001HV_{0.05}$, which is 1.53 times that of the roll substrate. This

IOP Conf. Series: Materials Science and Engineering **62** (2014) 012015 doi:10.1088/1757-899X/62/1/012015

paper aims to investigate the effect of tempering on microstructure, phases, FWHM (full width at half maximum) of presented peaks, microhardness and wear behavior of the alloyed layers on high-Ni-Cr infinite chilled cast iron rolls.

2. Experimental details

2.1 Sample preparation

Fabrication of the alloyed layer was from the reference [10]. Sample C was fabricated with the laser power of 7.2 kW, spot diameter of 1.5 mm, scanning speed of 11 m/min, and an overlap ratio of 33.3% was chosen to be tempered. Tempering of the alloyed layer was processed in a furnace with a ramping rate of 20℃/min, a holding temperature of 500℃ and a holding time of 60 min, followed by a furnace cooling to the room temperature in about 20 h.

The tempered sample was sectioned, mounted, grounded polished, and etched with an etchant containing 20 g $FeCl_3$, 50 mL HCl and 100 mL H_2O to reveal the microstructure by immersion technique.

2.2 Morphology and phase analysis

The morphology of the surfaces were observed using a scanning electron microscope (SEM) (JSM 7001F; JEOL, Ltd., Japan) with energy dispersive spectroscopy (EDS) (Oxford Inca Energy-350; Oxford Instruments plc). Different phases formed in the alloyed layers and FWHM of presented peaks were identified by X-ray diffractometer (D8 ADVANCE; Bruker Corporation, Germany) with Cu Kα radiation. The 2*θ was in the range of 30° to 120° with a scanning speed of 1°/min.

2.3 Microhardness analysis

The microhardness distribution in the cross section of the alloyed layer was measured using Vickers indenter (HXD-1000TMSC/LCD, Shanghai Taiming Optical Instrument Co. Ltd., China) at a load of 50g and a dwell time of 10s. The distance between indentations was in the range of 30-50 μm.

2.4 Wear experiments

Dry sliding wear test of the tempered layer and substrate was performed with a ball-on-disk apparatus (CETR UMT-2 tribometer) in ambient air using the straight line reciprocating wear form. Their surfaces were grounded and polished with 1200 grit paper and rinsed with alcohol before wear testing. The counter-body was a Co-WC hard alloy ball with a diameter of 9.5 mm. The applied load was 20 N. The workbench was set to rotate at a fixed speed of 300 rpm and the test time was 60 minutes. The reciprocating length of the wear track was 5 mm. Friction coefficient was recorded. The wear ball was changed prior to each test to ensure reproducibility in the wear conditions. The contour curves of the cross-section of the wear tracks were measured on 5 different positions for each specimen with a Surface Texture Measuring Instrument (SURFCOM 130A, ACCRETECH, Tokyo, Japan). The data measured was processed to get the area value through integrating with software Origin 8.0. The average of the five groups of area data was taken as the final value for each specimen. The wear volume loss is the product of the cross-section area and length of the wear track, which was used to evaluate the wear resistance. The microstructure of top surface of the wear track was observed with SEM.

3. Results and discussion

3.1 Phase and FWHM analysis

X-ray diffraction of the alloyed and tempered layer is shown in Fig. 1. Both layers show the presence of M23C6 and iron carbides (Fe3(C, B), Fe3C). Differences are that Fe3(C, B) and residual austenite changed to Fe3C and ferrite in the tempered layer. A point worth mentioning is that martensite is not

IOP Conf. Series: Materials Science and Engineering **62** (2014) 012015 doi:10.1088/1757-899X/62/1/012015

observed in the alloyed layer by XRD, due to the small volume fraction of it. A lot of hypereutectic carbides form initially from the liquid, followed by the eutectic reaction with the formation of ledeburite. Small proportion of austenite attributes to the small amount of transformed martensite. Residual austenite was always observed in laser treated steel layers or coatings due to the fast cooling rate [1, 11, 12]. Laser alloying is a fast heating and cooling process. Under non-equilibrium solidification, metastable phases are easily formed, such as the formation of M23C6 and Fe3(C, B), as the composition and thermal condition (gradient, cooling rate) of different areas varied enormously. Atoms (C, B) absorbing the energy offered by tempering started to diffuse, and stoichiometric carbide was formed in the tempered layer. Furthermore, residual austenite transformed into ferrite after a series of transformation. First, carbon in the martensite segregated. Second, martensite transformed into tempered martensite, with the precipitated tiny flaky carbides distributing in the martensite matrix. Third, residual austenite decomposed. Fourth, epsilon carbide dissolved into α phase and at the same time, cementite precipitated from α phase. Solubility of carbon in martensite decreased to the equilibrium composition of ferrite. Tempered martensite transformed into tempered troostite, with fine granular cementite distributing in ferrite. Fifth, cementites aggregated and grew up and ferrite recrystallized. These elaborations explained the existence of ferrite in the tempered layer.

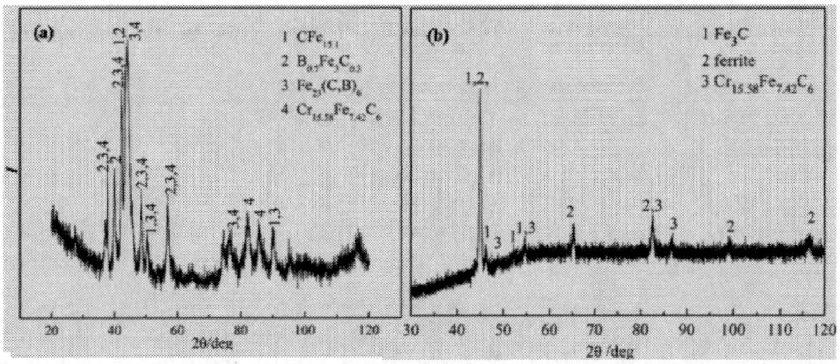

Figure 1. XRD pattern of (a) alloyed layer, (b) tempered layer.

The X-ray measurements allowed researchers to obtain important information about the surface state of material: the width of the diffraction peak at half the maximum value of diffraction (FWHM). This quantity is related to the grain distortion, to the dislocation density and to the so called type II micro residual stresses [13]. It is assumed as an index of the hardening of the material, which indicates that, the larger the FWHM is, the harder the measured surface. FWHM values of the first primary peak in the alloyed and tempered layer were 0.5196 and 0.2558 with the 2 theta position at 43.4832 and 44.8195 deg. Peak positions shifted to a relatively large Bragg's angle after tempering, indicating a decrease in lattice parameters. Narrowing of diffraction peaks testified a decrease of defects in structure and high strain values after laser treatment [14]. Furthermore, dense dislocations were reported in broadened austenite in laser melted layer. The narrowing of diffraction peaks testified a decrease of defects in structure and high strain values after tempering. This confirmed that stress release reduced the solid solubility of dissolved alloying elements kept in the solid solution after laser alloying due to the precipitation of carbides. Colaco et al reported a reduction of the super-saturation of martensite induced by post heat treatment followed by laser surface melting [15], which was similar to our results.

3.2 Microstructure analysis

Microstructure of the tempered layers is shown in Fig. 2. It is composed of hypereutectic carbides and eutectic ledeburite. However, compared with those in Fig. 2 [10], the heat and oxidation effect are observable.

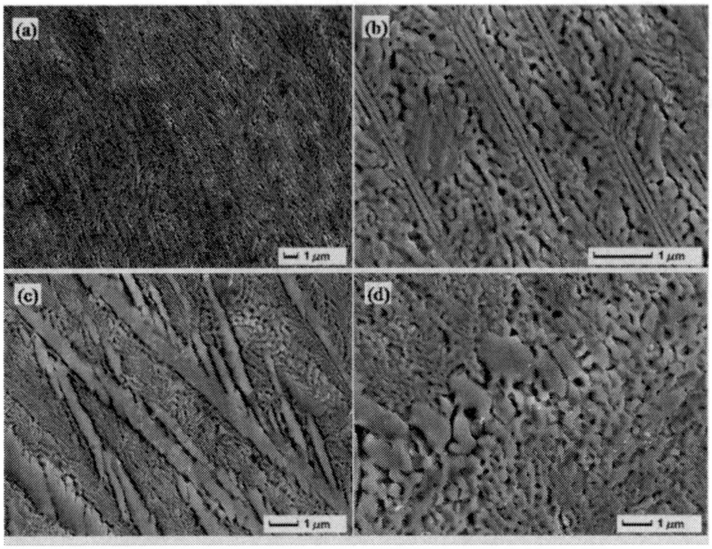

Figure 2. Microstructure of the tempered layer.

3.3 Microhardness analysis

Microhardness distribution in the tempered layer is shown in Fig. 3 . The average value decreased from 1001 HV0.05 of the alloyed layer to 558 HV0.05 of the tempered layer, decreasing by 44%. The average value decreases from 656 HV0.05 of the roll substrate to 410 HV0.05 of the tempered substrate, decreasing by 37.5%. Heat treatment decreased the surface hardness of the shot-peened nitride steel due to the residual stress release [16]. Tempering decreased residual stress and hence the microhardness in this investigation. In addition, diffusion of carbon and precipitation of carbides decreases the degree of the super-saturation of solid solution, resulting in the decrease of microhardness.

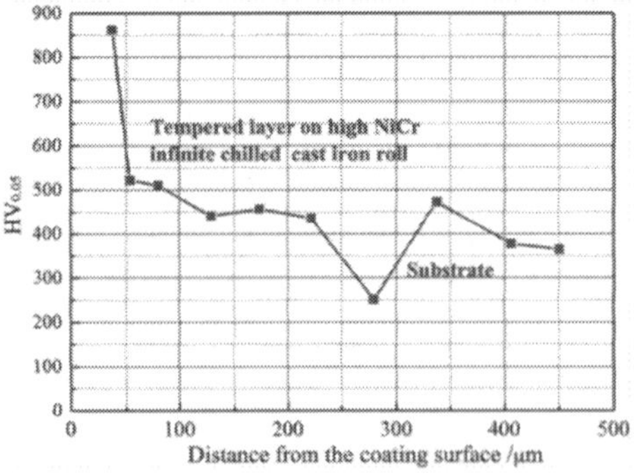

Figure 3. Micro-hardness distribution in tempered layer.

3.4 Friction and wear behavior

Friction coefficients of the tempered layer and substrate were close after sliding for 3600s as shown in Fig. 4. The running-in time for the tempered layer and substrate were 80s and 300s, respectively. The running-in wear mechanisms were very complicated, simultaneously accompanied by plough, adhesive, fatigue and oxidative wear, etc. It was fast for the tempered layer to run into steady wear process.

Contour curves of the cross-section of the wear tracks of the tempered layer and substrate are shown in Fig. 5. Wear volume losses were 1.735 ×10-3 and 1.92×10-3 mm3, respectively. Relative wear resistance of the tempered layer was 1.11 times that of the substrate.

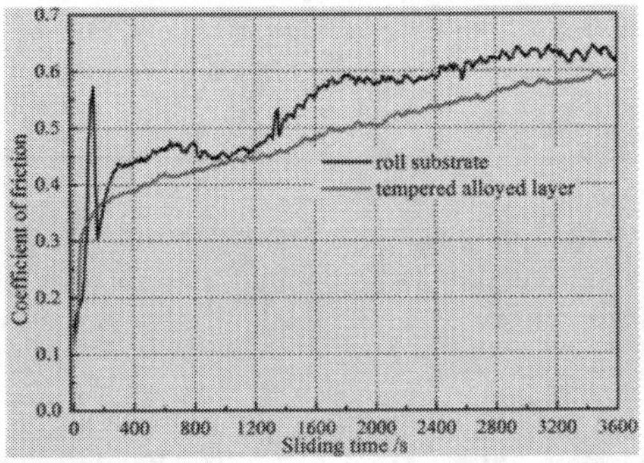

Figure 4. Friction coefficients of the roll substrate and tempered layer.

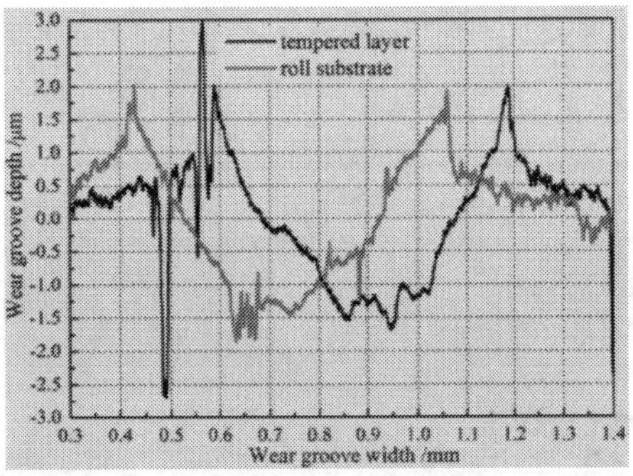

Figure 5. Contour curves of the cross-section of the wear tracks of the roll substrate and tempered layer.

Worn surface morphology of the tempered layer is shown in Fig. 6(a) and (b). Thick oxide films covered the tempered layer. No ploughing grooves can be seen on it. However, spalling of oxide films were observed due to the shear force during wear process. Shallow ploughing grooves, detached wear debris, and shrouded oxide films can be seen on the worn surface of tempered substrate shown in Fig. 6(c) and (d), indicating abrasive and adhesive wear. The surface films were the results of complex physical and chemical processes, including material transfer, debris generation and oxidation. These films have unique lubrication and electrical properties [17, 18]. The temperature rise caused by heating increased the wear of the material and favors oxidation [19, 20]. These oxide films served as the interface between the friction pair, avoiding the formation of large friction heat and energy between two sliding metallic components. However, the integrity and the bonding strength with the matrix of oxide films was one important factor to determine the wear resistance of matrix [21].

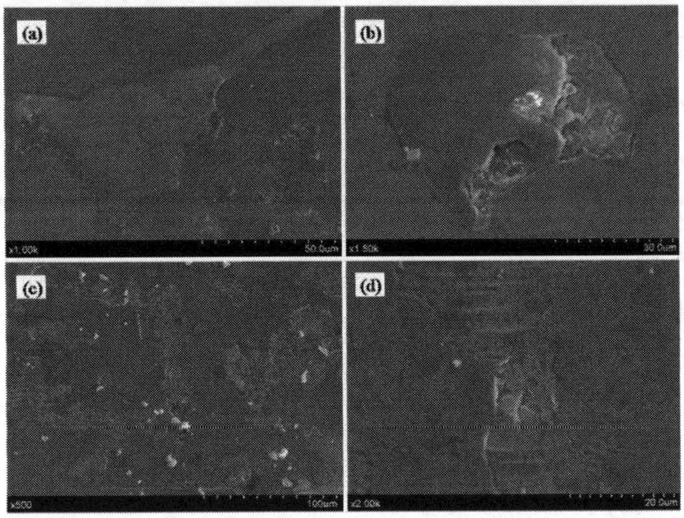

Figure 6. Surface morphology of (a), (b) the tempered layer, (c), (d) roll substrate.

4. Conclusions

Effect of tempering at 500℃ on the microstructure, phases, FWHM of peaks, microhardness and wear behavior of laser alloyed layer on high-Ni-Cr infinite chilled cast iron roll was investigated. The conclusions are as follows:

(1) Ferrite was formed in the tempered layer.
(2) Metastalbe carbides changed to stoichiometric carbides after tempering.
(3) FWHM (full width at half maximum) of the first primary peak was decreased by tempering.
(4) Microhardness decreased by 44%.
(5) Friction coefficient of the tempered layer exhibited gentle increase and a stable value of 0.6 was achieved for both the tempered layer and substrate.
(6) Relative wear resistance of the tempered layer was1.11 times that of the tempered roll substrate.
(7) Stress release, decreased solid solubility, improved toughness and easy formation of oxide films were positive to the wear resistance of the tempered layer.

Acknowledgments

Financial supports from the National Science Foundation of China (No. 51201070), Doctoral Fund of Ministry of Education of China (No. 20113227120006), Natural Science Foundation of Jiangsu Province of China (No. BK2012713), Special Research Foundation of Young teachers of Jiangsu University of Science and Technology (112110122) and the University Science Research Project of Jiangsu Province (11KJB460001) are acknowledged.

References

[1] Sun G F, Zhang Y K, Liu C S, Luo K Y, Tao X Q and Li P 2010 *Mater. Des.* **31** 2737-44
[2] Sun G F, Zhou R, Li P, Feng A X and Zhang Y K 2011 *Surf. Coat. Technol.* **205** 2747-54
[3] Zeinberger K H and Windhager M 2003 45th MWSP Conf. Proc. (Warrendale: Iron & Steel Society) P133-42
[4] Masanta M, Shariff S M and Roy Choudhury A 2011 *Wear* **271** 1124
[5] Jagdheesh R, Mudali U K and Nath A K 2007 *Surf. Eng.* **23** 93-8
[6] Sharma G, Awasthi R and Chandra K 2010 *Intermetallics* **18** 2124-27
[7] Klachuk S and Bamberger M 2010 *Mater. Sci. Tech.* **26** 1059-67
[8] Xin X Z, Chen J, Xiang N, Gong Y and Wei B 2014 *Dent. Mater.* **30** 263-70
[9] Takaichi A, Suyalatu, Nakamoto T, Joko N, Nomura N, Tsutsumi Y, Migita S, Doi H, Kurosu S, Chiba A, Wakabayashi N, Igarashi Y, Hanawa T 2013 *J. Mech. Behav. Biomed.* **21** 67-76
[10] Sun G F, Liu C S, Tao X Q and Chen S Y 2008 *J. Northeast. Univ. (Nat. Sci.)* **29** 845-8
[11] Colaco R and Vilar R 1998 *J. Mater. Sci. Lett.* **17** 563-7
[12] Colaco R and Vilar R 2005 *Wear* **258** 225-31
[13] Noyan I C and Chen J B. 1987 *Residual Stress Measurement by Diffraction and Interpretation* (New York: Springer-Verlag)
[14] Burg M Van Den and Hosson J Th M De 1993 *Acta Metall. Mater.* **41** 2557-64
[15] Colaco R, Gordo E, Ruiz-Navas EM, Otasevic M and Vilar R 2006 *Wear* **260** 949-56
[16] Pariente I F and Guagliano M 2008 *Surf. Coat. Technol.* **202** 3072-80
[17] Zaidi H, Chin K J and Frene J 2001 *Surf. Coat. Technol.* **148** 241-50
[18] Hu Z L, Chen Z H and Xia J T 2008 *Wear* **264**:11-7
[19] Ding T, Chen G X, Zhu M H, Zhang W H and Zhou Z R 2009 *Wear* **267** 1080-6
[20] Yi F, Zhang M and Xu Y 2005 *Carbon* **43** 2685-92
[21] Zhang Q, Jiang Z Y, Wei D B, Zhu H T, Chen Z X, Han J T and Xie G L 2013 *Wear* **301** 598-607

IOP Conf. Series: Materials Science and Engineering **62** (2014) 012016 doi:10.1088/1757-899X/62/1/012016

An experimental investigation on the three-point bending behavior of composite laminate

Azzam A [1, 2] and Li W [1]

[1]College of Textiles, Donghua University, Shanghai 201620, China
[2]Sudan University Science and Technology, School of Engineering and Technology Industries, Department of the Textile Engineering, Sudan

E-mail: liwei@dhu.edu.cn

Abstract. The response of composite laminate structure to three-point bending load was investigated by subjecting two types of stacking sequences of composite laminate structure by using electronic universal tester (Type: WDW-20) machine. Optical microscope was selected in order to characterize bending damage, delamination, and damage shapes in composite laminate structures. The results showed that the $[0/90/-45/45]_{2s}$ exhibits a brittle behavior, while other laminates exhibit a progressive failure mode consisting of fiber failure, debonding (splitting), and delamination. The $[45/45/90/0]_{2s}$ laminate has a highly nonlinear load–displacement curve due to compressive yielding.

1. Introduction

Composite materials have been competitive alternatives to traditional metallic materials for a while, due to their lower density, higher stiffness, higher strength, and better fatigue resistance when compared to steel or aluminum. Such properties enable composites to be an ultimate candidate for structural applications in aerospace and automotive products [1]. In order to understand more fully the process of damage initiation and propagation in carbon fiber composites, a series of tests have been performed by using a three-point beam test. Beam tests have an advantage over plate tests in that they allow a physical observation of the damage evolution process and information is not restricted to the post test examination of the failed specimens [2]. Quasi-static and dynamic loading of composite laminates result in complex damage mechanisms, among which multiple delamination and interfacial fractures are dominant ones. Damage initiation and growth in composite structures subjected to bending loads have been studied by numerous researchers, using various FE models [3]. The use of a three-point support flexural test to predict the stiffness of anisotropic composite plates in bending was studied by J.P. Nunes et al, and the results show that the flexural behavior of the composites depends on several factors, such as fiber orientation, laminate stacking, surface waviness and molding temperature [4]. Three-point beam impact tests on T300/914 carbon fiber composites have been performed by S.R. Hallett, and the results show that the consistent values are elastic deflection to failure and hence peak impact force. Once the peak force has been reached, failure initiates by fiber fracture at the back face [5]. Several methods have been proposed for determining mechanical properties of laminated composite materials. Those methods consider several specimen configurations with different laminate arrangements subjected to static loading. In the case of unidirectional

IOP Conf. Series: Materials Science and Engineering **62** (2014) 012016 doi:10.1088/1757-899X/62/1/012016

laminated composites, the mechanical properties can be divided into two main categories: normal and shear properties. Concerning in-plane or intralaminar shear properties, there exist a variety of different shear test methods depending on the specimen geometry, the type of applied load and the laminate configuration [6]. Many studies has been carried out on bending properties of composite laminate by considering several factors, such as fiber orientation, laminate stacking, and manufacturing conditions [7-13]. In this study, three point bending of the laminates composite structure was investigated. Two types of stacking sequence were chosen for bending test, $[0/90/\text{-}45/45]_{2s}$ and $[45/45/90/0]_{2s}$. To characterize the bending loads in composite laminate structures, optical microscope was used. The load-displacement curves were obtained to characterize the failure mechanisms in the composite laminate by considering the effects of stacking sequences used.

2. Materials and experimental

2.1 Materials

Unidirectional Prepreg carbon fiber/epoxy produced by Toray Company was used to manufacture composite laminates.

2.2 Laminates Fabrication

Laminates fabrication, which consist of two different stacking sequences, was prepared by heat pressure composite process. The pressure was 2 MPa, the cure was at 80℃ for 30 min, and the post-cure was at 120℃ for 2 h. The dimension of the entire sample was 16 cm x 16 cm (length*width) Table 1 represents the calculated mechanical properties of the produced lamina.

Table 1. Mechanical properties of composite lamina

Property	value
Young modulus(longitudinal)	E_1= 135.0 GPa
Young modulus (transverse)	E_2 = 8.91 GPa
Shear modulus	G_{12} = 98 GPa
Poisson ratio	0.3
Layer thickness	t = 2 mm

The testing equipment used in the experiments included flat-panel vulcanizer, electronic universal tester (Type: WDW-20) machine (Figure 1), and optical microscope (Nikon eclipses E200), as listed in Table 2.

Table 2. Testing equipment

Number	Name	Model
1	Vulcanizing machine	XLB
2	electronic universal tester	WDW-20
3	Optical microscope	Nikon eclipses E200

Figure 1. Electronic universal tester (Type: WDW-20) machine

2.3 Three-point bending test

The three-point bending tests were conducted with an electronic universal tester (Type: WDW-20) machine. The applied velocity of the bending load was 2 mm/min. Figure 2 shows the load configuration for a beam in three-point bending test. Different stacking sequence laminate types were tested, firstly quasi-isotropic and secondly unbalance stacking sequence. Load–displacement plots were obtained for each test specimen, and three specimens of each type of composite laminate were tested. Table 3 shows the lay-up, number of ply and dimensions for each specimen. The dimensions and all the procedures of the three-point bending test were in accordance with standard ASTM D7264 / D7264M – 07 [14]. The flexural stress Q_f is given by the following equation:

$$Q_f = \frac{3FL}{2bh^2},$$

Table 3. Specimen characteristics and specifications

Laminate type	Specimen name	Lay-up	N of plies	Span L (mm)	Width b (mm)
Quasi-isotropic stacking sequence	Qu1	$[0/90/\text{-}45/45]_{2s}$	16	40	15
	Qu2	Same	16	40	15
	Qu3	Same	16	40	15
Unbalance stacking sequence	Un1	$[45/45/90/0]_{2s}$	16	40	15
	Un2	Same	16	40	15
	Un3	Same	16	40	15

(Qu means quasi-isotropic, Un means unbalance)

where

Q_f is the flexural stress, in megapascals (MPa);

F is the load in newton's (N);

L is the span, in millimeters (mm);

h is the thickness of the specimen, in millimeters (mm); and

b is the width of the specimen, in millimeters (mm).

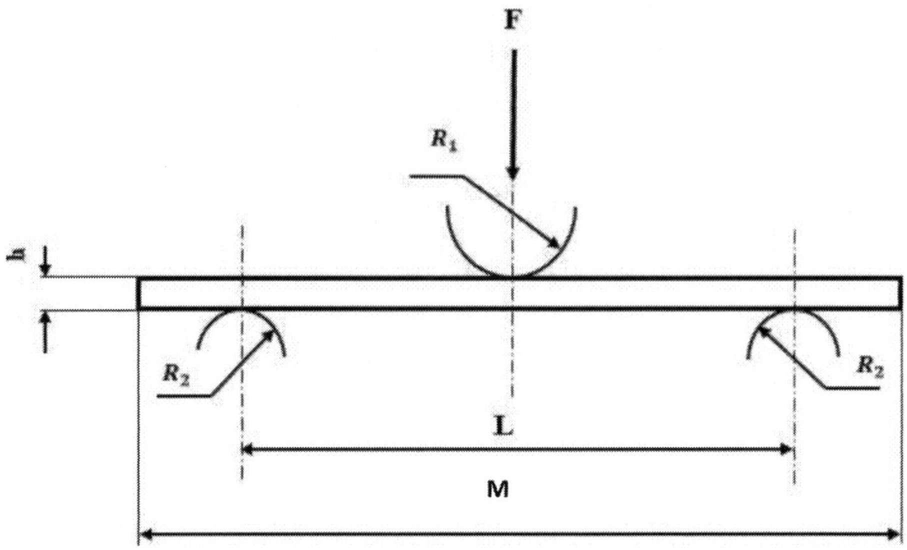

Figure 2. Load configuration for a beam in three point bending

In Figure 2, F = applied force, R_1= indenter radius, R_2= fixed support radius, h = specimen thickness, L = support span and M = specimen length.

3. Results and discussion

3.1 Three-point bending test

This section describes the experimental results after the various composite samples were broken under three-point bending (initial delamination, matrix crack, fiber crack). All results were plotted in terms of applied load versus center displacement of the sample under the crosshead of the electronic universal tester machine. All the samples for two stacking sequences had the same span length, thus making it possible to superimpose the load/displacement plots for each group of samples. This allowed a more accurate comparison of the resulting curves. The two stacking sequences were considered to evaluate the better stacking sequence and their mechanical properties $[0/90/\text{-}45/45]_{2s}$ call quasi-isotropic, $[45/45/90/0]_{2s}$call unbalance stacking sequence and s means symmetric. In each case, there is reasonably good correlation in the results. For these samples, the failure mechanism observations were reproducible.

Laminate: $[0/90/\text{-}45/45]_{2s}$

Figure 3 illustrates the load-displacement plots for laminate composite structures of $[0/90/\text{-}45/45]_{2s}$ consisting of 16 layers. The observations of the curve was bending stiffness to peak load due to no delamination observed and no oscillations before peak load. The curves can be divided into three regions. The first region, linear in appearance, can explain the elastic deformation of the composite laminate. The deformation represents when the displacement of the indenter was affected in the upper surface, due to the presence of some imperfections on the surface of the laminate. The second region after the load reached a peak value, significant drop of about 30–60 % of the peak load in the load-displacement curve was observed in composite laminate structures, and this sudden drop was due to the fiber cracking (Figure 4). After the crack initiated in tensile side, it propagated to the compressive side within all types of specimens before the final failure occurred. After the load drop, the specimen continued to sustain the load but never exceeded the previous peak load as only the reinforcement carried the load, which is reflected in the third region. In this region, plateau region was observed until reaching the final failure.

Laminate: $[45/45/90/0]_{2s}$

As seen in Figure 3, the behavior of composite laminate in this situation was bending stiffness, delamination and oscillations during the test until peak load in the laminate. This stacking sequence showed a little oscillations before peak load, which may result from vibrations of the supports, some defects in composite laminates, dents and delamination in the top and bottom faces of laminate composite structures, and initiation of damage in the specimens. The curves were divided into three regions. The first region, linear in appearance, can explain the elastic deformation of the composite laminate. The second region after the load reached a peak value, significant drop of about 20–50 % of the peak load in the load-displacement curve was observed in composite laminate structures, and this sudden drop was due to the fiber cracking and delamination. After the load drop, the specimen continued to sustain the load but never exceeded the previous peak load as only the reinforcement carried the load, which is reflected in the third region. In this region, plateau region was observed until reaching the final failure (Figure 5). The maximum fiber stress at failure on the tension side of a flexural specimen is considered the flexural strength of the material. Comparing the two types of stacking sequences used in this test, $[0/90/-45/45]_{2s}$ had superior advantages, such as high peak load, small damage area, and no delamination, to $[45/45/90/0]_{2s}$, due to different lay-ups, fiber orientations, and difference in their respective fiber modulus. The $[0/90/-45/45]_{2s}$ exhibited a brittle behavior, but other laminates exhibit a progressive failure mode consisting of fiber failure, debonding (splitting) , and delamination. The $[45/45/90/0]_{2s}$ laminate had a highly nonlinear load–displacement curve due to compressive yielding.

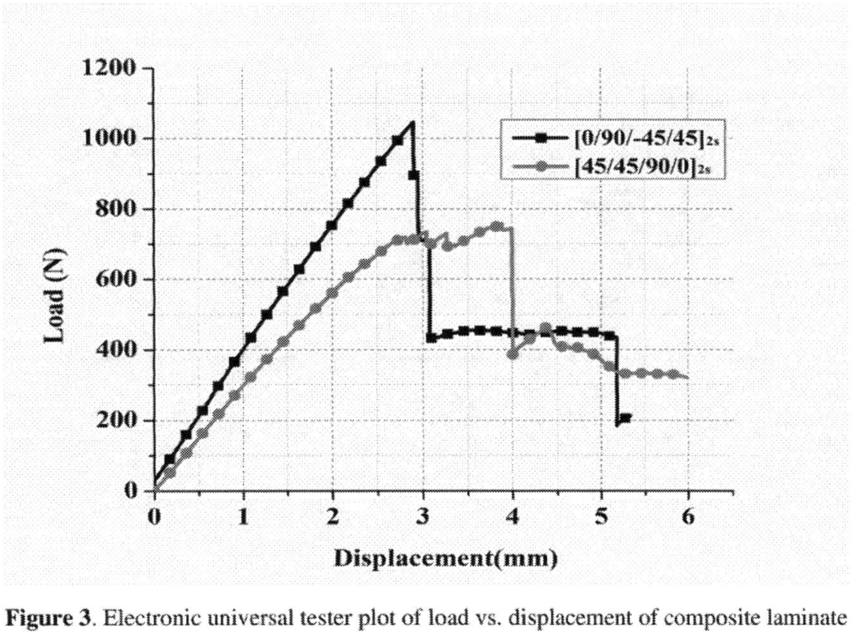

Figure 3. Electronic universal tester plot of load vs. displacement of composite laminate

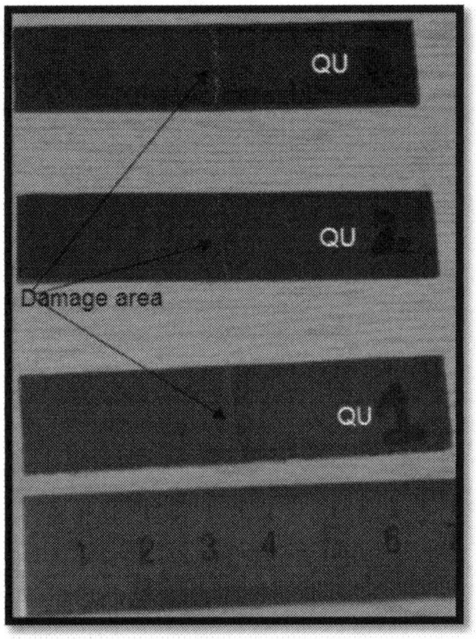

Figure 4. Crack shape on quasi-isotropic, Qu1, Qu2, Qu3: $[0/90/-45/45]_{2s}$

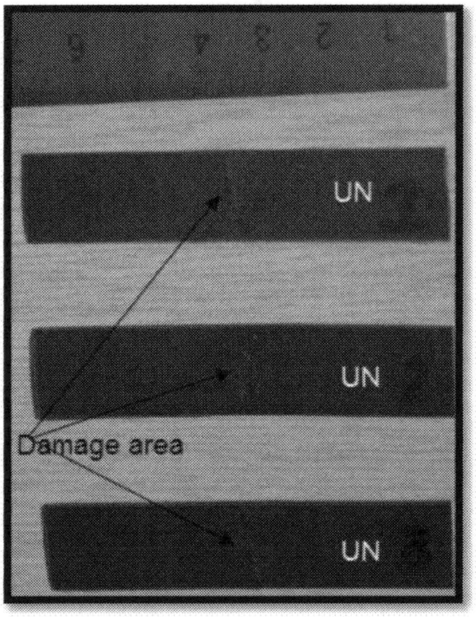

Figure 5. Crack shape on unbalance laminate, n1, Un2, Un3: $[45/45/90/0]_{2s}$

3.2 Optical microscopic images

Optical microscope results show shapes of the failures occurred in the samples, it was observed that the highly delaminated area occurred in $[45/45/90/0]_{2S}$ unbalance stacking sequence due to the orientation of the layer, and fiber orientation, as seen in Figures 6 and 7, which present the photos taken in transverse direction of the composite laminates for two types of the stacking sequences.

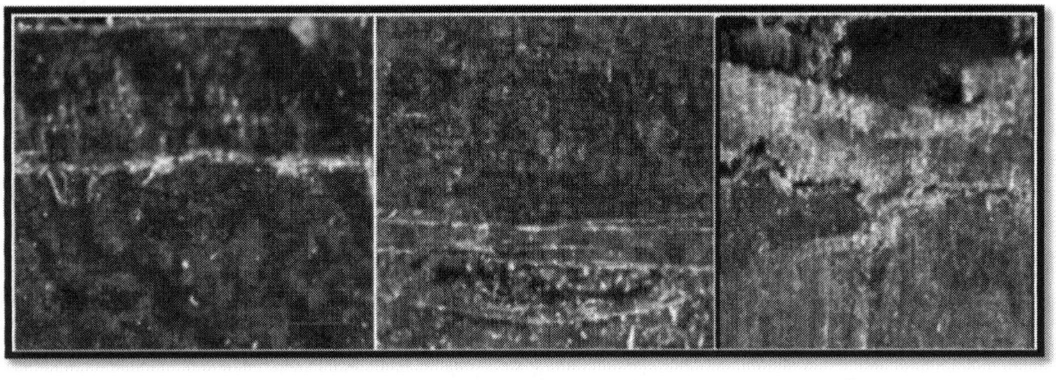

Figure 6. Optical microscope photos for Qu1, Qu2, Qu3: $[0/90/-45/45]_{2s}$

Figure 7. Optical microscope photos for Un1, Un2, and Un3: $[45/45/90/0]_{2s}$

As seen in Figure 6, the photos show a small delaminated area, small damage area, which almost has the same damage shape in upper surface of the laminate. However, in Figure 7, bigger delaminated area, big damage indentation, various damaged shapes, fiber failure, and debonding (splitting) are observed. These observations could be attributed to the important role and effects of fiber orientation on composite laminates.

4. Conclusion

Three point-bending test behavior of the composite laminates was investigated. Three point-bending results showed that $[0/90/-45/45]_{2s}$ exhibits a brittle behavior, but other laminates exhibit a progressive failure mode consisting of fiber failure, debonding (splitting), and delamination. The $[45/45/90/0]_{2s}$ laminate has a highly nonlinear load–displacement curve due to compressive yielding. Comparing the two types of stacking sequences used, $[0/90/-45/45]_{2s}$ in upper face has lower fail area, delamination, indentation in damage location, and damage size sensitivity than $[45/45/90/0]_{2s}$. This behavior is attributed to variation on stacking of the layers in composite laminate, fiber orientation, and mechanical property of lamina.

References

[1] Q Shen, M Omar and S Dongri Journal of materials science research, **1** (2012).
[2] S Hallett, C Ruiz and J Harding Composites science and technology, **59** (1999) 749-758.
[3] H Ullah, A R Harland, T Lucas, D Price and V V Silberschmidt Computational Materials Science, **52** (2012) 147-156.
[4] J Nunes, A Pouzada and C Bernardo Polymer testing, **21** (2002) 27-33.
[5] S Hallett Composites science and technology, **60** (2000) 115-124.
[6] G Vargas and F Mujika Polymer Testing, **30** (2011) 204-215.
[7] I D G Ary Subagia, Y Kim, L D Tijing, C S Kim and H K Shon Composites Part B: Engineering, **58** (2014) 251-258.
[8] G Belingardi and M P Cavatorta International Journal of Fatigue, **28** (2006) 815-825.
[9] N Carbajal and F Mujika Polymer Testing, **28** (2009) 618-626.
[10] I De Baere, W Van Paepegem and J Degrieck International Journal of Fatigue, **31** (2009) 1095-1101.
[11] K Fujihara, Z M Huang, S Ramakrishna and H Hamada Composites Science and Technology, **64** (2004) 2525-2534.
[12] Y Tomita, K Morioka and M Iwasa Materials Science and Engineering: A, 319–321 (2001) 679-682.
[13] G Vargas and F Mujika Polymer Testing, **30** (2011) 204-215.
[14] Standard test method for flexural properties of polymer matrix composite materials. ASTM 2007, D 7246M-2007.

IOP Conf. Series: Materials Science and Engineering **62** (2014) 012017 doi:10.1088/1757-899X/62/1/012017

Preparation and Properties of Graphene Oxide Modified Nanocomposite Hydrogels

Sihang Liu[1], Mei Huang[2*]

Department of Chemical & Biochemical Engineering, Zhejiang University, Hangzhou, China

E-mail: [1]igorbarton@zju.edu.cn, [2]huangm@zju.edu.cn

Abstract. Nanocomposite hydrogels with graphene oxide as chemical cross-linker were synthesized after graphene oxide being pretreated by methacryloyl chloride. Moreover, the mechanical behavior of nanocomposite hydrogels based on acrylamide (AAm) and graphene oxide (GO) was studied with different compositions. Experimental results of the swollen state properties of the nanocomposite hydrogels indicated that the addition of GO could effectively enhance the strength but lowers the swelling degree of nanocomposite hydrogels.

1. Introduction

The design and preparation of hydrogels have attracted a great deal of interest in biomedical engineering and biomaterials science because of their tunable chemical and physical structures, high water content and biocompatibility. Since hydrogels are prepared from hydrophilic polymer matrices that are cross-linked by physical or chemical cross-linking, hydrogen bonds, crystallized domains, hydrophobic interactions and chemical cross-linkers have been utilized for the synthesis of different types of hydrogels.

In recent years, nanomaterial has been explored for the preparation of nanocomposite hydrogels. For example, Nano materials, such as Laponite [1-5] and HNTs [6], have been used as the basic framework to obtain enhanced mechanical strength. And with the addition of multiple walls nanotube [7], hydrogels show increased electrical properties with 1.04 ms/cm.

In this study, the preparation of nanocomposite hydrogels based on acrylamide and graphene oxide was investigated by using methacryloyl chloride as pretreated agent for graphene oxide. And the properties of the nanocomposite hydrogels were characterized by swelling degree and mechanical tests.

2. Experimental

1.1. Chemicals

Graphene oxide was prepared by a modified Hummers method [8]. Acrylamide (AAm), N,N'-methylene bisacrylamide (BIS), Potassium persulfate (KPS), Dimethyl sulfur dioxide (DMSO), isopropanol, HCl and triethylamine were supplied by Aladdin Industrial Co. (Shanghai, China). Methacryloyl chloride (purity 99%) was purchased from Sun Chemical Technology (Shanghai, China) Co. Ltd. and used as received.

IOP Conf. Series: Materials Science and Engineering **62** (2014) 012017 doi:10.1088/1757-899X/62/1/012017

1.2. Synthesis of modified graphene oxide

After 0.5 g of GO was dispersed in 200 ml of DMSO by ultrasonic treatment for 10 min, 4.8 ml of triethylamine was gradually added and the mixture was stirred for another 30 min. Subsequently, 3.4 ml of methacryloyl chloride was added dropwise in an ice-salt bath at 0℃. After stirring for 2 h, the temperature of the suspension was raised to 45℃, which was maintained for another 24h. Then, the suspension was further diluted by adding 100 ml of isopropanol. The mixture was filtered and washed with HCl and deionized water until the pH was 7. Finally, the resulting solid was dried under freeze drying at -50℃ for 48 h to obtain modified graphene oxide. The as-prepared graphene oxide was named as GOM.

1.3. Synthesis of nanocomposite hydrogels

0.3 g of GOM in 15 ml of deionized water was ultrasonically dispersed for 10 min in ultrasonic cleaning bath at room temperature. 1.8 g of AAm monomer was added with stirring and maintained there for another 10 min. Then, the initiator (0.5% KPS) was added into the mixture and bubbled with high purity argon for 10 minutes. The resultant sample solution was poured into the space between two glass plates with a 2 mm silicone rubber as a spacer. Heat-initiated polymerization was carried out at 60 ℃ for 24 h.

For comparison, AAm solution with BIS as cross-linker was also prepared with the same initial monomer and initiator concentration.

1.4. Characterization and measurements

The water content of nanocomposite hydrogels was evaluated in terms of the swelling ratio. The swollen nanocomposite hydrogels were cut into small pieces and weighed at regular intervals until equilibrium was reached. Then, these hydrogels were carefully deswollen in a series of water-alcohol mixture with increasing alcohol contents until all samples were dried under vacuum to constant weight. The swelling degree (SD) of the nanocomposite hydrogels was determined by the equation:

$$SD = \frac{m_t - m_o}{m_t} \times 100\% ,$$

where mt and m0 are the weights of swollen and dry hydrogels, respectively.

The compressive stress-strain measurements were performed to the equilibrium swollen hydrogels, using a tensile-compressive tester (Regger 3010, Orientec Co.). The average values of stress-at-break and strain-at-break of the nanocomposite hydrogels were then determined from the mechanical stress data.

3. Results and Discussion

3.1 Graphene oxide modified nanocomposite hydrogels

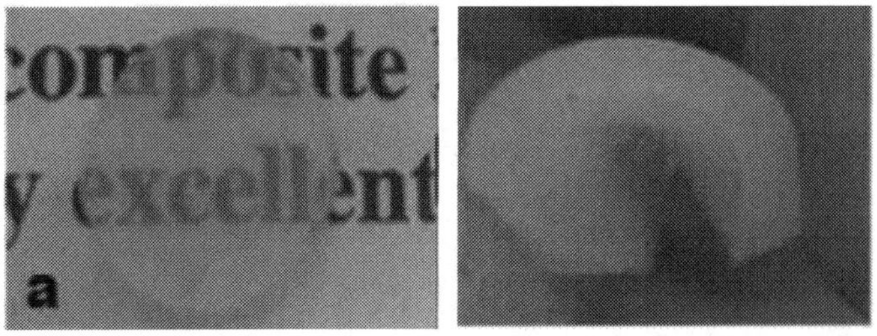

IOP Conf. Series: Materials Science and Engineering **62** (2014) 012017 doi:10.1088/1757-899X/62/1/012017

Figure 1. Photographs of nanocomposite hydrogels and PAAm hydrogels.(a) PAAm hydrogels (b) graphene oxide modified nanocomposite hydrogels.

Nanocomposite hydrogels based on AAm and GO were investigated by using methacryloyl chloride as pretreated agent for graphene oxide. Since GO sheet is a single-layer graphite sheet with several hydrophilic oxygenated functional groups, the carboxyl groups were chosen to interact with methacryloyl group in the pretreated reaction. It is thought that the presence of the double bond on GO sheets can act as a good chemical cross-linker for the synthesis of different types of hydrogels. In the GO-based hydrogels photographs shown in Figure 1, it can be seen that graphene oxide modified nanocomposite hydrogels are black with good flexibility and recoverability.

3.2 Swelling property of nanocomposite hydrogels

In order to have insights into the water transport process through the GO-based hydrogels, the dependence of the swelling degree on the increase of time was given in Figure 2. It can be seen that comparing with hydrogels using BIS as cross-linker, graphene oxide modified nanocomposite hydrogels exhibit a decreased swelling capacity in pure water. Moreover, the maximum values of swelling degree for graphene oxide modified nanocomposite hydrogels are only about 80%.

3.3 Compression tests

To explore the effect of nanoparticles on mechanical strength of hydrogels, the Stress-at-break of hydrogels was plotted against different cross-linkers of hydrogles, respectively (Figure 3). It is clear that with the addition of graphene oxide as cross-linker, nanocomposite hydrogels exhibit the toughness without obvious fracture even at 90% strain under increased compresses, while BIS crosslinked PAAm hydrogels fractured with the stress of 25%. It seems that in the nanocomposite hydrogels, graphene oxide not only works as the chemical cross-linker, but also greatly improves the elasticity of hydrogels by tangling with PAAm chain.

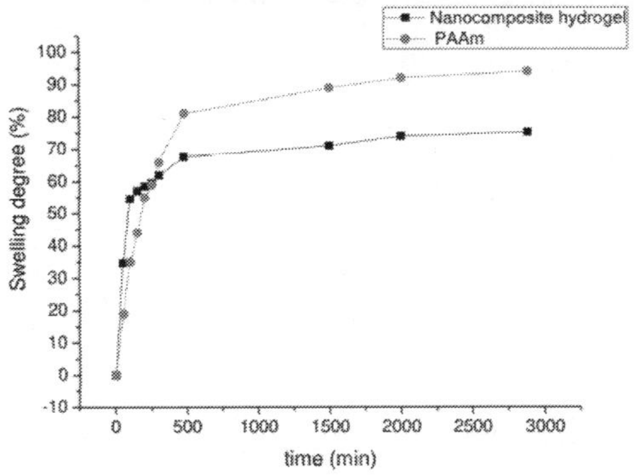

Figure 2. Swelling degree vs. time on nanocomposite hydrogels

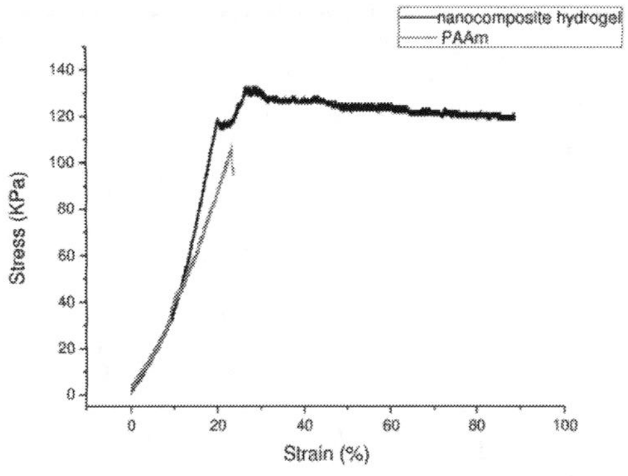

Figure 3. Influence of graphene oxide on the stress-at-break of nanocomposite hydrogels and PAAm hydrogels

4. Conclusion

Nanocomposite hydrogels cross-linked by graphene oxide was synthesized by one-step process. Experimental results of nanocomposite hydrogels based on AAm and GO indicated that graphene oxide could be modified and used as chemical cross-linker. Meanwhile, in the compression tests, the results that no obvious fracture appeared even at 90% strain indicated that the existence of graphene oxide might contribute to the physical entanglement of unbonded PAAm chain, which greatly improves the elasticity of hydrogels.

Acknowledgements

This work was financially supported by the Innovation Program of Zhejiang Environmental Protection Bureau (2011B09), the Program for Zhejiang Leading Team of S&T Innovation (2012).

References

[1] T. Huang, P(NIPAM-co-AA)/clay nanocomposite hydrogels exhibiting high swelling ratio accompanied by excellent mechanical strength, Appl Phys A. 107 (2012) 905–909.

[2] S. Sharifi, S.B.G. Blanquer, T.G. van Kooten, D.W. Grijpma, Biodegradable nanocomposite hydrogel structures with enhanced mechanical properties prepared by photo-crosslinking solutions of poly(trimethylene carbonate)–poly(ethylene glycol)–poly(trimethylene carbonate) macromonomers and nanoclay particles, Acta Biomaterialia. 8 (2012) 4233–4243.

[3] Y.R. Wang, Y.F. Dai, J.H. Gong, J.H. Ma, Preparation and properties of PDMAA/clay nanocomposite hydrogel, Journal of Donghua University(Natural Science). 37 (2012) 671-676.

[4] C.J. Wu, A.K. Gaharwar, B.K. Chan, G. Schmidt, Mechanically tough Pluronic F127/laponite nanocomposite hydrogels from covalently and physically cross-linked networks, Macromolecules. 44 (2011) 8215–8224.

[5] P. Zhang , X.Y. Wang , S.P. Li , H.L. Dai, Nanocomposite hydrogels with high mechanical strength and high swelling ratio by RAFT polymerization, International Journal of Polymeric Materials, 62: 10–16

[6] M.X. Liu ,W.D. Li , J.H. Rong , C.R. Zhou, Novel polymer nanocomposite hydrogel with natural

[7] clay nanotubes, Colloid Polym Sci 290 (2012) 895–905.

[8] A.F. Zhu , Z. Shi , J.H. Jin , G. Li , J.M. Jiang, Synthesis and properties of polyacrylamide-based conducting gels with enhanced mechanical strength, Journal of Macromolecular Science R, Part B: Physics. 51 (2012) 2183–2190.

[9] Y.W. Huang, M. Zeng, J. Ren, J. Wang, L.R. Fan, Q.Y. Xu, Preparation and swelling properties of graphene oxide/poly(acrylic acid-co-acrylamide) super-absorbent hydrogel nanocomposites, Colloids and Surfaces A: Physicochem. Eng. Aspects. 401 (2012) 97– 106.

IOP Conf. Series: Materials Science and Engineering **62** (2014) 012018 doi:10.1088/1757-899X/62/1/012018

3D printing of textile-based structures by Fused Deposition Modelling (FDM) with different polymer materials

R Melnikova[1], A Ehrmann[1] , K Finsterbusch[1]
[1] Faculty of Textile and Clothing Technology, Niederrhein University of Applied Sciences, 41065 Mönchengladbach, Germany

E-mail: andrea.ehrmann@hsnr.de

Abstract. 3D printing is a form of additive manufacturing, i.e. creating objects by sequential layering, for pre-production or production. After creating a 3D model with a CAD program, a printable file is used to create a layer design which is printed afterwards. While often more expensive than traditional techniques like injection moulding, 3D printing can significantly enhance production times of small parts produced in small numbers, additionally allowing for large flexibility and the possibility to create parts that would be impossible to produce with conventional techniques. The Fused Deposition Modelling technique uses a plastic filament which is pushed through a heated extrusion nozzle melting the material. Depending on the material, different challenges occur in the production process, and the produced part shows different mechanical properties. The article describes some standard and novel materials and their influence on the resulting parts.

1. Introduction

Rapid prototyping and rapid manufacturing systems are often based on solids, e.g. in the form of thin laser-cut layers sequentially bonded to each other, inkjet-like 3D printing with thermoplastics, selective laser sintering (SLS) from a powder, or the fused deposition modelling (FDM) process firstly established by Stratasys [1]. Most inexpensive 3D printers, developed for private use or small companies trying their first steps in the area of 3D printing, use the FDM process.

As a preparation for 3D printing, a geometric model created by CAD software is exported as STL (Surface Tesselation Language), OBJ (Wavefront) or – depending on the CAD program and the printer – other interchange formats and sliced afterwards (i.e. split into a number of layers). During the construction, care must be taken that the 3D model is "watertight" (i.e. contains no holes), manifold (i.e. edges must not be shared between more than two faces), and that the surface normals point outwards [2]. Corrections of the respective errors can be performed, e.g., by "netfabb" [3].

Depending on the printing process, such as SLS or FDM, additional requirements have to be fulfilled. For FDM models, it is important to take into account the position of the model in the printer – with the largest and necessarily most exact plane at the bottom –, the problem of strongly tilted faces (> 45°) necessitating support material, or the infill pattern inside a closed object. For SLS printing, instead, large volume models should contain so-called escape holes through which the polyamide powder can be removed after printing. Additionally, spaces between moving parts must be large enough to allow for the powder to flow away when the design is taken out of the 3D printer. In both processes, care must be taken that wall thicknesses are large enough, especially if the model is planned to be polished afterwards.

2. Experimental

While 3D printers are already used in several fields by companies and individuals, printing of textile-like patterns is still rare. In technical textiles, 3D printing has already been used to create applications such as flexible heating systems or wearable technology [4,5]. In design, garments such as bikinis [6], complete dresses [7] or shoes [8] are 3D printed. These garments, however, are clearly produced for show, not for everyday wearing. Thus, the project described here concentrates on more textile-like patterns which can be combined with traditional textiles or even replace them.
After designing different textile or textile-based structures in the free 3D graphics software Blender™, the models have been checked and – if necessary – repaired using netfabb. Afterwards, the repaired STL file has been imported in the "repetier host" and sliced into a layer model before the resulting file is sent to the 3D printer.
Most experiments depicted in this article have been performed using the FDM printer X400 produced by German RepRap with PLA (polylactic acid, elongation at break ~ 4 %, tensile modulus 1968 MPa [9,10]) or soft PLA (PLA + softener, elongation at break up to ~ 200 % [9]), BendLay (butadiene; harder than soft PLA, but not as brittle as ABS; impact strength 3kJ/m2, tensile modulus 1550 MPa [11]) and other traditional and experimental materials. ABS (Acrylonitrile butadiene styrene) has been excluded after first tests showed the insufficient mechanical properties of the resulting textile-based models [12].
For comparison with the FDM process, an SLS printer employed by Shapeways [2] has been used with the material "White Strong & Flexible" (nylon).

3. Results

First experiments aimed at reproducing textile structures, especially weft knitted single face structures, before the experiments concentrated on developing lace patterns and finally multi-material models. All structures were printed without using a base substrate, although 3D printing on flexible textile materials is principally possible, depending on material combinations and printer settings.

3.1. 3D printing of weft knitted structures

Creation of 3D printed weft knitted structures is based on the development of a CAD model (figure 1, upper left panel). For this, Blender™ has been used to import a model of a knitted stitch created in Adobe Illustrator, to duplicate the stitch and to add the 3D form by deforming the Beziér curve depicting the stitch. It must be taken into account that the material thickness, as well as the distance between neighbouring stitches, are large enough for the SLS or the FDM technology, respectively. With the program "netfabb", the printability of the design has been verified.
The developed CAD model has been used for 3D printing by SLS process (figure 1, upper right panel) with 0.8 mm material thickness and a minimum distance between the different stitches of 0.4 mm. The resulting model reproduces in principle the look of a single face weft knitted fabric; however, it should be mentioned that the lack of flexibility of the material itself leads to distinctly different mechanical properties of the model, compared with knitted structures created from traditional textile yarns. Additionally, the SLS process can normally only use one material per model; multi-material models have to be printed separately and joined afterwards.
For the FDM process, it is principally needed to use support structures – from a different or the same material as used for the model itself – to achieve the short distances between the stitches as depicted in figure 1 (upper left panel). As figure 1 (lower left panel) shows, however, small support structures from the same material are hard to realize. Additionally, the design had to be magnified to reach the minimum material thickness (here 1.88 mm) necessary to avoid breakage of the structures. Nevertheless, the support structures were still too fine to be produced by the printer as desired; instead undesirable clots were created which partly destroyed the model. In this way, it is not possible to build such a relatively fine model by FDM.
The same structure has been realized using soft PLA without support structures. Unexpectedly, this experiment gave a significantly better result (figure 1, lower right panel). While there are still fine

IOP Conf. Series: Materials Science and Engineering **62** (2014) 012018 doi:10.1088/1757-899X/62/1/012018

undesirable connections between the stitches, following the trace of the printing nozzle which sometimes extrudes material when this process should be stopped, the basic structure is clearly visible. The stitches are mostly separated, and the flexibility of the material itself leads to significantly more similar properties compared to a textile knitted structure than for the design produced by SLS.

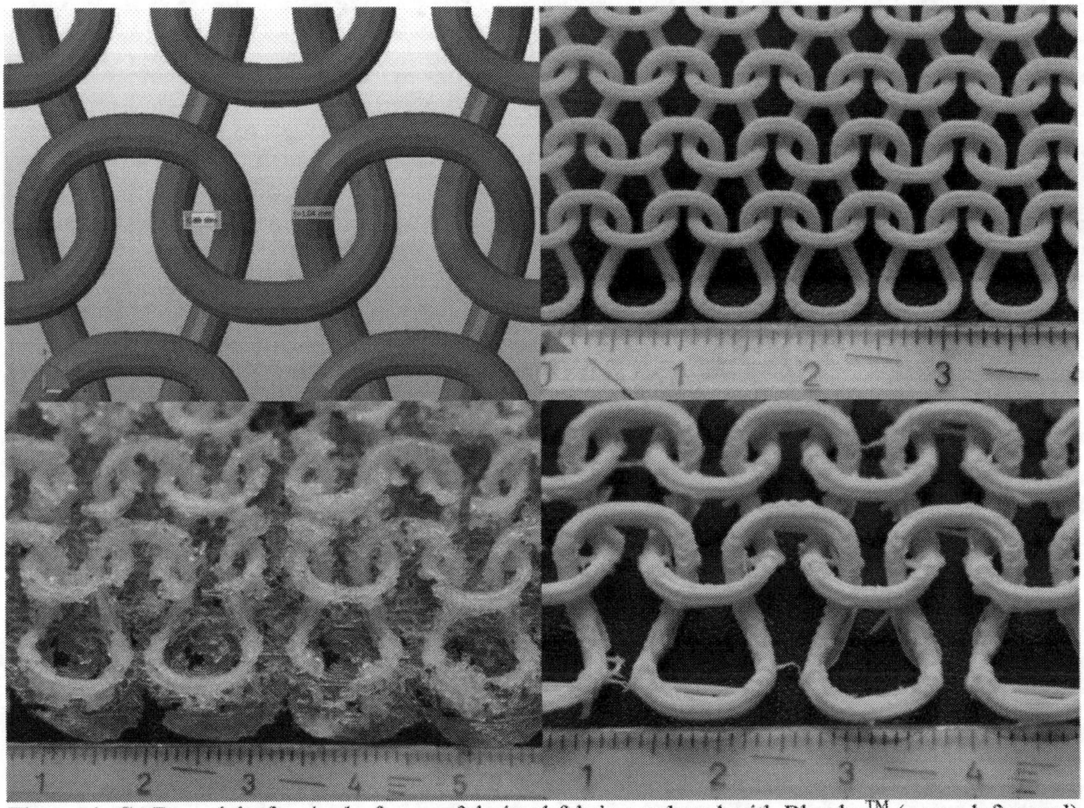

Figure 1. CAD model of a single-face weft knitted fabric produced with Blender™ (upper left panel), 3D printed with SLS (upper right panel), FDM with BendLay and support structures (lower left panel), and FDM with soft PLA without support structures (lower right panel). Dimensions depicted by ruler.

Nevertheless, it should be mentioned that the surface of the soft PLA model still exhibits roughness on a macroscopic scale, contrary to the microscopic roughness of man-made or natural textile fibres.
Nevertheless, these experiments already show that simple replication of common textile structures may not be the ideal way to produce 3D printed textile materials. Due to the layer-by-layer construction of the FDM technique, models that are built up from layers can be expected to be printed more successfully and with fewer problems occurring. Thus, the next sub-chapters deal with such thin layered structures.

3.2. 3D printing of layered structures

Several layered structures have been realized by FDM technology. Figure 2 shows a test structure, used to examine whether single strings can be deposited on top of relatively open structures without any support structures.

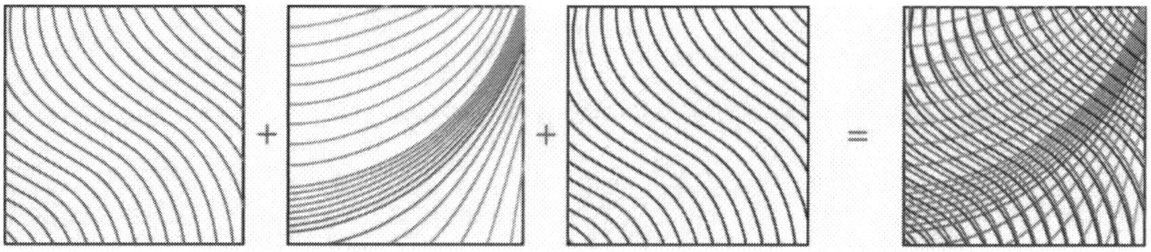

Figure 2. Test pattern for a layered structure, composed of three stacked layers.

The resulting FDM print (figure 3) shows that while this approach is mostly successful, the strings may sometimes break, even if a minimum diameter of 0.4 mm is maintained. Thus, as the next approach to create textile-based 3D-printed patterns, lace-like structures have been created which start with a partly open base layer, too, but do not include free-floating areas.

Figure 3. Resulting FDM printed 3-layer structure (dimension 8 cm x 8 cm).

3.3. 3D printing of lace patterns

Lace patterns have been created, inspired by the well-known Plauen lace, containing mostly floral and round elements on a base layer connecting these parts. Such a design is depicted in figure 4 (left panel). Due to the absence of free-floating areas, printing by the FDM process is unproblematic, if all connection lines have large enough diameters (figure 4, right panel).

Figure 4. Multi-layer lace pattern, depicted in "netfabb" (left panel), and the resulting 3D print (right panel). Detail dimension ~ 4 cm x 7 cm.

Additional to soft PLA, which has been used for most lace patterns to avoid possible mechanical problems with harder material resulting in breaks of the base connections, an experimental polymer has been used: Lay Tekkks, one of four different types of porous filament from the line PORO-Lay. Lay Tekkks is a hard filament, produced by Kai Parthy from CC-Products (Cologne), which is transported to the nozzle without the transport problems that too soft filaments may generate in several FDM printers. After finishing the printing process, however, it can be put into warm water for a period of minutes to hours, leading to the hard part of the material being dissolved and thus softening the resulting sample more and more, the longer it is immersed in water (figure 5).

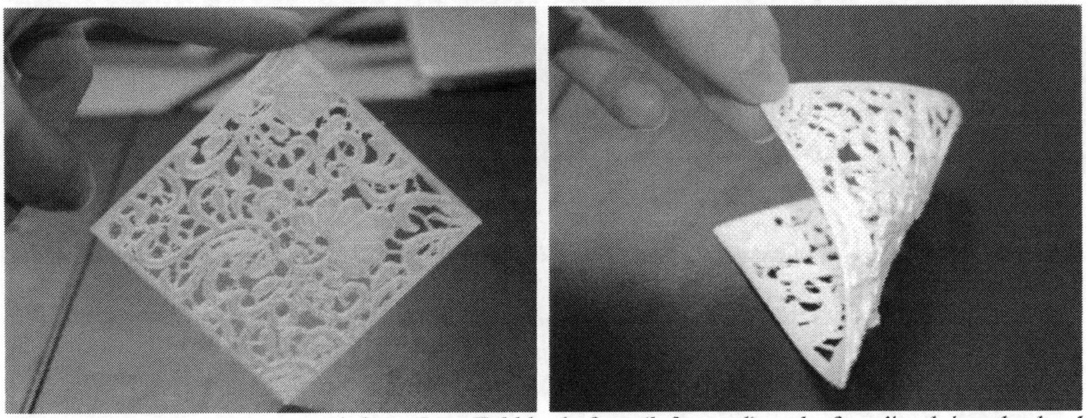

Figure 5. Lace pattern created from Lay Tekkks before (left panel) and after dissolving the hard material fraction in warm water for three hours (right panel). The dimension of the design is 8 cm x 8 cm.

3.4. 3D printing of multi-material models

FDM printers with two or more nozzles offer the possibility to mix different materials in one printed part. Since garments often contain harder parts, such as eyelets, buttons or other reinforcements, such material mixes have also been developed and printed.

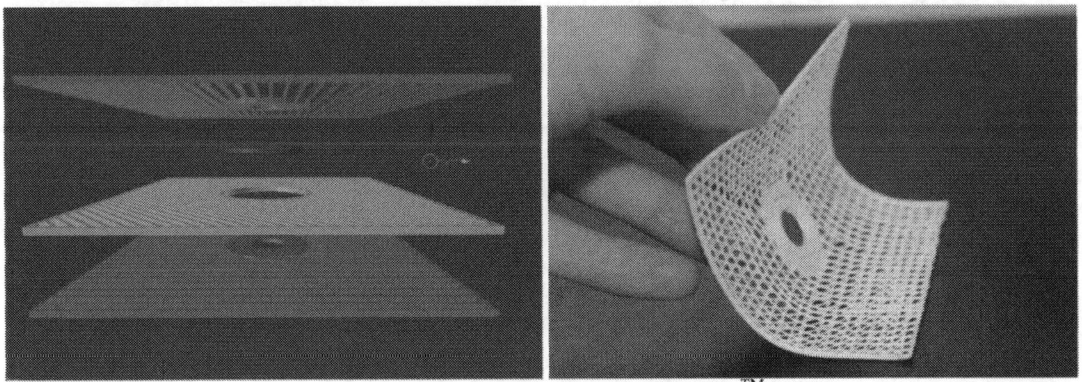

Figure 6. Multi-layer multi-material structure, depicted in Blender^{TM} as three separate layers plus separate ring from different material marked red (left panel), and the resulting two-material print with hard BendLay ring in soft PLA layer construction (right panel). The dimension of the design is 8 cm x 8 cm.

Figure 6 (left panel) shows exemplarily the construction of a three-layer base, planned to be printed with soft PLA, and a harder ring (depicted red) which fits in the gap of the soft middle layer, thus being partly fixed by the top and the bottom layer after finishing the design. In this way, it is indeed possible to create such a multi-material model (figure 6, right panel). Similarly, lace models with three or more layers of soft PLA can contain reinforced parts constructed from hard PLA or BendLay.

4. Conclusion

In a recent project, textile-based structures have been 3D printed using SLS and FDM technology. While ABS has turned out to be often too brittle for the desired fine structures, and hard PLA, as well as the nylon used in SLS printing, can be too hard to be used in typical textile applications such as garments, soft PLA – also in combination with less flexible materials such as BendLay – has proven to be able to reproduce some textile-based structures.

These materials enable the development of novel patterns and construction methods for clothing and related areas, allowing for new designs as well as new functionalities that cannot be achieved by conventional textile fabrics.

5. References

[1] Chua C K, Leong K F and Lim C S 2003 *Rapid Prototyping: Principles and Applications, 2nd Edition* (Singapore: World Scientific Publishing Co. Pte. Ltd.)

[2] Shapeways 2014 Things to keep in mind when designing for 3D printing http://www.shapeways.com/tutorials/things-to-keep-in-mind

[3] netfabb® – Software for 3D Printing http://www.netfabb.com/de/

[4] Toeters M, ten Bhomer M, Bottenberg E, Tomico O, Brinks G, Vincenzini P, Carfagna C 2013 Smart and interactive Textiles, Book Series: Advances in Science and Technology Vol. 80, pp. 112-117 (Switzerland: Trans Tech Publications Ltd.)

[5] Treadeway, C 2009 Smart Clothes and wearable technology, Book Series: Woodhead Publishing in Textiles, pp. 300-318 (England: Woodhead Publ. Ltd.)

[6] dezeen magagzine 2011 N12 3D-printed bikini by Continuum Fashion and Shapeways http://www.dezeen.com/2011/06/07/n12-3d-printed-bikini-by-continuum-fashion-and-shapeways/

[7] Francis Bitonti Studio 2014 Dita's Gown http://www.francisbitonti.com/ditas-gown/

[8] Mashable 2012 Bikinis? Shoes? This Duo Will 3D Print Your Wardrobe http://mashable.com/2012/07/17/3d-printing-continuum-fashion/

[9] Gérard T, Noto T and Budtova T 2012 11th European Symposium on Polymer Blends - Polyblends 2012, Donostia – San Sebastian / Spain

[10] Bijarimi M, Ahmad S, Rasid R 2012 International Conference on Agriculture, Chemical and Environmental Sciences (ICACES'2012) Oct. 6-7, 2012 Dubai (UAE)

[11] 1.75 mm Bendlay, MyThings4U.nl http://www.mythings4u.nl/product_info.php/manufacturers_id/12/products_id/83

[12] Passlack B, Ehrmann A and Finsterbusch K 2013 *Melliand Textilberichte* **94** 224

2014 Global Conference on Polymer and Composite Materials (PCM 2014) IOP Publishing
IOP Conf. Series: Materials Science and Engineering **62** (2014) 012019 doi:10.1088/1757-899X/62/1/012019

Experiment on the concrete slab for floor vibration evaluation of deteriorated building

S. U. Hong[1], J. H. Na[1], S. H. Kim[1] and Y. T. Lee*[1]

Department of Architectural Engineering, Hanbat National University, 125 Dongseodero, Yuseong-gu, Daejeon 305-719, Korea

E-mail: ytlee@hanbat.ac.kr

Abstract. Damages from noise and vibration are increasing every year, and most of which are noises between floors in deteriorated building caused by floor impact sound. In this study, the floor vibration of the deteriorated buildings constructed with the concrete slabs of thickness no more than 150 mm was evaluated by the vibration impact sound. This highly reliable study was conducted to assess floor vibration according with the serviceability evaluation standard of Reiher / Meister and Koch and vibration evaluation standard of ISO and AIJ. Designed pressure for the concrete slab sample of floor vibration assessment was 24MPa, and the sample was manufactured pursuant to KS F 2865 and JIS A 1440-2 with size of 3200 mm× 3200 mm × 140 mm. Tests were conducted twice with accelerometers, and Fast Fourier Transform was performed for comparative analysis by the vibration assessment criteria. The peak displacement from Test 1 was in the range of 0.00869 - 0.02540 mm; the value of peak frequency ranged from 18 to 27 Hz, and the average value was 22Hz. The peak acceleration value from Test 2 was in the range of 0.47 - 1.07 % g; the value of peak frequency was 18.5 - 22.57 Hz, and the average was 21Hz. The vibration was apparently recognizable in most cases according to the Reiher/Meister standard. In case of Koch graph for the damage assessment of the structure, the vibration was at the medium level and causes no damage to the building structure. The measured vibration results did not exceed the damage limit or serviceability limit of building according to the vibration assessment criteria of ISO and residential assessment guidelines provided by Architectural Institute of Japan (AIJ).

1. Introduction

The weights of the buildings at present were reduced in order to secure the large space by using high-strength lightweight concrete, design technology development, long span and high strength steel development. Therefore, the damping ratio and natural frequency of the building structure is decreasing. This phenomenon has caused the problem of dynamic load on the floor plate due to the vertical vibration. Thus, in the case of high-rise residential building, required to the effective standard for controlling the floor vibration that large space and long spanned is caused by the steel material, steel frame, high strength concrete [1]. For the modern architectures, the building spans are gradually increasing and the weight of the buildings become lighter with the development of structural analysis techniques and the improved strength of the conctruction materials. The rigidity of the floor plate reduces the natural frequency and damping capacity. to the problems of vibration comes from

[1] Department of Architectural Engineering, Hanbat National University, 125 Dongseodero, Yuseong-gu, Daejeon 305-719, Korea

'serviceability', which causes residents to feel anxious and uncomfortable (Ad Hoc Committee on Serviceability Research, 1986) [2]. The recent high-rise building has serviceability problems that caused by the wind-induced vibration acceleration. Therefore, accurate prediction of the dynamic characteristics is very important in the calculation of the reliability acceleration such as damping capacity and natural frequencies. In general, the natural frequency of the high-rise buildings can be calculated through the eigenvalue analysis at the design stage, but the prediction of the damping rate is difficult as the exact natural frequency is unpredictable. Detectable building vibration can cause anxiety of the residents. Such degree of building vibration that ordinary people cannot withstand is defined as 'serviceability limit state [3]. For the passengers of the automobiles and airplanes which have the moving frequency of 5-8 Hz, their bodies suffer the most unpleasant experience due to the fact that the natural frequencies of the human ribs and abdomen overlap with this frequencies and thus create resonance. A vibration may not damage the structure, but it can cause discomfort for people and disturb working process of the sensitive machines. (Ellingwood, 1989) [4]. In case of the horizontal member(beams, slabs) of the structure to induce the vertical is no problem in the structural safety in the design of the stress examine and limits by static deflection, it is possible for the residents to generate anxiety and discomfort due to the vibration. Buildings are exposed to vibrations on a daily basis, and the effect of vibrations on structural safety of buildings differs according to various factors such as amplitude and cycle of vibration. Also in general, there is a substantial difference between human perception and safety of building. In case of Korea, the standard of heavy-weight impact noise interruption performance for standard floor structures was reinforced because of the damages from the noises between floors, and the slab thickness of wall type and mixed structure buildings was increased from 150 mm and below to 210 mm and above [5]. Regular inspections are performed on buildings that using slabs with thickness of 210mm or above such as newly constructed apartment houses, but the old apartment houses that using slabs with thickness of 150mm and below has not been inspected yet. In this study, the floor vibration of the deteriorated buildings constructed with the concrete slabs of thickness no more than 150 mm was evaluated by the vibration impact sound. This highly reliable study was conducted to assess floor vibration according with the serviceability evaluation standard of Reiher / Meister and Koch and vibration evaluation standard of ISO and AIJ.

2. Methodology

Sources of vibration impact are generated in the form of shock wave and converted to elastic wave with amplitude and cycle. Here, the elastic wave generated inflicts damage on nearby structures and facilities. Ground vibration is described with displacement, velocity, acceleration and frequency. When a typical form of ground vibration is assumed to be a simple harmonic motion with constant cycle, the correlation shown in Eq. (1) - (4) is established among displacement, velocity and acceleration at maximum amplitude [10].

$$\mu = \mu_0 \sin(\omega t + \theta)\ ;\ \int \dot{\mu}\, dx \tag{1}$$

$$\mu = \frac{d\mu}{dt} = \omega \sin\left(\omega t + \theta + \frac{\pi}{2}\right);\ \dot{\mu} = \int \ddot{\mu} dx \tag{2}$$

$$\ddot{\mu} = \dot{d}\frac{\mu}{dt} = d^2 \frac{\mu}{dt^2} = \omega^2 \mu_0 \sin(\omega t + \theta + \pi) \tag{3}$$

$$\omega = 2\pi f = 2\pi\left(\frac{1}{T}\right) \tag{4}$$

Where, μ, μ_0 and θ are displacement, maximum displacement (mm) and phase angle (rad) respectively. $\dot{\mu}$ and $\ddot{\mu}$ are particle velocity (mm/sec, cm/sec, kine) and acceleration $(gal, cm/sec^2)$. t and ω are time (sec) and circular frequency (angular velocity). f and T are frequency (Hz) and +ycle (sec).

Since the absolute value of maximum particle motion is important in most situations related to characteristics of ground vibration, it can be represented as an absolute value as shown in the following two equations where $\mu max = \mu_0$, $\dot{\mu} max = v, \ddot{\mu} max = a$ (better have real equations here instead of an image). As they have a functional relationship, the variable of maximum particle velocity can be computed by Eq. (5) - (7) once vibration frequency is known and the vibrational wave is approximately a simple sinusoidal wave.

$$v = \omega\mu_0 = 2\pi f\mu_0 \; ; \; \mu_0 = \frac{v}{2\pi f} \tag{5}$$

$$\alpha = \omega^2\mu_0 = (2\pi f)^2\mu_0 = 2\pi f v \; ; v = \frac{a}{2\pi f} \tag{6}$$

Also, the relationship among wavelength (need a real symbol) λ, propagation velocity of elastic wave c, and cycle T was presented in Eq. (7).

$$\lambda = cT = c(\frac{1}{f}) \tag{7}$$

The functional relationship of the attenuation relation for the ground vibration was shown in Eq. (8).

$$V = V(D)^n \tag{8}$$

Where V is the ground vibration velocity (mm/sec, cm/sec) and D is the distance between station point and vibration source and n is the location constant by soil conditions.

3. Experimental

Designd pressure of the concrete slab sample for floor vibration assessment was 24MPa, and the sample was manufactured pursuant to KS F 2865 and JIS A 1440-2 with the size of 3200 mm×3200 mm×140 mm[12-13] . D13 rebar was arranged for 28 days of atmosphere curing. In this study, a 70kg adult male was allowed to jump for every second at the center of the floor to inflict impacts that were measured at different points. The tests were conducted twice with the accelerometers, and Fast Fourier Transform was performed for comparative analysis according with the vibration assessment criteria. The concrete slab sample was shown in Fig. 1. 1(a) was the sample of floor vibration and 1(b) was the measurement points. The conceptual diagram of the acceleration vibration measurement of Test 1 was shown in Fig. 2(a). 2(b) was the process of this experiment. Two tests were run to evaluate the respective measurements at each point with different standard. Test 1 was analyzed with standard of Koch and Reiher / Meister, Test 2 was analyzed with standard of ISO and AIJ. Fig. 3 showed the measured data from the accelerometer [6-9].

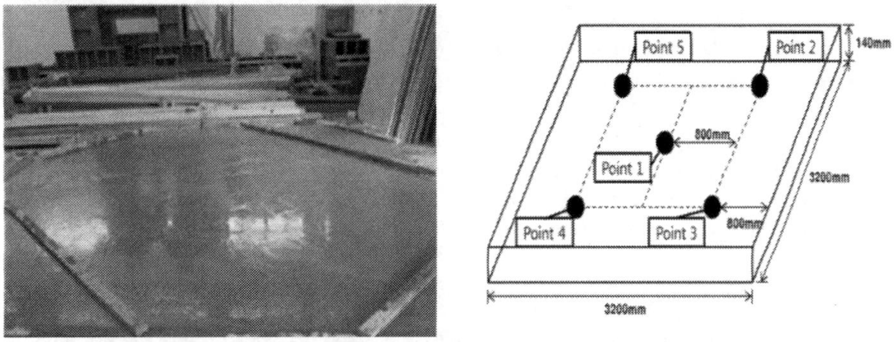

(a) Specimen of floor vibration (b) Measurement point

Figure 1. Concrete slab specimen

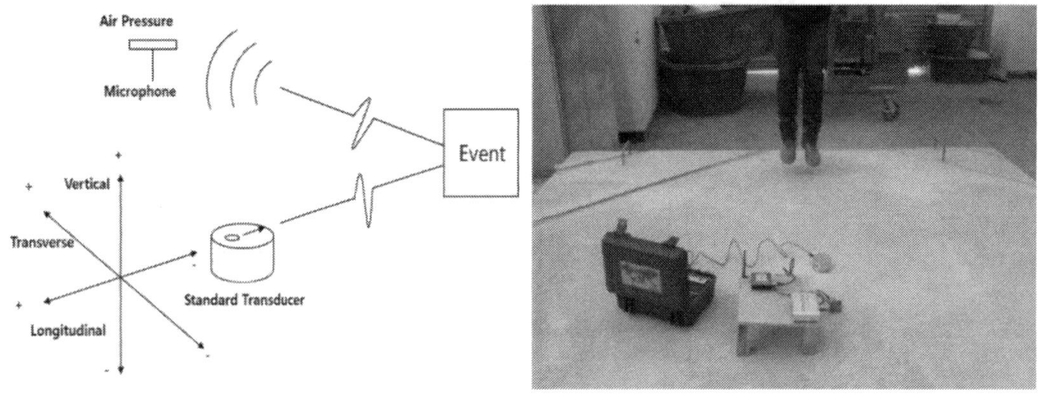

(a) Vibration measurement concept of accelerometer (b) Process

Figure 2. Experiments

(a) Test 1 (b) Test 2

Figure 3. Data measurement using accelerometer

4. Results and Discussion

4.1 Test 1

The experimental data in Table 1 was compared with the Meister and Reiher serviceability evaluation standard for building. The peak displacements ranged from 0.00869 to 0.02540 mm, and the peak frequency of vertical vibration had a range of 18 - 27 Hz and an average of 22Hz. The vibration correspond with most of the measured data can be identified as clearly sensible. The measured vibration was at the medium level and would cause no damage to the building structure according to the Koch Standard. The analysis result of the vibration measurement, the vibration about damage to the structure was not measured. However, the vibration was measured vibration enough to detect in general. The vibration measurement and the damage to the structure was not analysed in details. The vibration can be detected in general however.

4.2 Test 2

Serviceability value at which people feel discomfort was based on the recommended value of ISO 2603. As in Table 2, peak frequencies at points 1 to5 were 19, 20, 23, 20 and 22Hz, and the average was 21Hz. Serviceability limit for office and residential spaces at predominant frequencies of 18.5, 19.6, 22.6, 20.0 and 22.3Hz for Points 1~5 are 0.7 %, 0.5 %, 0.9 %, 1.1 % and 0.9 % g of gravitational acceleration. Accordingly, 0.7 %, 0.5 %, 0.9 %, 1.1 % and 0.9 % g were set as reference values. The highest acceleration amplitudes measured were 0.73 %, 0.47 %, 0.87 %, 1.07 % and 0.87% respectively. These values were not close to the recommended values for office and residential spaces. The range of vibration was 19% - 23% g, which was within the limit of serviceability. The results from Test 1, the serviceability standard of Reiher/Meister and the damage assessment by Koch were shown in Fig. 4(a) and 4(b). The data from Test 2, the serviceability standard of ISO and residential performance assessment of AIJ were shown in Fig. 4(c) and 4(d).

Table 1. Measurement result of Test 1

Test 1	Peak Frequency (Hz).	Peak Displacement. (mm)	PVS (mm/s)	Serviceability standard	Structure damage standard
				Reiher/Meister[1)]	Koch[2)]
Point 1	24	0.01080	2.52	Surely recognition	Moderate vibration (Nothing)
Point 2	18	0.02540	3.30	Surely recognition	Moderate vibration (Nothing)
Point 3	27	0.00869	1.78	Pretty recognition	Moderate vibration (Nothing)
Point 4	21	0.01570	2.66	Surely recognition	Moderate vibration (Nothing)
Point 5	21	0.01830	2.77	Surely recognition	Moderate vibration (Nothing)

1) Standard level : Incognizance - Pretty recognition - Surely recognition – Slightly unpleasant - Extremely unpleasant - Unbearable
2) Standard level : Light vibration (Nothing) - Moderately vibration (Nothing) - Wildness vibration (plastering crack) - Serious vibration (wall damage) - Extremely serious vibration (destruction)

Table 2. Measurement result of Test 2

Test 2	Peak Acceleration (%g)	Peak Frequency (Hz)	Time (sec)	Serviceability	Residential performance assessment
				ISO	AIJ
Point 1	0.73	19	36	Baseline	Baseline
Point 2	0.47	20	29	Baseline	Baseline
Point 3	0.87	23	47	Baseline	Baseline
Point 4	1.07	20	19	Baseline	Baseline
Point 5	0.87	22	60	Baseline	Baseline

5. Conclusion

In this study, an experiment was conducted to evaluate the floor vibration of deteriorated buildings. The impact sound of the concrete slab sample was measured and the results were stated as follow:

The peak displacement from Test 1 was in the range of 0.00869 - 0.02540 mm; the value of peak frequency ranged from 18 to 27 Hz, and the average value was 22Hz. The peak acceleration value from Test 2 was in the range of 0.47~1.07 % g; the value of peak frequency ranged from 18.5 to 22.57 Hz, and the average was 21Hz. Ranges of the vibration frequency from Test 1 and Test 2 were relatively consistent. When it is seen that the natural frequency values of floor slabs are concentrated in relatively the same range, it is known that occur temporary and intensive vibration on the specimen.

IOP Conf. Series: Materials Science and Engineering **62** (2014) 012019 doi:10.1088/1757-899X/62/1/012019

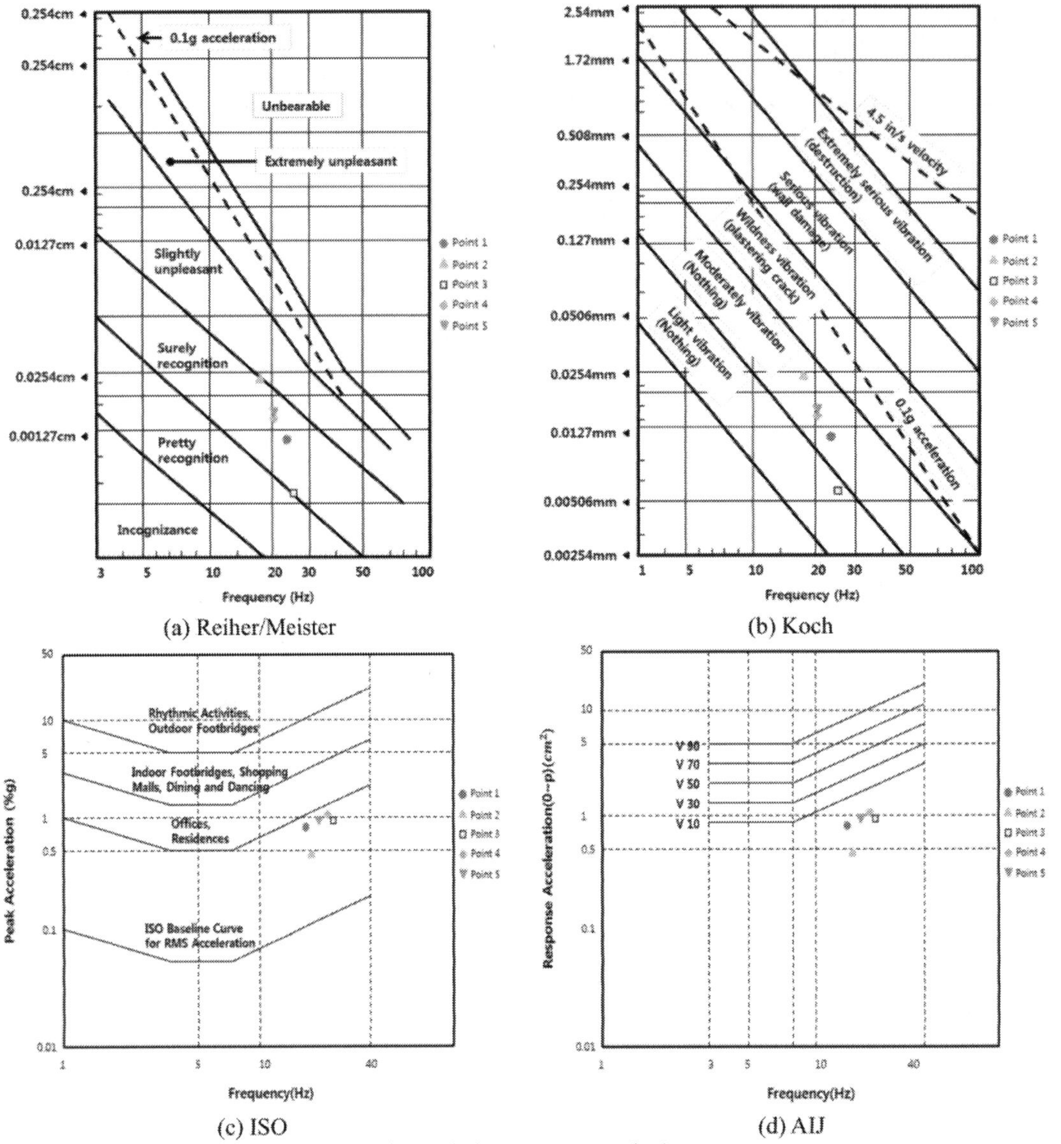

(a) Reiher/Meister (b) Koch

(c) ISO (d) AIJ

Figure 4. Compare to standard

Concrete slab floor that was used in deteriorated building with thickness lower than the latest standards creates living noise at level that does not exceed the standard range, but the noise is sufficient enough to make people feel uncomfortable. The vibration was apparently recognizable in most cases according to the Reiher/Meister standard. In the case of Koch graph for the damage assessment of the structure, the vibration was at the medium level and cause no damage to the building structure. The measured vibration results did not exceed the damage limit or serviceability limit of building according to the vibration assessment criteria of ISO and residential assessment guidelines provided by Architectural Institute of Japan (AIJ). . Vibration currently does not exceed damage limit or serviceability limit of building, but there is a problem of serviceability for temporary vibration. As the natural frequencies were similar to the frequencies measured at the test points from the experiment, buildings can possibly be damaged by vertical vibration and by the amplification of longitudinal vibration. Further study was expected to confirm the experimental data for serviceability and to assess the applicability of floor vibration assessment criteria for aged buildings.

Acknowledgements

This work was supported by the National Research Foundation of Korea (NRF) grant funded by the Korea government (MEST) (No. 2012R1A2A2A01012013)."

Reference

[1] Colin Hansen, Scott Snyder, Xiaojun Qiu, Laura Brooks, Danielle Moreau, "Active Control of Noise and Vibration" second edition volume 1, 2013
[2] Ad Hoc Committee on Serviceabillity Research, "Structural serviceability : A critical appraisal and research needs." J. Struct. Engrg., 112(12), 2646-2664, 1986
[3] H. Reiher and F.J Meister, "Human Sensivity to Vibrations", Forsch. aufdem Geb. Des Ingen, vol. 2, no. 11, pp.381-386, 1931
[4] B. Ellingwood, "Probability-based codified design: past accomplishments and future challenges", Structural Safety, 13, pp. 159–176, 1994
[5] Ministry of Land, Infrastructure and Transport, "Management standards and accreditation of floor impact sound of apartment building", Republic of Korea, 2013
[6] Reiher W, Meister FJ. The effect of vibration on people (translated from German, originally published in 1931). Report, F-TS-616-RE, Headquarters Air Material Command, Wright Field, OH, 1946
[7] F. Koch, V. Petrova-Koch, T. Muschik, J. Lumin., 57, p. 271, 1993
[8] ISO, "ISO 2631 Mechanical vibration and shock-evaluation of human exposure to whole-body vibration", 1997
[9] AIJ, "Residential performance assessment criteria guideline explanation"", Architectural Institute of Japan, 2004
[10] Anil K. Chopra., "Dynamics of Structures", Theory and Applications to Earthquake Engineering.
[11] BlastMate III Operation Manual, software version 4.3, Instantel, ISO 9001, 2001
[12] KS F 2865, "Laboratory measurements of the reduction of transmitted impact sound by floor coverings on a heavyweight standard floor", 2012
[13] JIS A 1440-2 : Acoustics - Laboratory measurement of the reduction of transmitted impact sound by floor coverings on a solid standard floor - Part2 : Method using standard heavy impact sources, 2007References to preprints. *For preprints there are two distinct cases:*

IOP Conf. Series: Materials Science and Engineering **62** (2014) 012020 doi:10.1088/1757-899X/62/1/012020

A detailed bond stretch model of thermal decomposition

Jia Fu[1], WeiGuo Sun[1, 2*], QunChao Fan[2] and Yi Zhang[1]

[1] School of Physics, Sichuan University, Chengdu, Sichuan P.R.C. 610065

[2] School of Physics and Chemistry, Research Center for Advanced Computation, Xihua University, Chengdu, Sichuan P.R.C. 610039

weiguosun@x263.net

Abstract: Thermal decomposition is caused by the breaking of the bonds of polymer chains after they are stretched to a length threshold. Any factors that affect the bond stretch can affect the thermal decomposition process. The physical processes like glass-transition changes and the physical configuration of the polymer chains affect the stretch mechanism apparently. Thus, glass-transition and molecular weight may be reflected in the thermal decomposition process characterized by the thermogravimetric (TG) or differential thermogravimetric (DTG) curves. This idea was confirmed by the thermal decomposition of solid poly-α-methyl styrene (PαMS). The experiment results showed that the DTG curves change suddenly near the glass-transition temperature, and the activation energy goes up with an increase of molecular weight, which indicates that the physical process is vital in the thermal decomposition process.

1. Introduction

In the thermal decomposition process, solid polymer may go through complex physical and chemical processes. Usually, only the chemical processes like bonding and breaking are considered and studied by traditional chemical dynamic method used for homogeneous system [1-3]. However, physical processes tend to be crucial [4] in the solid polymer decomposition processes.

One of the important physical processes that polymer goes through is glass-transition, which causes remarkable chain configuration changes in a short time. Some polymer materials have a relatively high glass-transition temperature under which the thermal decomposition is already underway. The glass-transition temperature of Poly-α-methyl styrene (PαMS) is up to $192^{O}C$ [5] and an observable weight loss during thermogravimetric analysis starts around 170 ^{O}C according to Ye [6]. This makes the PαMS a possible material for the investigation of the impact of glass-transition and hence the chain configuration of polymer thermal decomposition. Such effect is seldom studied. On the other hand, the thermal decomposition is widely studied by thermogravimetric (TG) or differential thermogravimetric (DTG) method. If the impacts of glass-transition on thermal decomposition are evident, there may be some sudden change in TG or DTG curve around glass-transition temperature, which makes TG or DTG method a potential useful experimental way to locate the glass-transition temperature.

Another important factor that affects the physical configuration of the polymer chains is molecular weight, which on the one hand affects the glass-transition temperature and on the other causes more complex entanglement between chains. In the thermal decomposition study of PαMS [7], the activation energy goes up with an increase in molecular weight from 188 to 272 kJ/mol. The influence is evident.

In section 2, we propose two physical models to describe the different bond breaking mechanisms for glass and high elastic phase polymers. In section 3, this work is summarized.

2. Theory and application

For an amorphous polymer, as the temperature goes up, the polymer leaves the glassy phase and turns into high elastic phase suddenly. In glassy phase, the movement (like rotation) of polymer chain segments is forbidden, while in high elastic state some chain segments that get enough energy are moveable. This change causes a large expansion of the phase space of the polymer chain and happens in a tiny temperature range. Therefore, a small temperature change during glass-transition can cause apparent change of polymer physical performance (specific volume, refractive index, diffusion coefficient, etc.) [8]. On the other hand, the influence of molecular weight is evident in thermal decomposition process [7]. Thus, it is worth discussing whether and how the thermal decomposition performance is affected by the physical processes and finally establishing a model to describe them uniformly.

2.1. The thermal decomposition model for glassy phase polymer

In glassy phase, the movement of chain segments is forbidden; all atoms are vibrating near equilibrium position. The polymer material is like a three-dimensional wire entanglement. The structure configuration of every polymer chain is fixed. For a specified main chain with n units, the energy corresponding to a particular structure can be simplified as a function of bond length r and bond orientation (in three-dimensional) $\boldsymbol{\theta}$ noted as $U(r_1,\ldots r_{n-1},\boldsymbol{\theta}_1,\ldots,\boldsymbol{\theta}_{n-1})$. A small structure change compared to equilibrium position will create an energy change of

$$\Delta U=\sum_{i=1}^{n-1}\frac{\partial U}{\partial r_i}\Delta r_i+\sum_{i=1}^{n-1}\frac{\partial U}{\partial \boldsymbol{\theta}_i}\Delta\boldsymbol{\theta}_i+\ldots, \tag{1}$$

which means the energy change is strongly affected by the original structure. It is worth noting that the structure change $(\Delta r_1,\ldots\Delta r_{n-1},\Delta\boldsymbol{\theta}_1,\ldots,\Delta\boldsymbol{\theta}_{n-1})$ caused by thermal excitation is not independent. The bond and orientation coordinates are not canonical coordinates so the particular form of U with initial condition is needed to determine the correlation of $(\Delta r_1,\ldots\Delta r_{n-1},\Delta\boldsymbol{\theta}_1,\ldots,\Delta\boldsymbol{\theta}_{n-1})$. However, one can infer that the changes of various bond lengths will differ from one another when stretching the bonds of the main polymer chain by thermal excitation. Each change meets a certain distribution according to the equilibrium structure configuration

2.2. The thermal decomposition model for high elastic phase polymer

In high elastic state, some chain segments that get enough energy are moveable. The movement makes some chain segments softer and reduces the differences among bonds. An energy change caused by thermal excitation can spread over the bonds of the segment more easily and therefore more equally, which differs notably from the case of glassy phase. On the other hand, the segment movements (like torsion and slippage) are correlated with the chains around it; when the temperature goes up the interactions between surroundings and chains are weakened. The movements like torsion and slippage become easier. Thus, even in the high elastic phase, the physical configuration is still in evolution.

It is worth noting that the movements of chain segments (torsion, slippage, stretching, etc.) are correlated. The existence of torsion and slippage may affect the stretching apparently. An explicit proof of the correlation of slippage and stretching is found in the fracture process and the mechanics of polymer bearing a macroscopic applied force. The fracture can be caused by two reasons, one is the bond breaking of main chain, and the other is slippage [9]. The former happens in the low temperature range while the latter happens in the high temperature range. The same applied force applied on different slippages may cause different stretching results (with or without bond breaking). In the thermal decomposition process, the thermal excitation can be treated as random applied forces on the polymer chain, thus the bond breaking mechanism in the macroscopic situation is a very good reference for its thermal decomposition counterpart. In the glass-transition process, the free volume effects and hence the torsions could have something to do with the slippage and stretching correlation.

2.3. The thermal decomposition during glass-transition

During glass-transition, the number of moveable chain segments increases suddenly. Thus, the thermal decomposition behaviour changes suddenly according to the mechanism change as described above. This sudden change is observed in the DTG experiment as shown in the Fig. 2 of [6]. One can figure out that there is an inflection at around 185 OC, which is very near the glass-transition temperature of PαMS (180OC or 192 OC). Another observation comes from the DTG study of polyvinyl chloride (PVC) as illustrated in the Fig. 8 of [2]. There is a bump in the DTG curve at around 90°C, which is near the glass-transition temperature of 85°C [10] of polyvinyl chloride. It is noteworthy that the bumps we found in the DTG curves may due to some other processes, including thermal decomposition process. For example, the free volume effects due to glass-transition may open many new escape channels for the impurity molecules and cause a fluctuation in the TGA study. A lot of works have to be done to confirm which reason is predominant. However, the situations in poly-α-methyl styrene and polyvinyl chloride show that the glass-transition does affect the DTG curve and the influence is observable. The DTG method could be used to detect the polymer glass-transition temperature and the mechanisms are definitely worth exploring. The bond stretch model is an interesting possibility.

2.4. The molecular weight dependence in thermal decomposition

The fact that activation energy goes up with molecular weight [7] indicates that the thermal decomposition process is highly affected by the conformation of chain structure (physical process), since the bond strength difference may not be apparent in polymers with different molecular weights. This can be explained by our bond stretch model since any factors that affect the bond stretch can affect the thermal decomposition process. The interchain interactions affect the segment motions like torsion, slippage and stretching, and are more complicated as the molecular weight goes up. The correlation of these motions may finally affect the bond breaking (caused by random stretching) of thermal decomposition process.

3. Conclusion

The bond stretch model gives a clear picture and hence qualitative analysis of the physical processes involved in thermal decomposition process. Any factors that can affect the bond stretch (like glass-transition, molecular weight) can affect the thermal decomposition process. Base on this model, this paper explains the DTG observation that the structure and movement of chain segments can affect the bond stretch and hence the decomposition process, and suggests that the DTG method could be used to detect the polymer glass-transition temperature when polymer's decomposition and glass-transition process coexist in a certain temperature range.

References

[1] Bockhorn, H.; Hornung, A.; Hornung, U. *Mechanisms and kinetics of thermal decomposition of plastics from isothermal and dynamic measurements*, *Journal of Analytical and Applied Pyrolysis* 1999, 50, 77-101

[2] Yang, J.; Miranda, R.; Roy, C. *Using the DTG curve fitting method to determine the apparent kinetic parameters of thermal decomposition of polymers*, *Polymer Degradation and Stability* 2000, 73, 455-461.

[3] Hsiang, Y. H.; Yung, C. C. *The effect of end groups on the thermal degradation of poly (methyl methacrylate)*, *Polymer Degradation and Stability* 2003 82, 81–88.

[4] House, J. E. In *Principles of Chemical Kinetics*, 2nd ed.; Elsevier Inc. 2007; p 230.

[5] Ren, P. Z. In *Polymer Chemistry*(in chinese), 5th ed.; Chemical Industry Publishing House. 2011; p 253.

[6] Na, Y. L.; Fen, C. S.; Huang, Y.; Bo, L.; Wen, Z. Z.; Xiao, J. Y. *Thermal degradation kinetics of poly-alpha-methyl styrene, High power Laser and particle beams* 2012, 24, 2400-2405.

[7] Wall, L. A. and Florin, R. E. *Pyrolysis of Monodisperse Poly-a-Methylstyrene, JOURNAL OF RESEARCH of the Notional Bureau of Standards* 1978, 83, 4, 371-380.

[8] Jun, H. M.; Dong, Z. H.; Xiao, C. W.; Xia, D. X. In *Polymer Physics*(in chinese), 3rd ed.; Fudan University Publishing House. 2006; p 108.

[9] Jun, H. M.; Dong, Z. H.; Xiao, C. W.; Xia, D. X. In *Polymer Physics*(in chinese), 3rd ed.; Fudan University Publishing House. 2006; p 205.

[10] Kent, J. A. In *Handbook of Industrial Chemistry and Biotechnology: Volume 1*; Springer. 2012 ; p 577.

IOP Conf. Series: Materials Science and Engineering **62** (2014) 012021 doi:10.1088/1757-899X/62/1/012021

The Effect of Ultraviolet Light on the "Liquid Wood"

Nedelcu Dumitru[1], Zăgan Sabina[2], Negreanu-Pirjol Ticuta[2], Zăgan Remus[2], Cărausu Constantin[1]

[1]Technical University "Gheorghe Asachi" of Iasi, Romania, Blvd. Mangeron, No. 59A, 700050 Iasi, Romania
[2]Constanta Maritime University, Naval Electromechanical Faculty, Mircea cel Batran Street, No. 104, 900663, Constanta, Romania

Corresponding author: Dumitru Nedelcu, nedelcu1967@yahoo.com

Abstract. To preserve resources, the goal is to use biobased materials containing the maximum possible amount of renewable biomass-based derivatives to secure a sustainable future. Bioplastics, biocomposites, biological fibres and related biomaterials will serve as substitutes for materials and products traditionally made from petroleum resources. To support this need, in 1998, Fraunhofer Institute of Chemistry and Tecnaro GmbH Company (founded by Jurgen Pfitzer and Helmut Nagele in Germany) investigated and developed a new compound made of wood components that can be processed as a thermoplastic material. It is well known that bioplastic and biocomposite materials represent another important group of bio-materials that include plastics reinforced with natural fibers and wood-plastic composites (WPCs). The aim of this research is to identify the behavior of "Liquid Wood" after UV irradiation. The materials used were Arbofill Fichte, Arboblend V2 Nature and Arboform L, V3 Nature. The samples were obtained by injection and the experimental study plan followed the Taguchi method with six input parameters and two levels for each of them. Three samples from each material were tested in an ultraviolet environment using different time of exposure in order to establish the material characteristics. After the irradiation process the material did not turn to yellow, which suggests that the composition of the sample liquid timber inhibitors have stability and the number of α-carbonyl (C = O) groups is sufficiently low. After the graphs analyzing can be inferred relatively similar performance of the samples in the first stage both UV and the VIS, indicating that the activity is almost absent. In UV is clearly observed the peak amplitude (maximum absorbance) at different wavelengths (λ). So, in terms of peak intensity the samples follow the order: Arboform L, V3 Nature; Arbofill Fichte and Arboblend V2 Nature.

1. Introduction

Human society has benefited tremendously from the use of plastics, one of the most highly valued materials mainly because of their extraordinary versatility and manufacturability.

Almost all synthetic plastics are made from petroleum and its allied components. In addition, plastics derived from fossil sources are mostly non-biodegradable. The increased use of plastics, along with other materials, over the years has resulted in an increase of plastic waste, which often is dumped as municipal solid waste, [1]. Taking into account the samples properties from "liquid wood", this biodegradable material [2] is capable of replacing plastic and providing mankind with new materials, [3, 4, 5]. Norbert Eisenreich, a senior researcher at the Fraunhofer Institute for Chemical Technology in Germany (ICT), said that *Arboform*, the new material, is made of lignin, which can be derived from soft tissues of wood, [8]. Once mixed with several other ingredients, the substance turns into solid and non-toxic alternative for plastics.

The starting point for a new class of thermoplastic materials from renewable resources is lignin, a natural polymer that is the main constituent of biomass, in trees and other woody plants with an amount of up to 30%, [1]. This growing interest about a green and sustainable chemistry has also contributed to call attention to biomass and specifically on lignocellulosic feedstock as a promising, renewable and vast resource for biopolymers [6]. Lignocellulosic biomass is primarily composed of three biopolymers: cellulose, hemicelluloses and lignin [7]. Lignin plays a major role in woody plants, adding strength and structure to the cell walls, controlling fluid flow and protecting against biochemical stresses by inhibiting enzymatic degradation of other chemical components [6].

The absorption spectra represent the signal dependence by the wavelength λ. There are many possibilities of obtaining absorption spectra, but the most common used is the representation of absorbance according to wavelength: $A = f(\lambda)$. The other variants, less used - called all absorption spectra - are $T = f(\lambda)$, $\log A = f(\lambda)$, $\varepsilon = f(\lambda)$ or $\log \varepsilon = f(\lambda)$. The last two are used mainly to characterize the molecular species because it does not depend on the experimental conditions are the spectra measurements. Each material has a characteristic absorption spectrum, like general form, like spectral domains, like peaks numbers and the ratios between the intensities of the different peaks. The maximum number and the general shape of the curve represent the qualitative characteristic after that can be identifying substances.

2. Technology, Methodology and Research Plan

The samples were obtained using the injection molding process, [10]. "Liquid wood" pellets, (also known under the name of arboform) are mainly composed of lignin discarded during the paper-making process, combined with cellulose and wood waste from industrial plants. These pellets are put in a bunker where are subjected to hot-air drying for two-four hours. After that, they are transferred into the bunker of an injection machine, from where the material enters into an injection cylinder, electrically heated at 160-180°C, being afterwards injected into a mold, which is water cooled at a temperature of 23°C. The main parameters of the "liquid wood" during the injection process, which could be taken into account, are: melting temperature, molding temperature, injection rate, injection and holding pressure, cooling time, holding pressure time inside the mould and injection time.

The experiments were planned using the Taguchi methodology, [9] and the model proposed by Viger and Sisson was easy to be studied. Each experiment was conducted three times. The proposed matrix model takes into consideration six technological parameters with two levels (table 1). The experimental research determined the coefficients of a type (1) model:

$$Z_t = M + T_{top} + t_{inj} + t_r + S_s + P_{inj} + T_{mat} + P_{inj}\,T_{top} + P_{inj}\,t_{inj} + P_{inj}\,t_r + P_{inj}\,S_s + P_{inj}\,T_{mat} \quad (1)$$

where: M-general average; Ttop-melting temperature, [oC]; tinj-injection time, [s], tr-cooling time, [s], Ss-screw displacement, [mm], Pinj-injection pressure, [MPa], Tmat-matrix temperature, [oC]. The most significant influence on the process is exercised by injection pressure followed by melt temperature and matrix temperature. These are followed by screw speed, injection time and cooling time, which are though less significant. After the orthogonally and number of degrees of freedom conditions were analyzed, there were 16 experimental tests to be made.

In this study were used three samples from each of the three types of "liquid wood": Arbofill Fichte, Arboblend V2 Nature and Arboform L V3 Nature. All the samples were exposed to different exposure times to UV-VIS radiation, table 2.

Table 1. The levels values of input parameters

Input parameter Levels	T_{top} [°C]	t_{inj} [s]	t_r [s]	S_s [mm]	P_{inj} [MPa]	T_{mat} [°C]
First level	145	9	18	60	80	50
Second level	160	11	25	80	100	80

Table 2. Exposure time to UV-VIS radiation

	Exposure time, [min]		
	Arboform L V3 Nature	Arbofill Fichte	Arboblend V2 Nature
First Sample	300	315	315
Second Sample	300	600	600
Third Sample	140	760	760

The samples were irradiated with UV lamp Philips TUV 11W T5 4P-SE in a hermetically sealed steel enclosure. It was used ultraviolet light, with short range UV (UVC) considered that they are also germicidal UV, because UV-C radiation can break down chemical bonds. The value of wavelength used is of 2.537 Angstroms (254 nm). It is known that the exposure to light such as ultraviolet irradiation lignin in phenols are known to produce phenoxy radicals that can be readily oxidized to yellow o-quinones (figure 1).

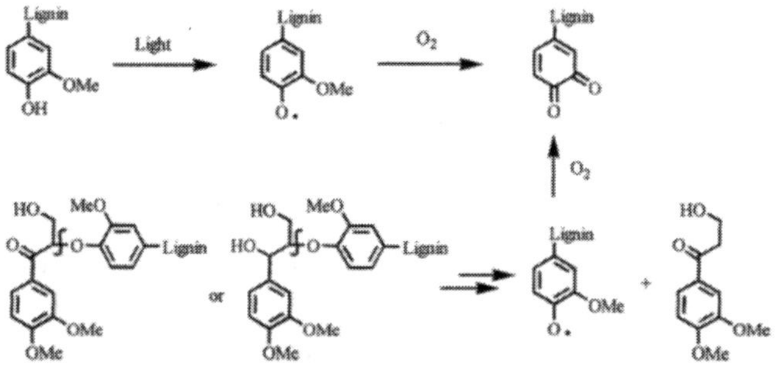

Fig. 1. Several lignin structures – formation and subsequent oxidation

Spectrophotometric analysis was performed using CECIL EC -3021 Super Aurius spectrophotometer in order to study the quality of the spectra in the UV -VIS (ultraviolet - visible) domain. Solid samples of each material were brought to fine-grained, homogeneous and then passed in liquid and the testing for homogeneity has been performed. Thus, the samples were subjected to a digestion process (mineralization) to be read by the equipment. For this purpose, there were used different mixtures of acids HNO_3 / H_2SO_4 and HNO_3 / H_2SO_4 / HCl. Second mixture was chosen for more efficient digestion. There have been used high concentrations of pure acids: HNO_3 65 % GR; H_2SO_4 96 % GR; HCl 37 %. The proportion was HNO_3: H_2SO_4: HCl = 3ml: 3ml: 3ml. After digestion, the samples were brought to volume in 100 ml volumetric flasks with distilled water as follows: mineralized sample was poured the contents over a quantity of distilled water in 100 ml volumetric flask and then brought the volumetric flask to the mark with distilled water. After calibrating the apparatus, the reference sample (blank sample) was measured. It has been used high quality quartz cuvettes with a path spectral length of 10 mm for both reference and sample.

3. Results and discussion

The qualitative spectra of the samples analyzed are shown in Figures 2, 3 and 4.

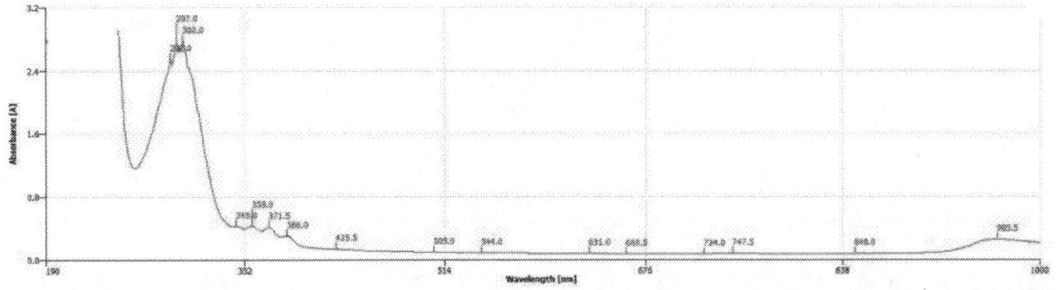

a

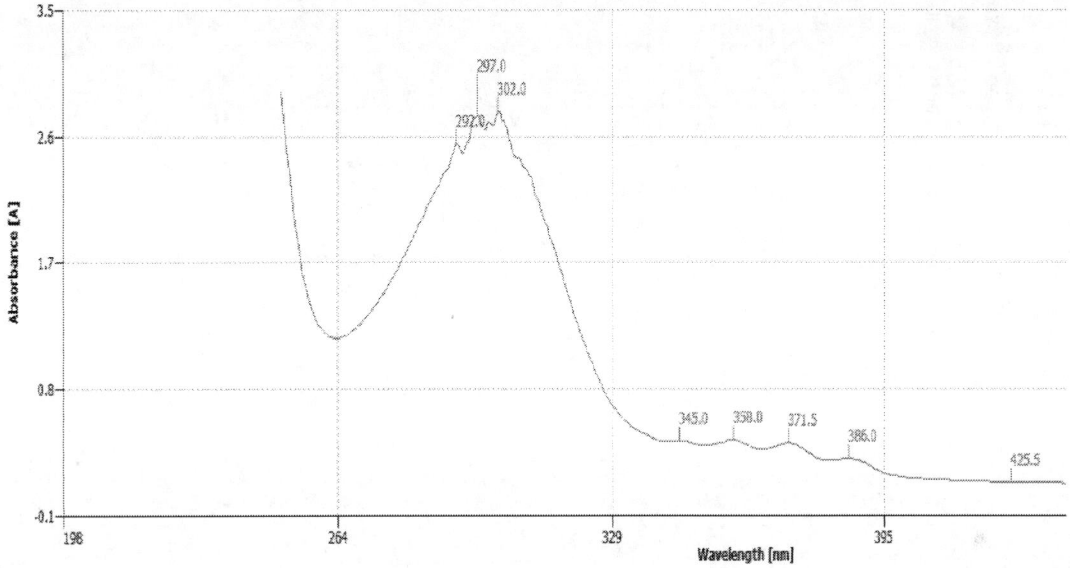

b

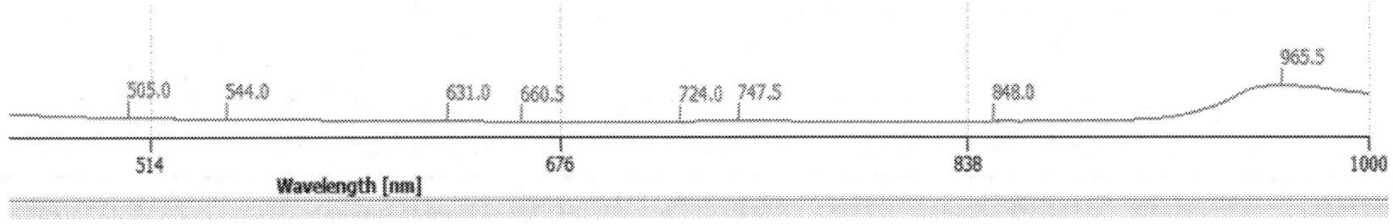

C

Fig. 2. The UV-VIS spectrum for Arbofill Fichte: a-UV-VIS full spectrum; b-UV spectrum; c-VIS spectrum

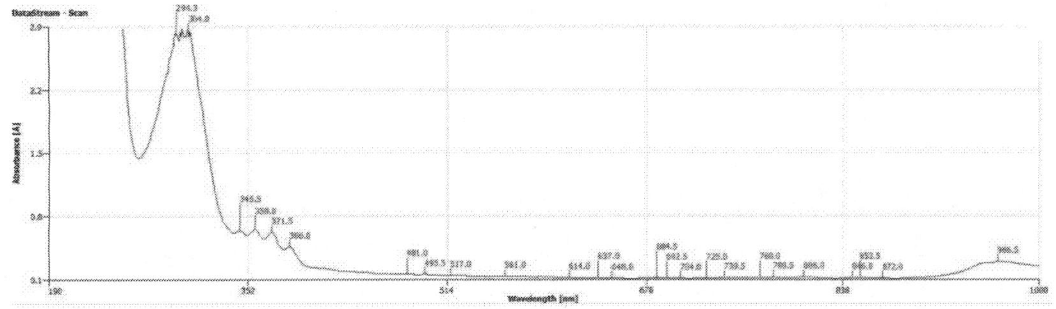

a

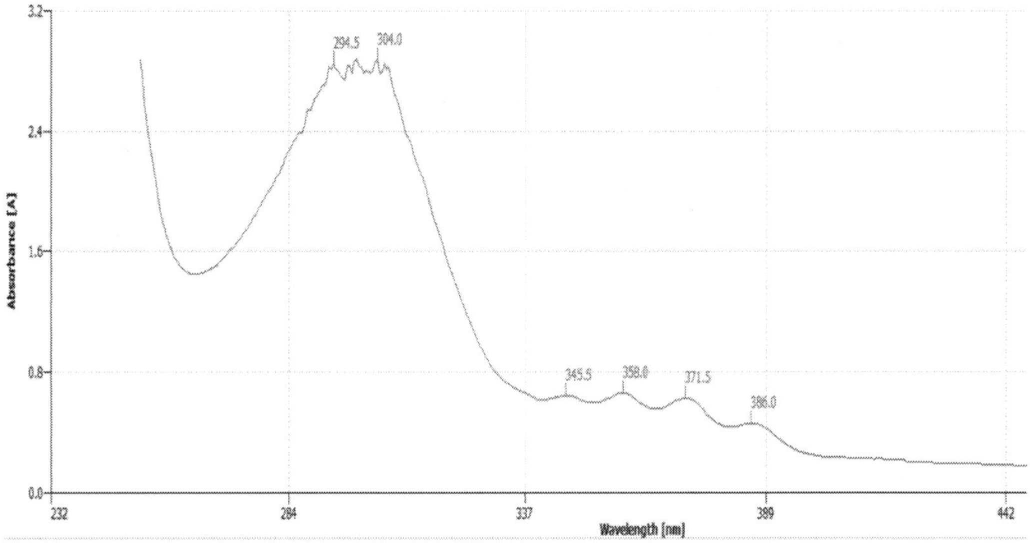

b

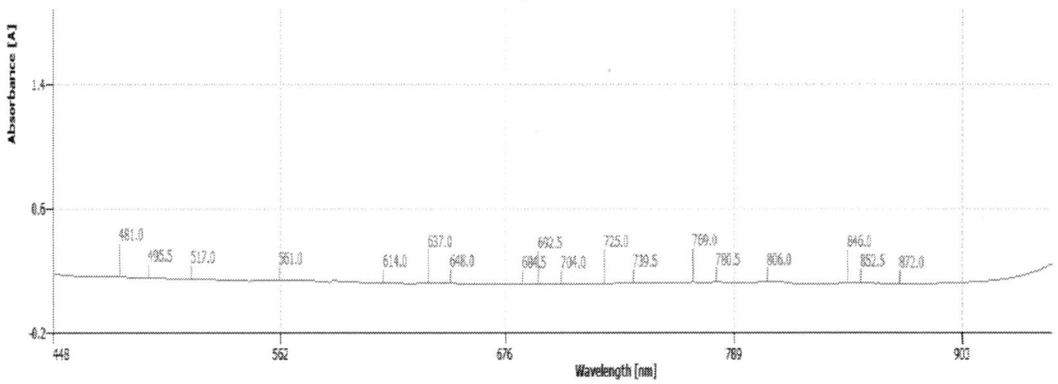

c

Fig. 3. The UV-VIS spectrum for Arboform L, V3 Nature: a-UV-VIS complete spectrum; b-UV spectrum; c-VIS spectrum

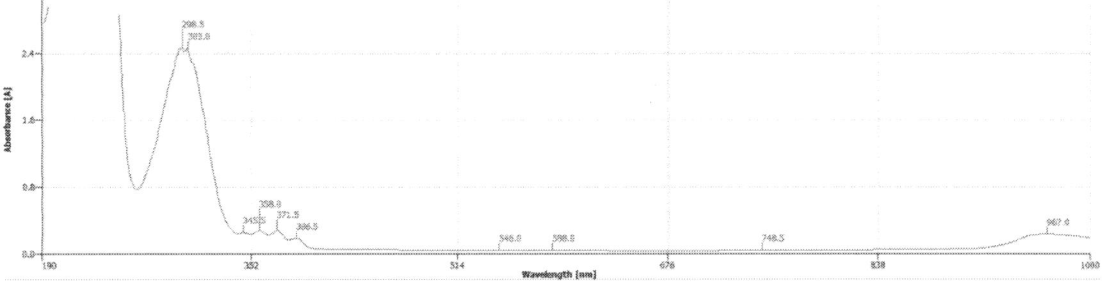

a

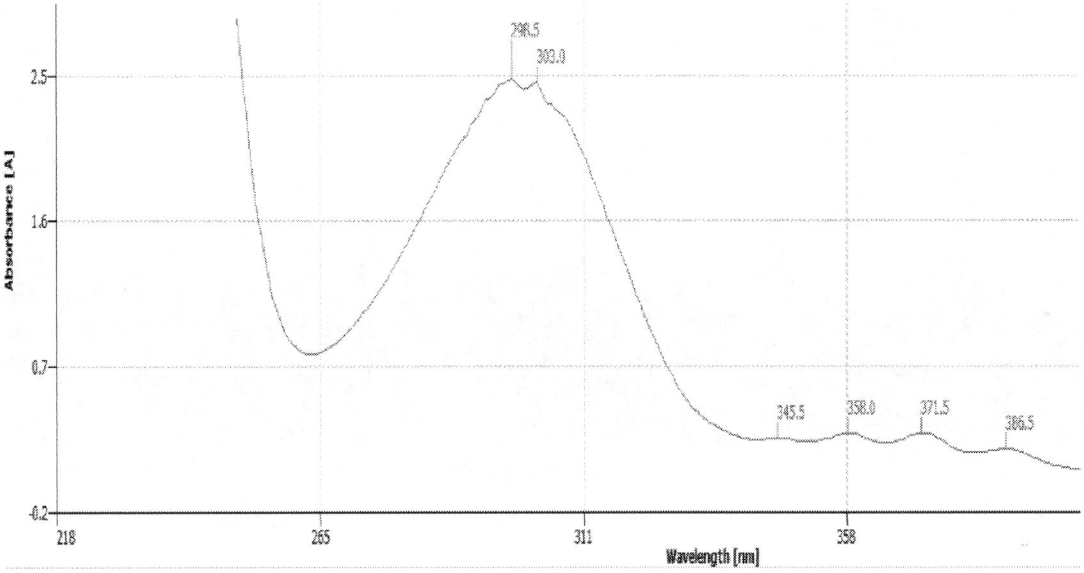

b

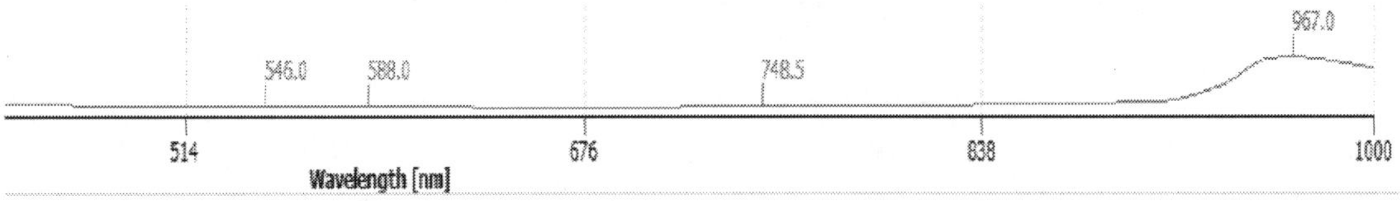

c

Fig. 4. The UV-VIS spectrum for Arboblend V2 Nature: a-UV-VIS complete spectrum; b-UV spectrum; c-VIS spectrum

Figure 2 shows the attenuation spectrum on the basis of UV. The spectrum that occur in the absorbance values read at 345, 358, 371, 386, 425, 505, 544, 631, 660, 724 and 747 nm, which is flatten and indicates that the tested material has absorbed UV and thus shows a good performance under the influence of UV irradiation.

From the figure 3, it can be seen a similar behavior as in the case of Arbofill Fichte in the sense of a flattening of the spectrum. In this case, the recorded values range from 345, 358, 371.5, and 386 up to 872 nm. It can also be concluded that this material absorbs UV radiation and has good efficiency under the UV irradiation.

From the graphs shown in Figure 4 is drawn the same conclusion, namely that this material has good performance under the UV radiation.

To achieve prominence between samples made of four materials was done in Figure 5, the spectra overlap, as follows: Figure 5a represents the UV-VIS spectra overlap completely; Figure 5b represents the UV spectra overlap and VIS spectra overlap in Figure 5c.

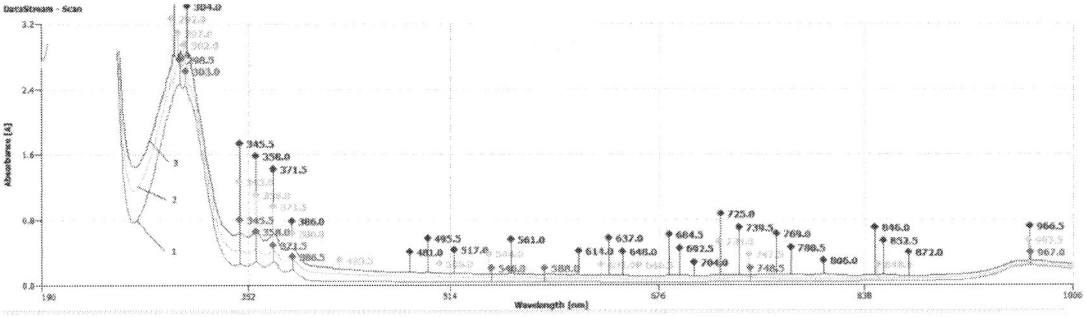

a

1-Arboblend V2 Nature; 2-Arbofill Fichte; 3-Arboform L, V3 Nature

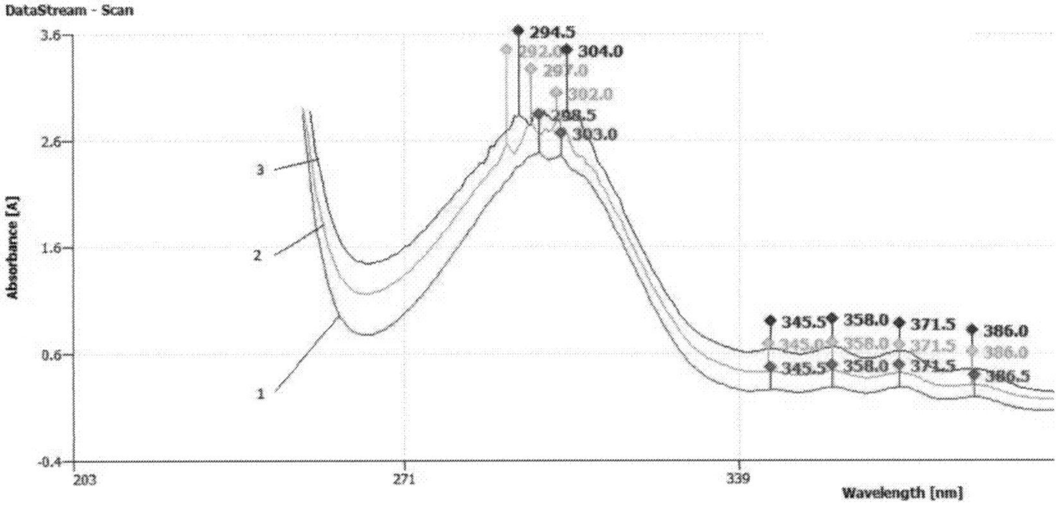

b

1-Arboblend V2 Nature; 2-Arbofill Fichte; 3-Arboform L, V3 Nature

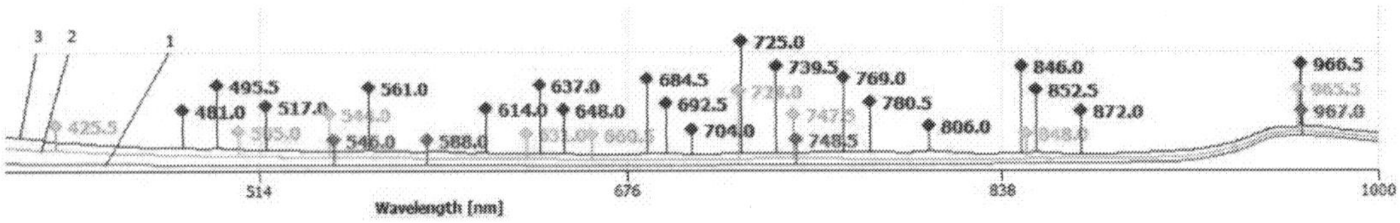

c

1-Arboblend V2 Nature; 2-Arbofill Fichte; 3-Arboform L, V3 Nature

Fig. 5. Spectra overlap

4. Conclusions

Digestion of the samples was performed with an array of acids. Was measured the final pH of the solutions prepared (9 ml acid / mineralized sample + 91 ml distilled water) and got the 0.5 value. The analyzing of graphs shows a relatively close performance of the samples in the first stage both for UV and VIS. During the last domain, the activity is almost nonexistent. In

UV was clearly observed the peak amplitude (maximum absorbance) to different wavelengths (λ). So, in terms of peaks intensity the studied materials are in the following order: Arboform L, V3 Nature, Arbofil Fichte and Arboblend V2 Nature. Absorbance peaks were seen in the UV into two groups lengths: first group: 291.5nm - 295.5nm; 297.0nm - 298.5nm; 302.0nm – 303.0nm; 304.0nm; 307.0nm, second group: 345.5nm; 358.0nm; 371.0nm -371.5nm; 385.0nm. If the study object in the next stage will be the acid addition influencing the spectrum of the sample characteristics, research can continue with another solvent matrix to maximize the effect of acids in the next stage, namely pH of samples. Future studies will focus on identifying these types of materials' resistance to the action of substances highly acidic (pH=1). The first data prove that these materials could play an important role in the manufacture of furniture specifically for areas with high acidity or specific use in chemical laboratories, using acidic substances. Based on actual research, it can be concluded that all materials studied subjected to such treatment will increase resistance to corrosive properties.

References

[1] SrikanthPilla (2011), Engineering Applications of Bioplastics and Biocomposites- An overview, Handbook of Bioplastics and Biocomposites Engineering Applications,

[2] BjörnWinberg, KAM Industries, Albis Plastic Scandinavia AB, Commercial biopolymers Properties and limitations, 2010

[3] Dumitru Nedelcu, Ciprian Ciofu, Nicoleta Monica Lohan, (2013), *Microindentation and Differential Scanning Calorimetry of "Liquid Wood"*, Composites Part B: Engineering, doi: 10.1016/j.compositesb.2013.05.024 , vol. 55, pp. 11-15.

[4] Dumitru Nedelcu, (2013), *Investigation on microstructure and mechanical properties of samples obtained by injection from Arbofill*, Composites Part B: Engineering, Volume 47, April 2013, pp. 126–129, doi: 10.1016/jcompositesb.2012.11.023.

[5] Dumitru Nedelcu, Radu Comaneci, (2013), *Microstructure, Mechanical Properties and Technology of Samples obtained by Injection from Arboblend*, Special Issue on *Evolution of Engineering, Technologies and Materials,* ModTech2013 International Conference, 27-29 June, Sinaia, Romania, Indian Journal of Engineering & Materials Sciences, manuscript accepted.

[6] Laurichesse S, Av´erous L, Chemical modificationof lignins: towards biobased polymers, *Progress in Polymer Science* (2013),http://dx.doi.org/10.1016/j.progpolymsci.2013.11.004

[7] Patanjali Varanasi, Priyanka Singh, Manfred Auer, Paul D Adams, Blake A Simmonsand Seema Singh, Survey of renewable chemicals produced from lignocellulosic biomass during ionic liquid pretreatment,Biotechnology for Biofuels 2013

[8] http://www.tecnaro.de/english/willkommen.htm?section=we, Accesed: 20.11.2011

[9] Nedelcu D.; Pruteanu, O. Aspecte ale formarii canelurilor exterioare prin deformare plastica la rece utilizand metoda Taguchi, Tehnica-Info Publishing House, Chisinau, 2000; pp. 243-261

[10] Ciprian Ciofu, Daniel Mindru, Injection and Micro Injection of Polymeric Plastics Materials, Int. J. of Modern Manufacturing Technologies, Vol. V, No. 1/2013, 49-68

IOP Conf. Series: Materials Science and Engineering **62** (2014) 012022 doi:10.1088/1757-899X/62/1/012022

Influence of External Forces on the Mechanical Characteristics of the a-IGZO and Graphene Based Flexible Display

H.-J. Kim[1], Youn-Jea Kim[2*]

[1]Graduate School of Mechanical Engineering, Sungkyunkwan University, Suwon 440-746, Republic of Korea
[2]School of Mechanical Engineering, Sungkyunkwan University, Suwon 440-746, Republic of Korea

*E-mail: yjkim@skku.edu, Tel: +82-31-290-7448, Fax:+82-31-290-5889

Abstract: Thin film transistors (TFTs) based flexible displays have optically transparent and m echanically flexible properties that are attractive for next-generation display technologies. In particular, "amorphous indium-gallium-zinc-oxide" (a-IGZO) and graphene have attracted much attention due to the advantages of their excellent unif ormity and compatibility with transparent and flexible substrates. To maintain these characteris tics, it is important to confirm the deformation characteristics of TFTs with applied external for ces, such as compressive or tensile stress, distortion effects, and temperature. The mechanical c haracteristics of modeled devices applied to different active layers on TFTs, such as a-IGZO and graphene, were investigated under various external conditions. The distributions of t he stress-strain curve on each active layer and the deformed shapes were assessed graphically.

Key words: thin film transistors (TFTs), flexible display, amorphous indium-gallium-zinc-oxide (a-IGZO), graphene, active layer

1. Introduction

Recently, many consumers want not only portable electronic devices with high-performance and high-efficiency, but also small and light ones. Accordingly, the display market has started to focus on flexible displays for compact arrangements in limited spaces in electronic devices. The flexible organic light-emitting diode (FOLED) has received much attention because of its advantages, such as lower materials costs, color purity, light-weight, and high light-emitting efficiency for applications in flexible displays and lighting.[1] The FOLED is a type of organic light-emitting diode (OLED) incorporating a flexible plastic substrate on which the electroluminescent organic semiconductor is deposited. The device can be bent or rolled while still operating. Currently a focus of research in industrial and academic groups, FOLEDs represent one method of fabricating a rollable display. Flexible displays built-in to commercial electronic devices are being released all the time. Thus, various researches are studying transparent thin-film transistor (TTFT), because there is growing interest in oxide active layers. TTFTs have mainly been used with transparent conducting oxides (TCOs), and amorphous indium-gallium-zinc-oxide (a-IGZO) [2] or graphene [3], which are being studied widely. The a-IGZO satisfies various requirements for such applications: it has very high transmittance in the visible range and high electron mobility so many researchers have studied the a-IGZO.[4-8]

Graphene has a large theoretical specific surface area (2,630 m^2g^{-1}), high intrinsic mobility (200,000 $cm^2v^{-1}s^{-1}$) [9,10], and thermal conductivity (~5,000 $Wm^{-1}K^{-1}$). [11]Moreover, its optical transmittance (~97.7%) and good electrical conductivity are notable advantages for many potential applications. [12] In this study, the mechanical characteristics of a-IGZO and graphene based flexible displays were investigated under the influence of external forces, such as tensile, compressive, and torsional forces at constant temperature. When an electronic device is operated, it is very difficult to maintain both the electrical and mechanical characteristics due to the generation of heat in the device. Additionally, the stress and strain energy density (SED) distributions at the active layer of the modeled device with TFTs were calculated using a commercial code, COMSOL multiphysics, and various deformed shapes were obtained.

2. Numerical Details

The schematic diagrams of the simplified TFTs for numerical analyses are shown in Figure 1. The modeled TFT is composed of a total of six different layers, and its detailed configurations are as follows: The top layer is the source/drain electrode (aluminum, Al). The second layer is the active layer (a-IGZO or graphene) as a target layer, and is located below the top layer. The third layer is an insulator (aluminum oxide, Al_2O_3). The fourth layer is an insulator (polyvinylphenol, PVP). Finally, the bottom layer is a flexible substrate (polyimide, PI). The material properties of each layer in the modeled TFTs are provided in Table 1. To assess the mechanical characteristics of the a-IGZO or graphene based flexible display, a simulation of the TFTs was performed with various applied external forces (5.7×10^8 N/m^2), [13] as shown in Figure 2. The surrounding temperature was set at room temperature, 20°C. The heating temperature of the modeled TFTs was set at 50°C to simulate the operating condition of the electronic device. Table 2 shows the cases of simulations with various external forces. The number of elements in the modeled TFT layer was generated approximately, from 16,000 to 19,000.

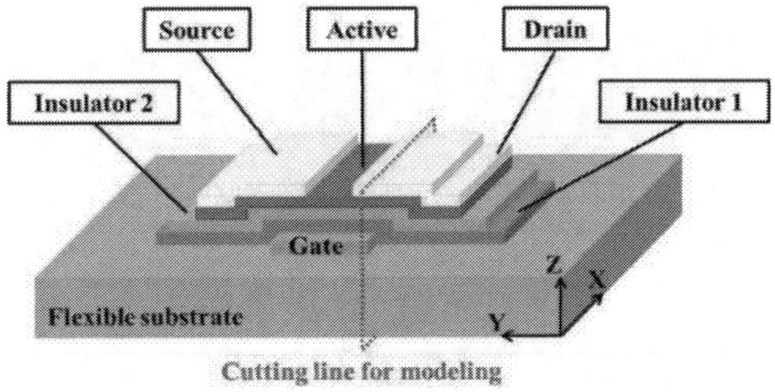

(a) Layer structure of the modeled TFT

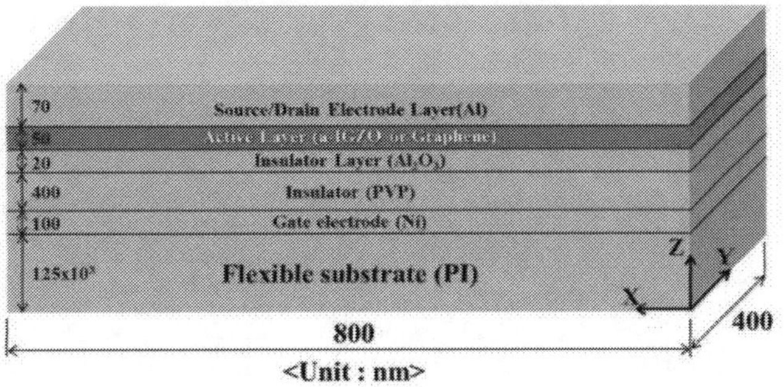

(b) Dimensions and materials of simplified model

Figure 1 Schematic diagrams of the modeled TFT for numerical analyses.

IOP Conf. Series: Materials Science and Engineering **62** (2014) 012022 doi:10.1088/1757-899X/62/1/012022

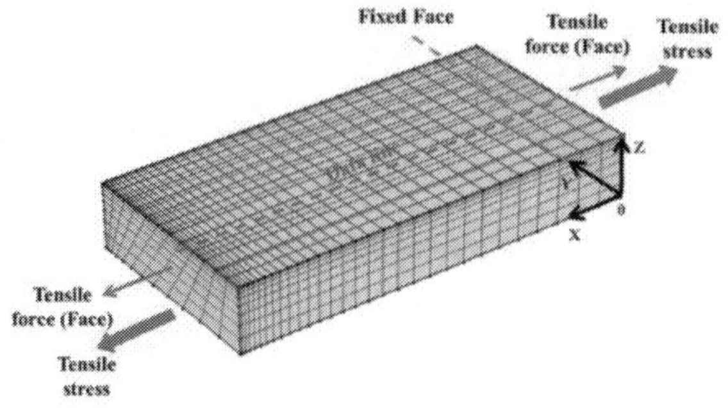

(a) Tensile force

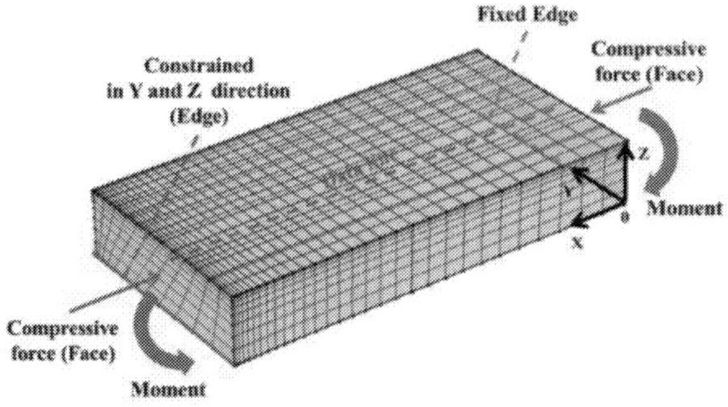

(b) Compressive force

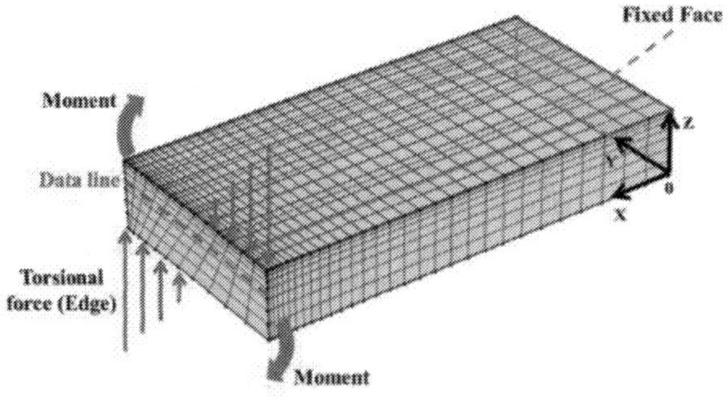

(c) Torsional force

Figure 2 Grid systems of the model for the applied external forces.

Table 1. Material properties of each layer of the modeled TFT.

Materials	Young's modulus [GPa]	Poisson's ratio	Density [kg/m^3]	Coefficient of thermal expansion [10^{-6}/K]
Al	68	0.36	2700	0.00099
a-IGZO	**137**	**0.36**	**5600**	**4.31**
Graphene	**1000**	**0.24**	**313008**	**0.000095**
Al_2O_3	370	0.22	3960	0.0022
PVP	0.8	0.34	1166	100
Ni	207	0.31	8880	0.048
PI	29	0.34	1880	0.23

IOP Conf. Series: Materials Science and Engineering **62** (2014) 012022 doi:10.1088/1757-899X/62/1/012022

Table 2. Simulations with various external conditions.

Active layer	External forces			
	Case 1	Case 2	Case 3	Temperature [°C]
a-IGZO	Tension	Compression	Torsion	50
Graphene				
Surroundings	-	-	-	20

3. Results And Discussion

The influences of external forces on the von Mises stress and SED distributions were investigated numerically to determine the mechanical characteristics of the a-IGZO and graphene based flexible displays.

The results of von Mises stress and SED in the active layers are presented in Figures 3 and 4. It can be seen from the results in Figure 3(a), when a tensile force was applied along the x-direction of the modeled TFTs, the von Mises stress in the a-IGZO was larger than that in graphene. Figure 3(b) shows similar stress tendencies in both the a-IGZO and graphene active layers, and the deformed shapes rose convexly in the z-direction when a bending moment was generated at the edge of the modeled TFTs by a compressive force. When a clockwise torsional force was applied to the nano layer structure in the x-axis, the graphene layer showed a greater value of von Mises stress than that of the a-IGZO one. In particular, as shown in Figure 3(c), stress was concentrated in the center of the active layer.

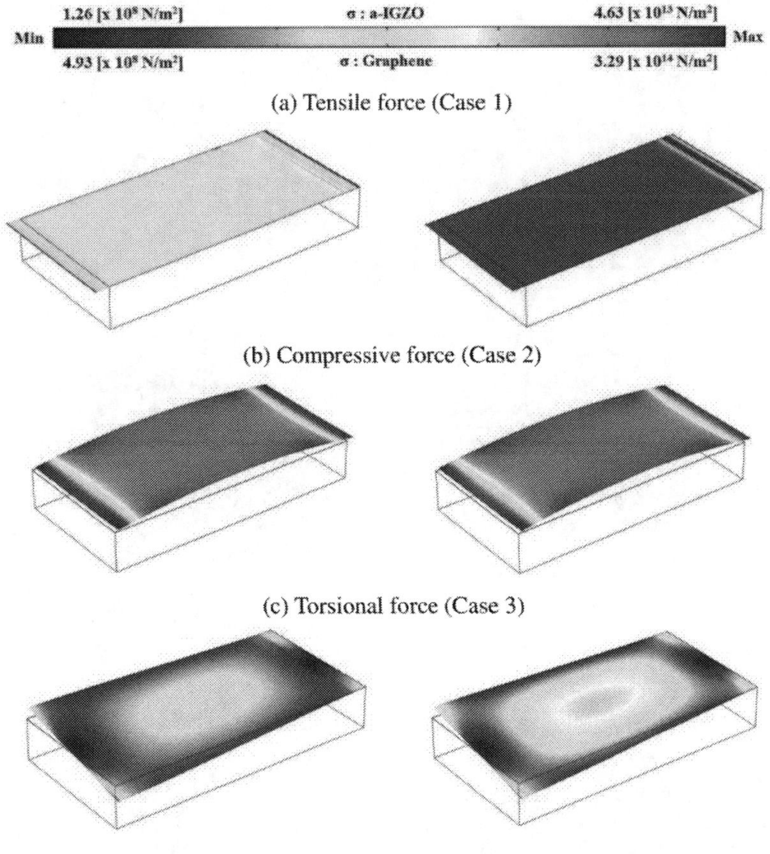

Figure 3 Influence of external forces on von Mises stress distributions in the active layers.

It is also seen from Figure 4(a) that the SED of the a-IGZO was greater than that of graphene in the x-direction when a tensile force was applied. When the bending moment was applied to the modeled TFTs by a compressive force, the deformation was small on both sides along the x-direction. The results also showed that the SED appeared largely at the center of active layer, and both active layers rose convexly in the z-direction. Additionally, the SED of graphene was higher than that of a-IGZO (Fig. 4(b)). A clockwise torsional deformation was generated on both sides along the x-direction when a clockwise torsional force was applied to the modeled TFTs. Furthermore, a larger value of SED was represented at the center of the graphene active layer (Fig. 4(c)), resulting from the stress concentration phenomenon.

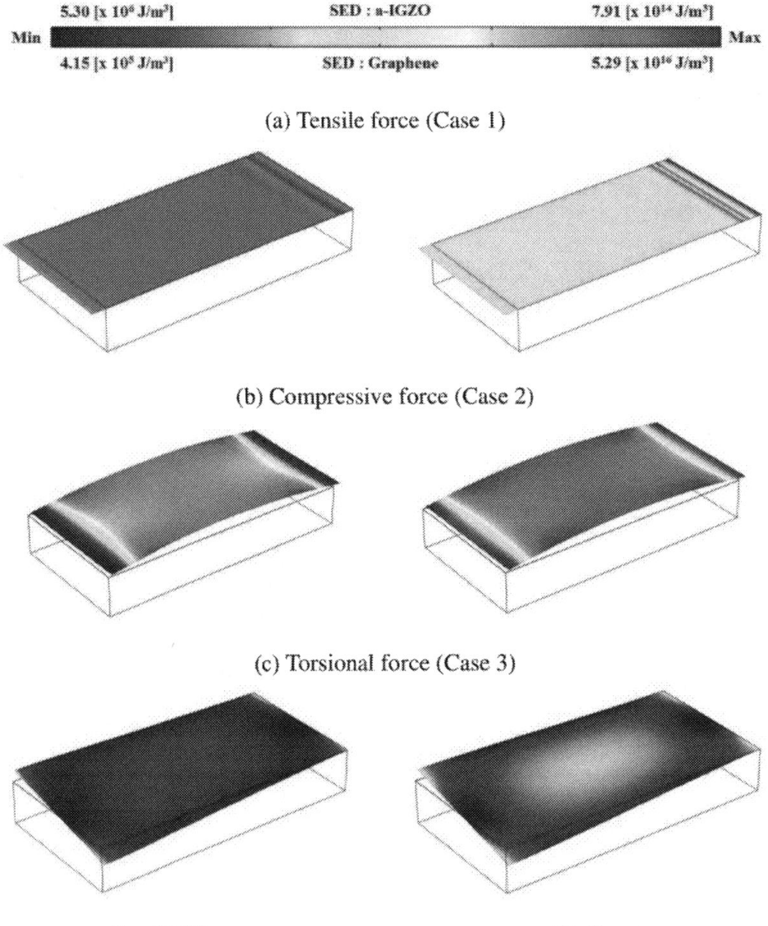

Figure 4 Influence of external forces on strain energy density distributions in the active layers.

As shown in Fig. 5, when the tensile force was applied to the TFTs structure, the values of von Mises stress and SED showed almost the same tendency in both active layers. However, the graphene layer showed slightly higher than those of a-IGZO layer.

Also, from the results of Figure 6(a), the value of von Mises stress in the graphene active layer was greater than that of a-IGZO when the compressive force was applied. The maximum value difference was 25.2630x1013 N/m2. Similarly, the SED of the graphene layer was also higher than that of a-IGZO (Fig. 6(b)), and the maximum value difference was 35.6456x1015 J/m3.

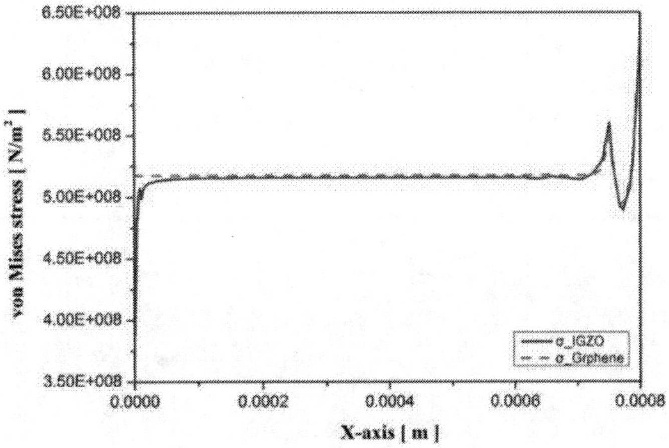

(a) von Mises stress

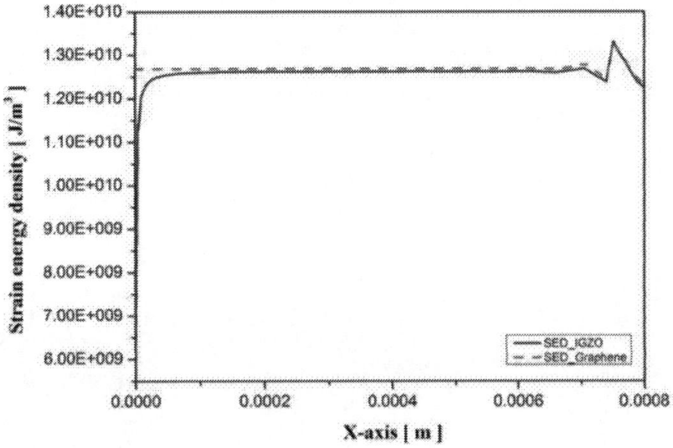

(b) Strain energy density

Figure 5 Comparison of von Mises stress and strain energy density distributions between the a-IGZO and graphene active layers with tensile force.

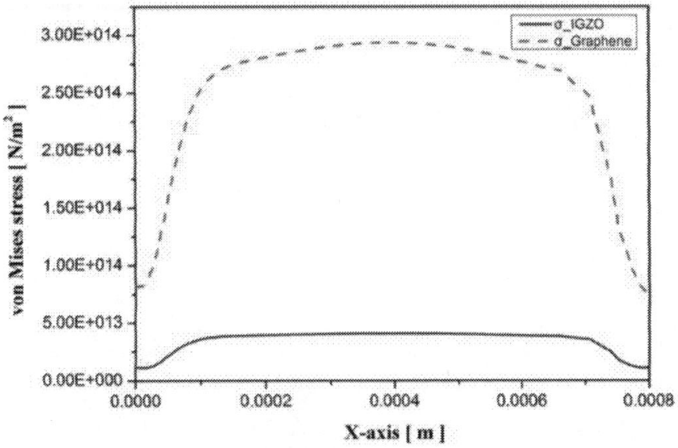

(a) von Mises stress

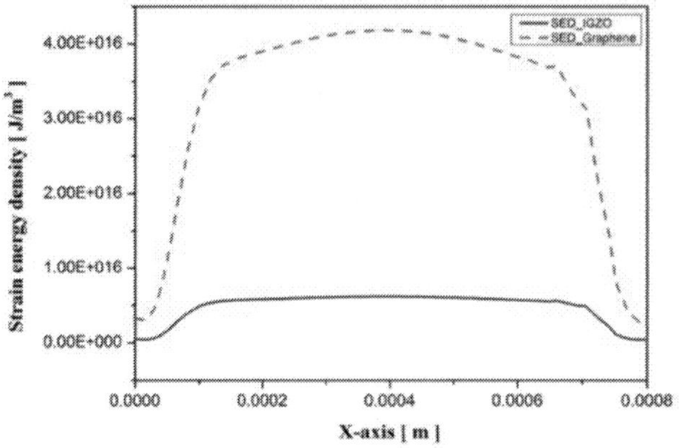

(b) Strain energy density

Figure 6 Comparison of von Mises stress and strain energy density distributions between the a-IGZO and graphene active layers with compressive force.

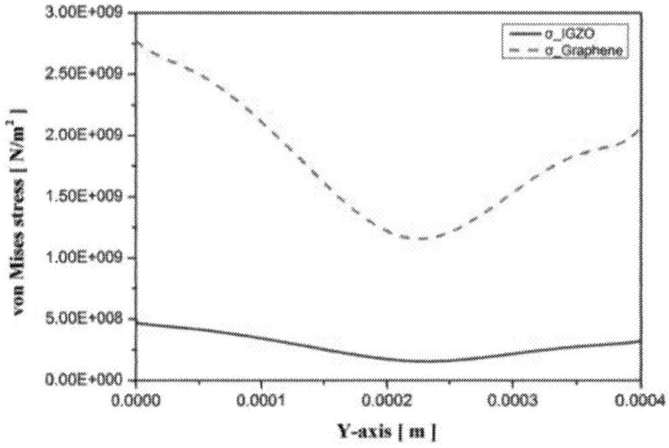

(a) von Mises stress

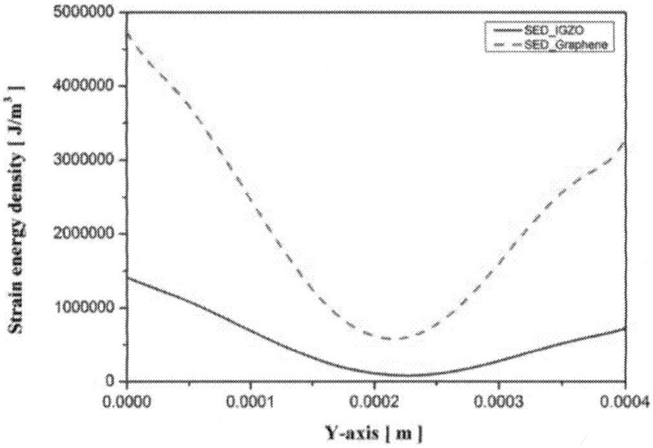

(b) Strain energy density

Figure 7 Comparison of von Mises stress and strain energy density distributions between the a-IGZO and graphene active layers with torsional force.

When the torsional force was applied to the modeled TFTs, the values of von Mises stress and SED of the graphene layer were higher than those of the a-IGZO active layer. As shown in Figure 7, the maximum value differences for von Mises stress and SED were 23.0498×108-N/m2 and 3.3233×106 J/m3, respectively.

4. Conclusions

In this study, the mechanical characteristics of a-IGZO and graphene based flexible displays were investigated numerically under the influence of external conditions, such as tensile, compressive, and torsional forces at a constant operating temperature of 50°C. The following conclusions were reached:

1) The a-IGZO and graphene active layers showed similar tendencies in mechanical characteristics; indeed there was no significant difference when a tensile force was applied to the TFT structures.

2) The values of von Mises stress and SED in the graphene layer were greater than those for the a-IGZO layer when a compressive or torsional force was applied. In particular, these mechanical characteristics were superior at the thin edges of the graphene layer.

3) The graphene active layer had better mechanical characteristics in terms of flexibility and strength than the a-IGZO layer when TFT multi-layers received external forces. Thus, it may be concluded that the graphene is worth investigating for use in next-generation flexible displays, that will be capable of bending or rolling.

Acknowledgements

This work was supported by the Agency for Defense Development through the Chemical and Biological Defense Research Center at Sungkyunkwan University, Korea.

References

[1] F. Li, Z. Lin, B, Zhang, T. Zhang, C. Wu, and T. Guo, Organic Electronics 14, 2139 **(2013).**

[2] T. Kamiya, K. Nomura, and H. Hosono, *Sci. Technol. Adv. Mater.* 11, 044305 **(2010).**

[3] A. K. Geim, *Science.* 324, 1530 **(2009).**

[4] D. Y. Cho, J. Song, C. S. Hwang, W. S. Choi, T. W. Noh, J. Y. Kim, H. G. Lee, B. G. Park, S. Y. Cho, S. J. Oh, J. H. Jeong, J. K. Jeong, and Y. G. Mo, *Thin Solid Films* 518, 1079 **(2009).**

[5] H. Hosono, K. Nomura, Y. Ogo, T. Uruga, and T. Kamiya, *J. Non-Cryst. Solids* 354, 2796 **(2008).**

[6] G. H. Kim, H. S. Kim, H. S. Shin, B. D. Ahn, K. H. Kim, and H. J. Kim, *Thin Solid Films* 517, 4007 **(2009).**

[7] D. H. Lee, K. Kawamura, K. Nomura, H. Yanagi, T. Kamiya, M. Hirano, and H. Hosono, *Thin Solid Films* 518, 3000 **(2010).**

[8] A. Suresh, P. Gollakota, P. Wellenius, A. Dhawan, and J. F. Muth, *Thin Solid Films* 516, 1326 **(2008).**

[9] K. I. Bolotin, K. J. Sikes, Z. Jiang, M. Klima, G. Fudenberg, J. Hone, P. Kim, and H. L. Stormer, *Solid State Commun.* 146, 351 **(2008).**

[10] S. V. Morozov, K. S. Novoselov, M. I. Katsnelson, F. Schedin, D. C. Elias, J. A. Jaszczak, and A. K. Geim, *Phys. Rev. Lett.* 100, 016602 **(2008).**

[11] A. A. Balandin, S. Ghosh, W. Z. Bao, I. Calizo, D. Teweldebrhan, F. Miao, and C. N. Lau, *Nano Lett.* 8, 902 **(2008).**

[12] D. Yoon, Y. W. Son, and H. Cheong, *Nano Lett.* 11, 3227 **(2011).**

[13] Y. G. Seol, J. S. Park, N. T. Tien, N. E. Lee, D. K. Lee, S. C. Lee, Y. J. Kim, C. S. Lee, and H. Kim, *J. the Electrochemical Society* 157, H1046 **(2010).**

IOP Conf. Series: Materials Science and Engineering **62** (2014) 012023 doi:10.1088/1757-899X/62/1/012023

Low-temperature baroplastic processing of graphene-based polymer composites by pressure-induced flow

Wei Tang[1], Cheng-en He[1], Yuanzhen Wang[1], Yingkui Yang[1,2] and Chi Pong Tsui[2]

[1] MOE Key Laboratory for Green Preparation & Application of Functional Materials, Faculty of Material Science & Engineering, Hubei University, Wuhan 430062, China
[2] Department of Industrial & Systems Engineering, The Hong Kong Polytechnic University, Hung Hom, Kowloon, Hong Kong, China

E-mail: yingkuiyang@gmail.com (YKY); gary.c.p.tsui@polyu.edu.hk (CPT)

Abstract: Two-stage emulsion polymerization was employed to synthesize nanoparticles consisting of a low glass transition temperature core of poly(*n*-butyl acrylate) (PBA) and a glassy poly(methyl methylacrylate) (PMMA) shell. Incorporation of graphene oxide (GO) into the PBA-PMMA latex produced GO/PBA-PMMA composites after demulsification and graphene/PBA-PMMA composites after chemical reduction of GO. The as-prepared powdery materials were processed into thin films by compression molding at room temperature as the result of a pressure-induced mixing mechanism of microphase-separated baroplastics. The presence of oxygen-containing groups for GO sheets contributed to better dispersion and stronger interface with the matrix, thereby showing greater reinforcement efficiency toward polymers compared to graphene sheets. In addition, both Young's modulus and yield strength for all materials increased with applied pressure and processing time due to better flowability, processability and cohesion at higher pressure and longer time. Low-temperature processing under pressure is of significance for energy conservation, recyclability and environmental protection during plastic processing.

1. Introduction

With extremely high values of Young's modulus (1.1 TPa) and fracture strength (125 GPa), thermal conductivity (~5×10^3 Wm/K), electrical conductivity (~6×10^5 S/m) and specific surface area (2630 m^2/g),[1, 2] graphene has thus great potential for improving the mechanical, thermal, electrical, and gas barrier properties of polymeric materials.[3] Over the past decade, considerable progress has been made for graphene-filled polymer composites.[4, 5] For instance, Chung *et al.*[6] prepared graphene/polystyrene (PS) composites by solution blending and subsequent compression molding, and obtained a low percolation threshold (0.19 vol.%) and an enhancement in storage modulus by ~28% at 1.94 vol.% of graphene loading. Macosko *et al.* [7] fabricated graphene/polycarbonate (PC) composites *via* melt compounding and observed better performance in suppressing gas permeability and tensile modulus relative to pure PC. We have recently employed an *in-situ* polymerization technology to fabricate graphene/poly(methyl methylacrylate) (PMMA) composites, which showed great increases in storage modulus (+58.3%) and glass transition temperature (T_g, +19.2 °C) as well as high electrical conductivity (13.37 S/m) at 2.08 vol.% loading.[8]

Graphene oxide (GO) traditionally serves as a precursor for graphene,[9] however, it is also considered as a chemically-functionalized graphene containing epoxy/hydroxyl groups on the basal plane and carbonyl/carboxyl groups at the edge.[10] GO has been increasingly attracting attention for its own characteristics.[11, 12] GO/poly(vinyl alcohol) (PVA) composite films could be simply fabricated by vacuum filtration and possessed high Young's modulus (4.8 GPa) and tensile strength (~110 MPa) for the film containing 3 wt.% GO. 132% and 36% of increases in tensile and compressive strengths have been reported for GO/PVA hydrogels upon addition of 0.8 wt.% GO, respectively.[13] Also, GO can effectively improve the compatibility of immiscible polyamide and polyphenylene oxide blends.[14] Recently, Liang *et al.* [15] have prepared GO/styrene-butadiene rubber (SBR) composites by aqueous-phase mixing of GO colloid with SBR latex in the presence of a small amount of butadiene-styrene-vinyl pyridine latex. The mechanical properties of GO/SBR composite with 2.0 vol.% GO was found to be comparable with those of the SBR composite reinforced with 13.1 vol.% of carbon black.

Till now, graphene- and GO-based composites have been frequently reported using various polymers including nylon, polyolefin, polyester, and polyimide.[4] Although these thermoplastic composites have been successfully prepared by solution mixing, melt compounding and *in-situ* polymerization methods,[3] they generally require high processing temperatures to enable compression and extrusion molding into final products or the desired forms for property measurement. However, the hot-melt processing causes not only energy consumption when heating the materials to form a melt, but also thermal degradation of polymers and organic additives, resulting in poor plastics' performance. In 2003, Mayes *et al.* [16] reported a novel class of "baroplastic" block copolymers consisting of a high-T_g shell of PS and a low-T_g core of poly(*n*-butyl acrylate) (PBA) or poly(2-ethylhexyl acrylate) (PEHA). These baroplastics could exhibit melt-like flow and hence be processed by compression molding and extrusion at 20-50°C under external pressure through an apparent semi-solid partial mixing mechanism.[17] Moreover, the baroplastics were easily shredded and remoulded for ten times without evident property degradation.

In the present work, core-shell PBA-PMMA nanoparticles were prepared by two-stage emulsion polymerization. To this polymer latex was added an aqueous GO colloid to produce GO/PBA-PMMA composites after demulsification and reduced graphene oxide (RGO)/PBA-PMMA composites after chemical reduction of GO, respectively. As expected, the obtained polymer composites with a characteristic of baroplastics could be compression-moulded under pressure at ambient temperature. Mechanical properties at different applied pressure and processing time were investigated for PBA-PMMA, GO/PBA-PMMA and RGO/PBA-PMMA composites.

2. Experimental Section

2.1. Materials

Methyl methacrylate (MMA, 99%) and *n*-butyl acrylate (BA, 99%) were purified by washing with aqueous solution of NaOH (5 wt.%) and drying with anhydrous calcium chloride. Water was distilled and deionized before use. Pristine graphite powder (>95%), sodium dodecyl sulfonate (SDS, 99%), hydrazine hydrate (N_2H_4, 50-60%), 1-dodecanethiol (98%), acetone (97%), potassium persulfate (KPS, 99%) and other chemicals were purchased from the Sinopharm Group Chemical Reagent Co., Ltd (Shanghai) and used without purification unless otherwise stated. GO was prepared from graphite using a modified Hummers method.[8, 11]

2.2. Synthesis of core-shell PBA-PMMA latex nanoparticles

Core-shell polymer nanoparticles were synthesized by two-stage emulsion polymerization according to the procedure in a literature with minor change.[17] In the first stage, a 250 mL glass flask charged with deionized water (60 mL), acetone (9 g) and SDS (0.9 g) was flushed with nitrogen for 20 min, and the resulting solution was heated to 65°C under vigorous stirring. To the above solution was slowly added 9 g of BA over a period of *ca.* 30 min and stirred for another 1 h. A solution of KPS

(0.18 g) in water (10 mL) was then added to initiate polymerization of BA and the reaction lasted for over 5 h at 65°C. In the second stage, MMA (9 g) was pre-emulsified by adding water (60 mL), acetone (9 g) SDS (0.27 g), and chain transfer agent of 1-dodecanethiol (0.18 g) under vigorous stirring for over 1 h. The pre-emulsified MMA solution was then added dropwise, with a flow rate of 2-3 drops/second, to the PBA emulsion, and the reaction was allowed to proceed at 65°C with constant stirring for an extra 12 h. The as-prepared copolymer emulsion was reserved for the fabrication of GO/PBA-PMMA and RGO/PBA-PMMA composites.

2.3. Preparation of GO/PBA-PMMA and RGO/PBA-PMMA composites

For the preparation of GO/PBA-PMMA composites, 81 mg of GO was ultrasonically dispersed into 300 mL of water to obtain a yellow-brown colloid. The stable suspension of GO was mixed with the PBA-PMMA latex for about 1 h with stirring, affording a GO/PBA-PMMA composite emulsion. The GO/PBA-PMMA emulsion was further treated by hydrazine hydrate for 25 min to obtain a RGO/PBA-PMMA emulsion. All as-prepared emulsions were demulsified by addition of an aqueous solution (5 wt.%) of aluminum sulfate. The obtained precipitates were purified by washing with a mixture of methanol/water (v/v, 50/50) for several cycles to remove emulsifier and demulsifier. Finally, the resulting powders were dried under vacuum at room temperature in the presence of phosphorus pentoxide, yielding GO/PBA-PMMA and RGO/PBA-PMMA composites as well as pure PBA-PMMA materials, respectively. The loading amount of GO and RGO were about 0.5 wt. % in the composites.

2.4. Room-temperature processing of baroplastic materials

Powders of GO/PBA-PMMA, RGO/PBA-PMMA, and PBA-PMMA were processed into films by compression molding at room temperature on a Minitype Thermo-compressor (Bl-6170-C, Dongguan Baolun Instrument Co., China). The processing was carried out under pressures of 10-60 MPa for 3-9 min. The resulting films were cut into rectangle samples for tensile tests.

2.5. Characterization

Particle size distribution of polymer latexes was determined at 25°C by dynamic light scattering (DLS) experiments on a Malvern HPPS5001 laser particle-size analyzer. Tg was measured on a TA differential scanning calorimeter (DSC, Q2000) at 10°C/min under nitrogen. Transmission electron microscopy (TEM) analysis was performed on a Tecnai G220 electron microscope operated at 200 kV. Two composites were also cryo-microtomed at -80°C to obtain thin films (~70 nm) for studying the dispersion state of GO and graphene sheets in the polymer matrix. Tensile tests were conducted on a universal testing machine (CMT-4104, SANS, China) at a crosshead speed of 5 mm/min. Five specimens were measured and averaged for each batch of material samples.

3. Results and Discussion

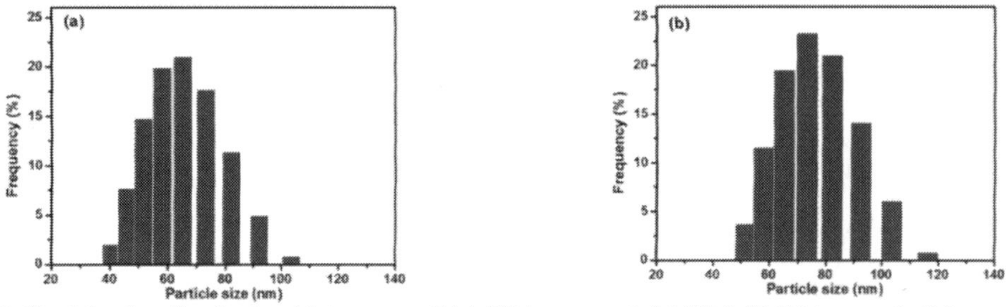

Figure 1. Particle size distribution histograms of (a) PBA core and (b) PBA-PMMA core-shell latex particles determined by DLS experiments.

Figure 1 gives particle size distribution histograms of the parent PBA core and PBA-PMMA core-shell particles. According to DSL data, the PBA core sizes obtained at the first stage of polymerization range from 35 nm to 110 nm in diameter (Fig. 1a), however, the PBA-PMMA core-shell particle sizes are in the range of 45-125 nm (Fig. 1b) after addition of MMA at the second stage. Their average particle sizes were calculated to be ~65 for core and ~76 nm for core-shell latexes, respectively. The increase of latex nanoparticles by ~11 nm in diameter implies an effective polymerization of MMA around PBA core. However, both PBA and PBA-PMMA latex particles show very large polydispersities due to the poor controllability of core and shell polymers during emulsion polymerization as described elsewhere.[17] TEM technique was further used to evaluate the particle sizes of PBA-PMMA latex. As shown in Figure 2, the average diameter of core-shell particles is approximately 75 nm, in good agreement with the DLS result (Fig. 1b).

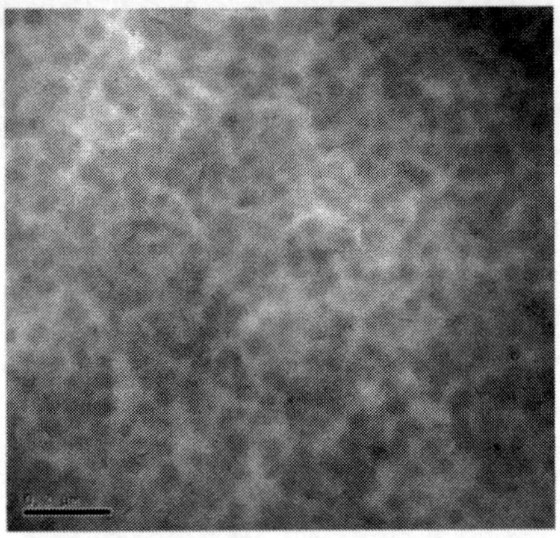

Figure 2. A typical TEM image of core-shell PBA-PMMA latex particles.

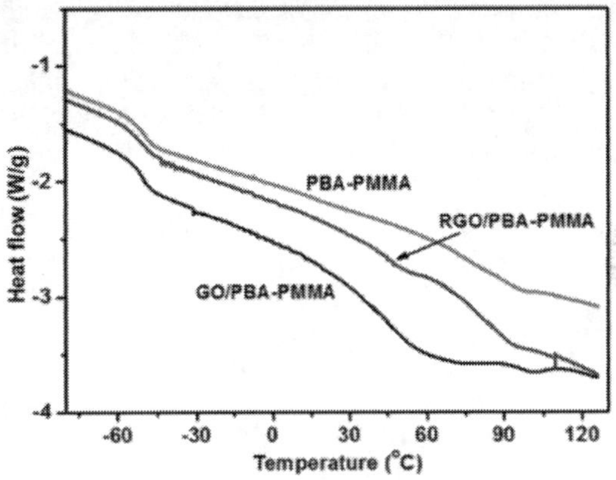

Figure 3. DSC curves of PBA-PMMA, GO/PBA-PMMA and RGO/PBA-PMMA composites.

Figure 3 shows DSC curves of GO/PBA-PMMA and RGO/PBA-PMMA composites and neat PBA-PMMA as dried. Two distinct T_gs for three samples occurs at about -45°C and 105°C corresponding to PBA core and PMMA shell, respectively, due to the microphase separation of PBA and PMMA components.[16, 17] Incorporation of GO and graphene into the matrix slightly increases T_gs of PBA-PMMA due to the mechanical interlocking and hydrogen bonding of crumpled sheets with polymer chains.[8] In addition, a new peak of T_g can be observed at around 55°C for GO/PBA-PMMA and RGO/PBA-PMMA composites, which is properly ascribed to the interphase of two components.[17]

The PBA core is generally soft due to its intrinsically low T_g, however, the PBA-PMMA latex becomes a powdery product after demulsification due to the coverage of PBA surrounded by a high-T_g hard shell of PMMA, as shown in Figure 4. As reported previously,[16, 17] baroplastic nanoparticles consisting of a low-T_g soft core and a high-T_g hard shell can be processed by compression molding or extrusion at temperatures as low as room temperature. The white PBA-PMMA powders changed to the transparent films upon compression molding at a pressure of 35 MPa and room temperature (Fig. 4a), and the GO/PBA-PMMA composite changed from yellow powders to brown films after processing (Fig. 4b). However, the RGO/PBA-PMMA composite films appeared to be black-colored from grey powders due to the chemical reduction of GO (Fig. 4c). The low-temperature processing mechanism of PBA-PMMA and its composites results substantially from a pressure-induced miscibility between two-polymer phases, in which the applied pressure creates a mobile phase around portions of the glassy phase. Therefore, the low-T_g PBA layer acts as a "solvent" for the PMMA component. This enables flow of baroplastic materials and facilitates cohesion to the processed products.[16, 17]

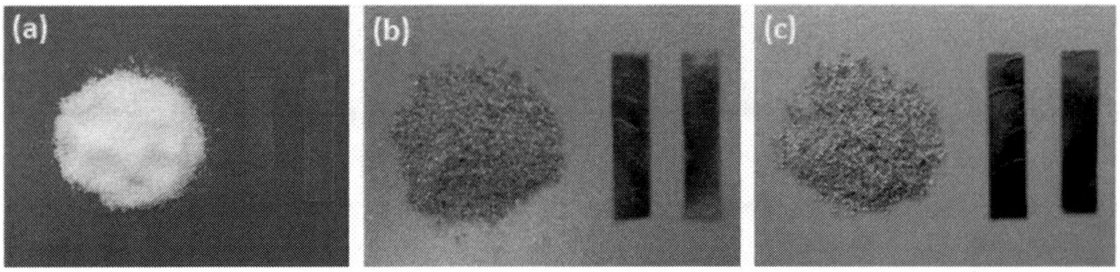

Figure 4. Original dried-powders of (a) neat PBA-PMMA, (b) GO/PBA-PMMA and (c) RGO/PBA-PMMA composites and their films processed by compression molding under 35 MPa for 5 min at room temperature.

Figure 5 shows TEM images of GO and graphene sheets dispersed in the composites processed by compression molding under 35 MPa for 5 min at room temperature. GO is readily dispersed in water or organic solvents because of polar groups,[18] while graphene sheets obtained by chemical reduction of GO tend to restack as a graphite-like structure.[19] Therefore, GO should show better dispersion than graphene in the latex and subsequent polymer matrix. As shown in Figure 5a, thin GO sheets with in-plane dimension (0.5~2 μm) are homogeneously dispersed in the matrix. In contrast, most of graphene sheets restack to form thick clusters in the matrix (Fig. 5b) due to the fact that the reduction of GO into graphene results in its hydrophobicity and re-agglomeration. It is noteworthy that these images imply that GO and graphene sheets might preferentially form segregated structures around the latex nanoparticles during mixing and precipitation, and the segregated sheets remain well after low-temperature compression molding. It is rationally accepted that these regions marked by white circles are supposed to be the latex particles before processing and the polymer phase after molding. Similar phenomena have been observed for Cu nanowire/PS [20] and graphene/PC [21] composites. Moreover, it should be noted that the core-shell nanostructure of PBA-PMMA could not be clearly discriminated in our TEM measurements because the PBA core or PMMA shell was not selectively stained by chemicals such as RuO_4 as described elsewhere.[17]

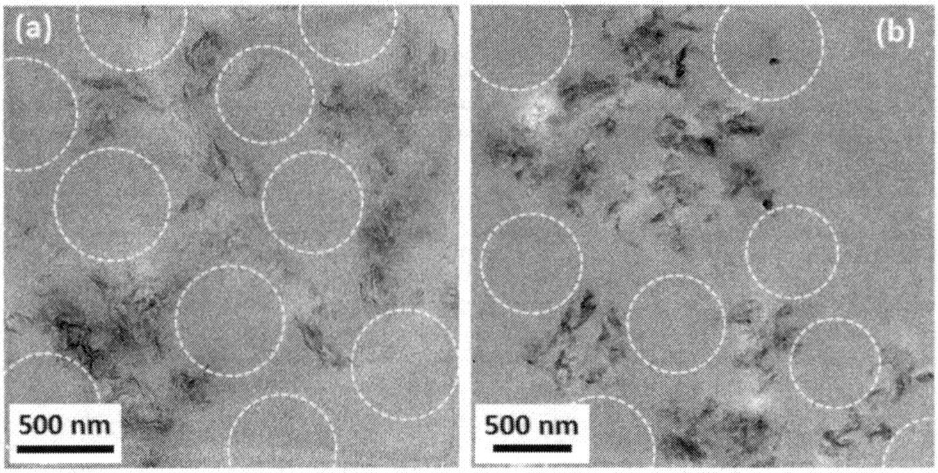

Figure 5. TEM images of (a) cryomicrotomed GO/PBA-PMMA, (b) RGO/PBA-PMMA composites, and white dash-circles correspond to the polymer phase coming from the core-shell latex nanoparticles.

Tensile tests were carried out on three kinds of baroplastic materials to examine their mechanical properties. Figure 6 shows the stress-strain curves for three materials processed by compression molding for 5 min with varied pressures (15-55 MPa) at room temperature. Tensile properties were found to be sensitive to the applied pressure, with Young's modulus and yield strength increasing with external pressure for each material. It can be also seen that both Young's moduli and yield strength of all baroplastic composites gradually increase with processing time (from 3 min to 9 min) at a fixed pressure (35 MPa) and room temperature, as shown in Figure 7. The improved tensile properties are presumably due to better flowability, processability and cohesion of baroplastics induced by higher pressure and longer time.[17]

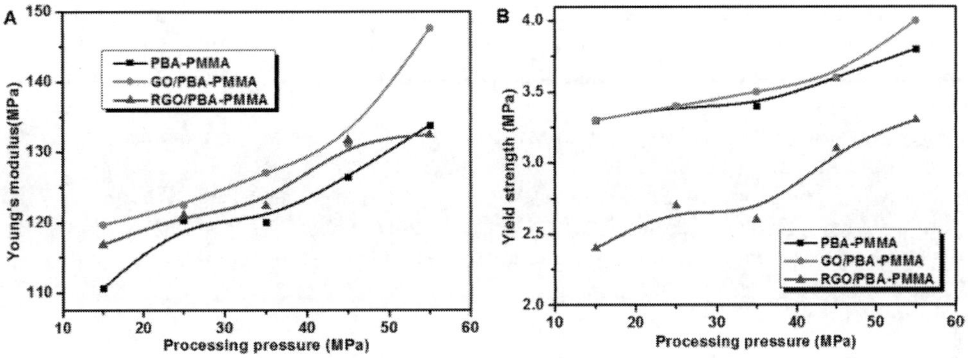

Figure 6. Young's modulus (A) and yield strength (B) of PBA-PMMA, GO/PBA-PMMA and RGO/PBA-PMMA composite films processed under different pressures for 5 min at room temperature.

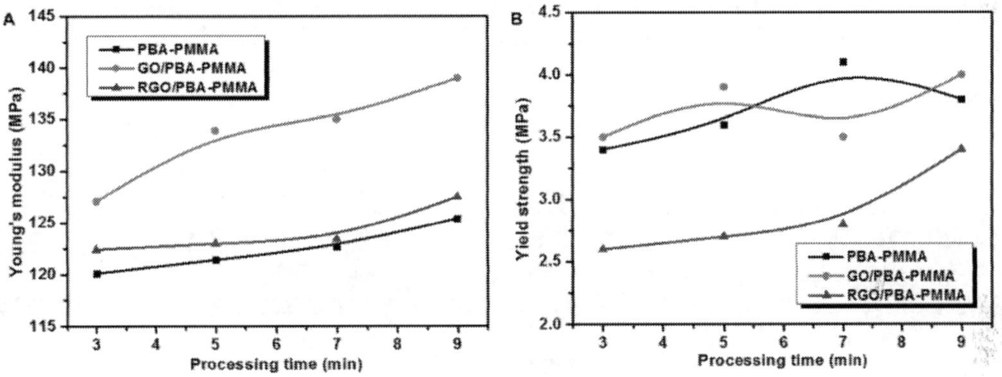

Figure 7. Young's modulus (A) and yield strength (B) of PBA-PMMA, GO/PBA-PMMA and RGO/PBA-PMMA composite films processed under 35 MPa at room temperature for different time.

Furthermore, incorporating graphene into the PBA-PMMA matrix slightly improves the composite's moduli but reduces yield strength because of the inefficient stress transfer and aggregated graphene sheets yielding high stress concentration sites in the polymer matrix.[22] However, both its modulus and strength are enhanced by adding the same loading of GO to PBA-PMMA. It is evident that the overall tensile properties of GO/PBA-PMMA composite show better improvements than RGO/PBA-PMMA relative to neat PBA-PMMA under identical processing conditions and comparable loading levels. This finding can be explained by three factors. First, GO shows better dispersion than graphene in the polymer matrix as shown in Figure 5. Second, oxygen-containing functional groups around GO can interact with pendant carbonyl groups of polymer chains *via* hydrogen bonding, forming a stronger interface with the matrix relative to graphene bearing less groups.[23] Third, GO is an insulator and not an optimum electrically-conductive filler for polymer composites, however, the moduli of GO sheets (208 GPa) [24] is very close to that of chemically-derived graphene (250 GPa).[25] This indicates a comparable efficacy of GO and reduced GO for the improvement of mechanical properties as previously reported in the PMMA composites.[26] Therefore, better dispersion and stronger interaction of GO relative to graphene in/with the polymer matrix achieve greater improvements in tensile properties for GO/PBA-PMMA composite.

4. Conclusions

Core-shell latex nanoparticles consisting of a PBA core and a PMMA shell were synthesized by a two-stage emulsion polymerization technique. Incorporation of GO aqueous solution into the PBA-PMMA latex successfully fabricated GO/PBA-PMMA and RGO/PBA-PMMA composites by *in-situ* reduction of GO. Coverage of a low-T_g soft core by a high-T_g hard shell resulted in powdery materials, which could be further processed into thin films by compression molding at room temperature due to the pressure-induced miscibility of microphase-separated baroplastics. GO showed better dispersion and stronger interaction in/with the matrix relative to graphene, thus achieving greater improvements in tensile properties for GO/PBA-PMMA composite. In addition, tensile properties for all materials were gradually improved with increasing external pressure and processing time because of better flowability, processability and cohesion between phases at higher pressure and longer time. The pressure-induced processing of baroplastic materials at low temperature reduces energy consumption, thermal degradation and processing time during plastic processing and also improves the recyclability.

Acknowledgments

We acknowledge the financial support from the National Natural Science Foundation of China (51073050, 51273057), the Program for New Century Excellent Talents in University (NCET-12-0709) and the Open Project of State Key Laboratory of Materials Processing and Die & Mould Technology (2012-P10) at Huazhong University of Science and Technology. CPT would like to thank for the support from Research Committee of The Hong Kong Polytechnic University (G-UB32).

References

[1] Allen M J, Tung V C and Kaner R B 2010 *Chem. Rev.* **110** 132
[2] Novoselov K S, Falko V I, Colombo L, Gellert P R, Schwab M G and Kim K 2012 *Nature* **490** 192
[3] Kim H, Abdala A A and Macosko C W 2010 *Macromolecules* **43** 6515
[4] Kuilla T, Bhadra S, Yao D, Kim N H, Bose S and Lee J H 2010 *Prog. Polym. Sci.* **35** 1350
[5] Peng RG, Wang YZ, Tang W, Yang YK and Xie XL 2013 *Polymers* **5** 847
[6] Pham V H, Cuong T V, Dang T T, Hur S H, Kong B-S, Kim E J, Shin E W and Chung J S 2011 *J. Mater. Chem.* **21** 11312
[7] Kim H and Macosko C W 2009 *Polymer* **50** 3797
[8] Yang YK, He CE, Peng RG, Baji A, Du XS, Huang YL, Xie XL and Mai Y-W 2012 *J. Mater. Chem.* **22** 5666
[9] Park S, Ruoff R S 2009 *Nat. Nanotechnol.* **4** 217
[10] Cai W, Piner R D, Stadermann F J, Park S, Shaibat M A, Ishii Y, Yang D, Velamakanni A, An S J, Stoller M *et al* 2008 *Science* **321** 1815
[11] Yang YK, He CE, He WJ, Yu LJ, Peng RG, Xie XL, Wang XB and Mai Y-W 2011 *J. Nanopart. Res.* **13** 5571
[12] Loh K P, Bao Q, Eda G and Chhowalla M 2010 *Nat. Chem.* **2** 1015
[13] Zhang L, Wang Z P, Xu C, Li Y, Gao J P, Wang W and Liu Y 2011 *J.Mater. Chem.* **21** 10399
[14] Cao Y W, Zhang J, Feng J C and Wu P Y 2011 *Acs Nano* **5** 5920
[15] Mao Y, Wen S, Chen Y, Zhang F, Panine P, Chan T W, Zhang L, Liang Y and Liu L 2013 *Sci. Rep.* **3** 2508
[16] Gonzalez-Leon J A, Acar M H, Ryu S-W, Ruzette A-V G and Mayes A M 2003 *Nature* **426** 424
[17] Gonzalez-Leon J A, Ryu S-W, Hewlett S A, Ibrahim S H and Mayes A M 2005 *Macromolecules* **38** 8036
[18] Paredes J I, Villar-Rodil S, Martinez-Alonso A and Tascon J M D 2008 *Langmuir* **24** 10560
[19] Stankovich S, Dikin D, Piner R, Kohlhaas K, Kleinhammes A, Jia Y, Wu Y, Nguyen S and Ruoff R 2007 *Carbon* **45** 1558
[20] Gelves G A, Al-Saleh M H and Sundararaj U 2011 *J. Mater. Chem.* **21** 829
[21] Yoonessi M and Gaier J R 2010 *Acs Nano* **4** 7211

[22] Yang YK, Yu LJ, Peng RG, Huang YL, He CE, Liu HY, Wang XB, Xie XL and Mai Y-W 2012 *Nanotechnology* **23** 225701

[23] Terrones M, Martin O, Gonzalez M, Pozuelo J, Serrano B, Cabanelas J C, Vega-Diaz S M and Baselga J 2011 *Adv. Mater.* **23** 5302

[24] Suk J W, Piner R D, An J and Ruoff R S 2010 *Acs Nano* **4** 6557

[25] Gomez-Navarro C, Burghard M and Kern K 2008 *Nano Lett.* **8** 2045

[26] Potts J R, Lee S H, Alam T M, An J, Stoller M D, Piner R D and Ruoff R S 2011 *Carbon* **49** 2615

The preparation of nanosized polyethylene particles via novel heterogeneous non-metallocene catalyst (*m*-$CH_3PhO)TiCl_3/CNTs/AlEt_3$

J Wang,[1] J P Guo,[1] J J Yi,[2] Q G Huang,[1,]* H M Li,[2] Y F Liu,[1,]* K J Gao[2] and W T Yang[1]

[1] State Key Laboratory of Chemical Resource Engineering, Key Laboratory of Carbon Fiber and Functional Polymers, Ministry of Education, Beijing University of Chemical Technology, Beijing 100029, People's Republic of China
[2] Lab for Synthetic Resin Research Institution of Petrochemical Technology, China National Petroleum Corporation, Beijing 100083, People's Republic of China

E-mail: huangqg@mail.buct.edu.cn

Abstract: This paper reports the preparation of coral-shaped topological morphology nascent polyethylene (PE) particles promoted by the novel heterogeneous non-metallocene catalyst (*m*-$CH_3PhO)TiCl_3$/carbon nanotubes (CNTs), with $AlEt_3$ used as a cocatalyst. Scanning electron microscope (SEM), high resolution transmission electron microscope (HR-TEM) and inductively coupled plasma (ICP) emission spectroscopy were used to determine the morphology of the catalyst particles and the content of (*m*-$CH_3PhO)TiCl_3$. The carbon nanotube surface was treated with Grignard Reagent prior to reacting with (*m*-$CH_3PhO)TiCl_3$. The catalyst system could effectively catalyze ethylene polymerization and ethylene with 1-hexene copolymerization, the catalytic activity could reach up to 5.8 kg/((gTi)h). Morphology of the obtained polymer particles by SEM and HR-TEM technique revealed that the nascent polyethylene particles looked like coral shape in micro-size. The multiwalled carbon nanotubes (MWCNTs) supported catalysts polymerized ethylene to form polymer nanocomposite in situ. The microscopic examination of this nanocomposite revealed that carbon nanoparticles in PE matrix had a good distribution and the cryogenically fractured surface was ductile-like when polymerization time was 2 min.

Keywords: Carbon nanotube; Non-metallocene catalyst; Nanocomposite; Particle morphology.

1. Introduction

An important feature of heterogeneous Ziegler-Natta catalyst is that the polymer particle shape can duplicate from the catalyst particle, which is useful for overcoming the fouling obstacle in industry procedure. Porous materials such as SiO_2 and $MgCl_2$ have been applied as supports for the preparation of heterogeneous Ziegler-Natta catalyst for olefins polymerization [1]· Basset [2] reported that metallacalixarenes anchored to mesoporous SiO_2 are efficient catalysts for propylene polymerization with a high initial activity. Copéret [3] prepared SiO_2- supported chloro alkyl aluminum co-catalyst which is a highly efficient co-catalyst for the Ni-catalyzed dimerization of ethylene. Our previous work [4] reported that $MgCl_2$-supported Ziegler-Natta catalysts with different structure ligands have efficient catalytic activity for the copolymerization of ethylene with 1-octene. The comonomer incorporation content of the copolymer is related to the catalysts' ligand structure. Carbon nanotubes (CNTs) have attracted much attention because of their excellent mechanical, thermal and electrical properties. Based on these distinctive advantages, they are considered to be promising fillers in various polymers. Kaminsky [5] reported that carbon black (CB), carbon nanofibers (CNFs) and different types of CNTs were first separated by ultra sound and then treated with methylaluminoxane (MAO). These catalytic active centers efficiently promoted ethylene/propylene polymerization and polyolefins nanocomposites obtained after adding zirconocenes or other transition metal complexes directly on the surface of the fillers. Recently, Wang [6] revealed that the active site $TiCl_4$ is directly bonded with the support oxidized nanosized carbon spheres through Ti-O bond. In this work, we report the novel coral-shaped and core-shell structure polymer particles produced by the non-metallocene catalyst (m-CH3PhO)$TiCl_3$, which was directly bonded on the oxidized multiwalled carbon nanotubes (MWCNTs) through Ti-O bond, with $AlEt_3$ used as a cocatalyst. The morphology of the catalyst particles and the obtained polymer particles including polyethylene and copolymer of ethylene with 1-hexene, as well as the performance of the catalyst for olefins polymerization, were investigated.

2. Experimental

All operations of air- and moisture-sensitive materials were performed using the rigorous repellency of oxygen and moisture in flamed Schlenk-type glassware on a dual manifold Schlenk line under a nitrogen atmosphere. CH_3MgCl with 22 wt % in tetrahydrofuran (THF) and triethylaluminum ($AlEt_3$) with 1.0 M in hexane were purchased from Acros Organics Agent in China. MWCNTs were in diameter of about 50 nm and length of about 15 - 20 μm. $AlEt_3$ (2.0 M in n-hexane) and *m*-cresol were purchased from Acros in China. 1-hexene were from Fluka in China. Other chemicals were from Beijing chemical agent company.

3. Results and discussion

Table 1 . Catalysts system (*m*-CH_3PhO)$TiCl_3$/CNTs/$AlEt_3$ for the (co)polymerization of ethylene with 1-hexene

Runs	**support**	**Common.**	**Time (min)**	**A**[a] **(kg/gTi. h)**	**C-Cont**[b] **(mol%)**	M_n[c] $(\times 10^{-4})$	M_w/M_n[c]	T_m[d]
1[e]	—	1-hex.	60	5.6	2.38	3.2	2.3	130.6
2	CNTs	—	2	5.2	—	3.2	4.8	134.2
3		1-hex.	2	5.8	2.64	2.8	4.9	128.4
4		1-hex.	10	5.7	2.44	2.8	5.2	129.5
5		1-hex.	120	5.5	1.55	2.9	5.3	129.8
6f		1-hex.	10	3.5	2.35	3.1	5.1	129.6
7g		1-hex.	10	2.5	1.23	0.9	5.6	126.3

Polymerization conditions: the catalyst was 65 mg(Ti: 5.4×10^{-5} M); ethylene pressure was 0.20 MPa, hydrogen pressure was 0.1 MPa; polymerization temperature was 80 °C; *n*(Al): *n*(Ti) = 100; *n*-hexane (as solvent) was 1000 mL; comonomer content in feed was 0.3 mol/L; and Ti content of the catalyst (*m*-CH_3PhO)$TiCl_3$/CNTs was 4.0 wt%.

[a]: catalytic activity.

[b]: comonomer molar content in copolymer determined by ^{13}C NMR.

[c]: numerical average molecular weight and molecular weight distribution of the polymers determined by GPC.

[d]: melting point of the polymers determined by DSC.

[e]: solution polymerization in *n*-hexane by catalyst system of (*m*-$CH_3PhO)TiCl_3/AlEt_3$.

[f]: polymerization temperature, 60 °C

[g]: polymerization temperature, 95 °C

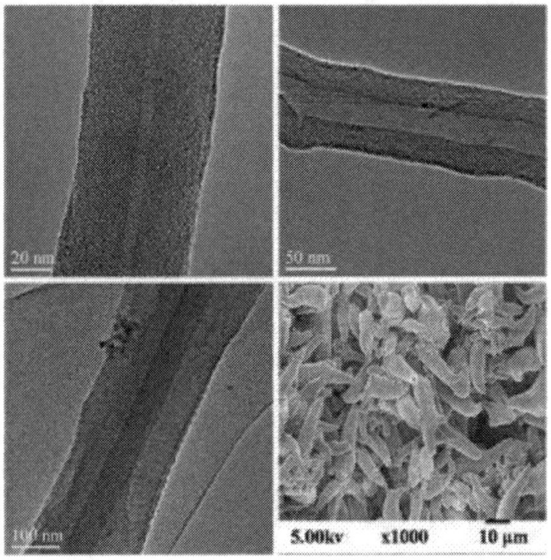

Figure 1. HR-TEM image of the heterogeneous catalyst particle of (*m*-$CH_3PhO)TiCl_3$/CNTs(a) HR-TEM and SEM images of the copolymer particles produced by the heterogeneous catalyst (*m*-$CH_3PhO)TiCl_3$/CNTs after polymerization time of (b) 2 min, (c) 10 min, and the SEM patterns of the copolymer particles (d) by catalyst (*m*-$CH_3PhO)TiCl_3$/ CNTs/$AlEt_3$

4. Conclusions

In this work, a non-metallocene catalyst system $TiCl_3$(*m*-$OPhCH_3$)/CNTs/$AlEt_3$ was prepared. The mass fraction of titanium in the catalyst $TiCl_3$(*m*-$OPhCH_3$)/CNTs was 4.0 wt %, and the catalytic activity was up to 5.8 kg/((gTi)h). The kinetics revealed that the product mass was increased linearly with increasing polymerization time. Coral-shaped copolymer particle and nanocomposites were formed *in-situ* utilizing $TiCl_3$(*m*-$OPhCH_3$)/CNTs/$AlEt_3$ for ethylene copolymerization. The ^{13}C NMR analysis confirmed that 1-hexene content of copolymer was up to 2.64 mol% when the concentration of 1-hexene in feed was 0.3 mol/L. This work indicated that the catalyst system $TiCl_3$(m-$OPhCH_3$)/CNTs/$AlEt_3$ could improve the incorporation of 1-hexene within the copolymer chain and fairly uniform nanocomposites could be made in situ.

Acknowledgment

We sincerely thank the National Natural Science Foundation of China (No. 21174011) and the Natural Science Foundation of Beijing, China (No. 2102036).

References

[1] Chung J S, Cho H, Ko Y and Young W 1999 Preparation of the Ziegler–Natta/ metallocene hybrid catalysts on $SiO_2/MgCl_2$ bisupport and ethylene polymerization *J. Mol. Catal. A: Chem.* 144, pp 61-67

[2] Espinas J, Pelletier J, Szeto K and Basset J 2014 Preparation and characterization of metallacalixarenes anchored to a mesoporous silica SBA-15 LP as potential catalysts *Microporous and Mesoporous Mater.* 188 , pp 77-85.

IOP Conf. Series: Materials Science and Engineering **62** (2014) 012024 doi:10.1088/1757-899X/62/1/012024

[3] Kermagoret A, Kerber R, Conley M, Callens E and Copéret C 2014 Chlorodiethylaluminum supported on silica: a dinuclear aluminum surface species with bridging μ^2-Cl-ligand as a highly efficient co-catalyst for the Ni-catalyzed dimerization of ethane *Catal.* 313, pp 46-54.

[4] Kong Y, Yi J, Dou X, Liu W, Huang Q G Gao K and Yang W 2010 With different structure ligands heterogeneous Ziegler–Natta catalysts for the preparation of copolymer of ethylene and 1-octene with high comonomer incorporation. *Polymer* pp 3859-3866.

[5] Wiemann K, Kaminsky W, Gojny F and Schulte K 2005 Synthesis and properties of syndiotactic poly(propylene)/carbon nanofiber and nanotube composites prepared by in situ polymerization with metallocene/MAO catalysts *Macromol. Chem. Phys. 206(15),* pp 1472-1478.

[6] Wang J, Yu M , Jiang W, Zhou Y, Li F, Cheng L, Yi J, Huang Q, Liu Y and Yang W 2013 The preparation of nanosized polyethyleneparticles via carbon sphere nanotemplates. Ind. Eng. Chem. Res. 52, pp 17691−17694.

IOP Conf. Series: Materials Science and Engineering **62** (2014) 012025 doi:10.1088/1757-899X/62/1/012025

Physical changes associated with gamma doses on Wood/ Polypropylene Composites

Diène NDIAYE[1]*, Adams TIDJANI[2]

[1]Université Gaston Berger de Saint-Louis, Sénégal
[2]Université Cheikh Anta Diop de Dakar, Sénégal

* *Corresponding author:* diene.ndiaye@ugb.edu.sn

Abstract: The effect of gamma- radiation on the morphology, thermal behavior and mechanical properties of wood polypropylene composites has been investigated. Simultaneous thermo gravimetric analysis (TGA) and differential scanning calorimetry (DSC) have been performed on WPC samples of (9.5 ± 0.1) mg. These samples were exposed to different gamma-dose in the range 10 to 100kGy. The results indicated that gamma radiation improves the mechanical properties while the thermal stability is decreased. With gamma radiation, the scanning electron microscopy (SEM) of the micrographs became smoother and we can notice an improvement of interaction between polymer and wood fibers.

Keywords: Wood polymer composites (WPC), morphology, gamma radiation, thermal stability, mechanical properties.

1. Introduction

The use of wood fibers as a filler is a choice offering an economical solution for the increasing costs of wooden products and construction materials. There is a considerable commercial interest in thermoplastic composites filled with wood fibers, due to the potential opportunities. The product has the aesthetic appearance of wood and the processing capability of thermoplastics and its performance in humid area. These composites have been widely studied. Properties of WPCs depend on the characteristics of matrix and fillers, chemical interaction between wood fibers and polymer, humidity absorption and processing conditions. The material can be considered as an easily attainable (natural) option, is competitive as far as price is concerned and convenient for a wide range of applications. Disadvantages associated to the use of natural fibers as reinforcement in thermoplastics are the result of a lack of a good interfacial adhesion and a poor resistance to humidity absorption. The first key point for the production of acceptable WPC is the compatibility between fibers and the polymer. In particular, the hydrophobic character of polymeric material (low surface tension) which contrasts with the hydrophilic character (high surface tension) of wood fibers [1]. Scanning the literature, one can find different surface treatments that have been experienced to improve wood/polymer adhesion in composites. Remind that the level of adhesion and/or the dispersion state of wood are the key points for the improvement of mechanical properties of the composites. Indeed, the wood particles which have high strength and modulus – with good adhesion and uniform dispersion – can impart better mechanical properties to the host polymer in order to obtain a composite with better properties than those of the unfilled polymer. The main solutions found to improve compatibility are the use of coupling agents, pretreatment of wood fiber and/or the polymer through surface coating treatment or graft copolymerization which improves mechanical properties, water absorption and dispersion [2, 3]. The most commonly used compatibilizer in WPC is usually polypropylene grafted with maleic anhydride (MAPP). It is proven in several studies that the nucleation density on the surface of the MAPP is very significant because the increase in the interfacial adhesion between the reinforcement and the MAPP. The thermal stability of composites is also improved by the addition of MAPP in the formulation compared to composites without MAPP. The compatibilizer also influences the uptake of water and leads to a decrease in water sensitivity of the composite from the same composite without MAPP. This reduction can be attributed to an increase in interfacial adhesion that reduces water accumulation in the interfacial areas and prevents water from penetrating into the fibers. This is what justifies the use of MAPP in this work to improve the dimensional stability of the irradiated material. Another limitation of using wood fibers in WPC outdoor applications is the high sensitivity to UV-light degradation [4]. Regardless of formulation, the introduction of wood fiber into a PP leads to a change in the morphology of the crystallizing polymer. The use of ionizing radiation on polymers continues to draw great attention because it modifies them significantly. Polymers exposed to ionizing radiation, even at low doses, often undergo structural changes accompanied by molecular crosslinking, grafting and chain scission reactions [5, 6, 7]. Using ionizing radiation represents an economic advantage, because it reduces the use of additives in the formulas. With suitable doses, it is the only technique that uses electromagnetic energy to generate significant changes. Ionizing radiation (gamma rays and accelerated electrons γ - emitted by a source of ^{60}Co or ^{137}Cs) were developed industrially since the 1960s for sterilizing medical and surgical equipment and preservation of food products. They also led to the development of chemistry radiation, radical type, primarily applied to polymers. By triggering a chemical process of formation of free radicals, ionizing radiation can then initiate a number of chemical reactions such as chains scissions, addition, polymerization, cures, etc.., which can lead to various applications like: degradation or crosslinking of plastics (including improving recyclability), wood, impregnated with resin or grafting on polymers. The objective of this study was to use the technique based on the γ irradiation of composites as an alternative to coupling or compatibilizing agents. It was used to improve the filler/matrix interface. This technique introduces energy into the material to generate favorable changes in its morphology. Scanning the literature, one can find a competition between scissions and crosslinking reactions in irradiated polymers. The prevailing effect brings about macroscopic changes in the molecular mass distributions, continuously affecting transition temperatures, diffusion coefficients and mechanical properties during polymer irradiations. Reyes et al. [8] studied the effect of gamma irradiation on the mechanical properties of polypropylene. The carbon- chlorine and carbon-fluorine bonds well are known radiation-sensitive, this severely limits the use of two industrial polymers, poly (vinyl chloride) (PVC) and poly (tetra fluoro ethylene) (PTFE) in radiation area. So in our study, we used PP. The results obtained showed that general behavior of the polymer blends was similar, being 10 to 50 kGy the most favorable irradiation range due to the remarkable improvement the studied mechanical properties experience within it. Adams et al. [9] examined the Gamma-oxidation of linear low-density polyethylene: The dose–rate effect of irradiation on chemical and physical modifications. Their results showed that the lower the dose rate, the higher the degree of oxidation in terms of γ-product formation, which were very dependent on the dose rate of initiation.

2. Experimentals

2.1 Materials

The basic materials used in this study are listed below: The wood fiber were kindly donated by American Wood fibbers (Schofield, WI) and are constituted predominantly with ponderosa pine, maple, oak, spruce, southern yellow pine, cedar. The isotactic polypropylene matrix (PP) has a density of 0.900 g/cm3 and a melt flow index of 2.5 g/10 min, it was provided by Solvay Co. - Polypropylene grafted with maleic anhydride (MAPP with an approximate maleic anhydride (MA) content of 3 wt. % and molecular weight of 40 000g/mol was purchased from Aldrich Chemical Company, Inc. (Milwaukee, WI). All ingredients were used as received.

2.2 Compounding and processing

Before compounding, the WF was dried in an oven for at least 48 h at 105°C to a moisture content of less than 1%. The dried WFs were stored in a sealed plastic container to prevent the absorption of water vapor. First, the PP was put in the high-intensity mixer (Papenmeier, TGAHK20, Germany), and the WF was added after the PP had reached its melting temperature. The mixing process took 10 min on average. After blending, the compounded materials were stored in a sealed plastic container. Several formulations were produced with various contents of PP, WF and MAPP. For the extraction of volatile and harmful gases, the hood was open. For the mechanical property experiments, test specimens were molded in a 33-Cincinnati Milacron reciprocating screw-injection molded (Batavia, OH). The nozzle temperature was set to 204°C. The extrudate, in the form of strands, was cooled in the air and pelletized. The resulting pellets were dried at 105°C for 24 h before they were injection-molded into the ASTM test specimens for flexural, tensile (Type I, ASTM D638), and Izod impact strength test. The dimensions of the specimens for the flexural tests were 120x3x12mm3 (length x thickness x width). The composition of the sample used in our study was in percent (PP/Wood/MAPP): 47/50/03. In our previous work 10, it was proven that the use of MAPP improved the adhesion and dispersion of the reinforcement in the matrix. So we used this composite to see the effect of gamma radiation on some properties.

2.3 Gamma irradiation

Specimens had been sealed in glass tubes filled with oxygen, prior to being irradiated integral doses at 80°C. The samples were irradiated with γ -rays at radiation doses of 10, 25, 50, 100 kGy using a dose rate of 4.8 kGy/h in air, at room temperature with a 60Co source. 60Co is the most common gamma emitter used because of its abundance in the market and the operational safety of devices that use it. The experimental measurements were taken after two days of storage under oxygen. The effect of gamma irradiation on the thermo-mechanical properties was then monitored. Ionizing radiation penetrates the wood structure and produces a uniform degree of polymerization. So, it can create radical sites on the cellulose chains from which branch vinyl polymer can grow to give graft copolymer [11]·

2.4 Scanning electron microscopy (SEM).

The state of dispersion of the wood inside the polymeric matrix was analyzed using optical microscopy on samples of 100–200 μm thick. The fractured samples were coated with a thin layer of gold to avoid electrostatic charging during examination. Scanning electron microscopy (SEM) was used to obtain microphotographs of the fracture surfaces of the wood composites. These fractures have been performed in liquid nitrogen to avoid any deformation. SEM has been performed using a Fei Quanta 400 microscope working at 30 kV. The polymer surface was examined with Leica optical microscope working in a transmission mode. Samples were thin enough that no special preparation was needed for their observations with the optical microscope.

2.5 Differential scanning calorimetry (DSC)

DSC is widely used to characterize the thermal properties of WPCs. DSC can measure important thermoplastic properties, including the melting temperature (Tm), heat of melting, degree of crystallinity χ (%), crystallization, and presence of recyclates/ regrinds, nucleating agents, plasticizers, and polymer blends (the presence, composition, and compatibility). Thermal analysis of the WPC samples was carried out on a differential scanning calorimeter (Perkin Elmer Instruments, Pyris Diamond DSC, and Shelton, Connecticut) with the temperature calibrated with indium. All DSC measurements were performed with powdered samples of about 9.5 ± 0.1 mg under a nitrogen atmosphere with a flow rate of 20 mL/min. Three replicates were run for each specimen. All samples were subjected to the same thermal history with the following thermal protocol, which was slightly modified from the one reported by Valentini et al. [12]:

1. First, the samples were heated from 40 to 180°C at a heating rate of 20°C/min to erase the thermal history.

2. Second, the samples were cooled from 180 to 40.00°C at a cooling rate of 10°C/min to detect the crystallization temperature (Tc).

3. Finally, the samples were heated from 40 to 180°C at a heating rate of 10°C/min to determine T_m. T_m and the heat of fusion (ΔH_m) were obtained from the thermograms during the second heating. The values of ΔH_m were used to estimate [χ (%)], which was adjusted for each sample.

2.6 Thermogravimetric analysis (TGA)

This analysis consists to record the weight of sample in dynamic conditions at a heating rate of 10 °C/min between 20°C and 700°C under an inert atmosphere of nitrogen. The experience was performed using Perkin equipment Elmer Pyris-1 TGA. The sample (masse of $(10 \pm 1)mg$) was placed in a little cup made of aluminium hanging from a microbalance. The variation of the mass of the sample allows drawing the TG (variation of the mass in function of the temperature) and TGD (derivative of loss of mass versus the time) thermograms. The combination of these two thermograms gives a clear indication of number of stages of the thermal degradation. This method allows determining the degradation temperature of the materials and thus the thermal stability of the composite.

2.7 Mechanical properties

Tensile tests (tensile strength and tensile strain) and three-point flexural tests (flexural modulus and flexural strength) were carried out on an Instron 5585H testing machine (Norwood, MA) with crosshead rates of 12.5 and 1.35 mm/min according to the procedures outlined in ASTM standards D 638 and D 790, respectively. Eight replicates were run to obtain an average value for each formulation. Before each test, the films were conditioned in a 50% relative humidity chamber at 23°C for 48 h. The impact strength is defined as the ability of a material to resist the fracture under stress applied at a high speed. The impact properties of composite materials are directly related to their overall toughness. The Izod impact strength test is the total energy required to break a notched sample. It was measured with an Instron impact pendulum tester (model PW5) according to ASTM D 256 with acutely notched specimens (notch depth 2 mm) at room temperature.

3. Results and Discussions

3.1SEM of the composites

Fig.1 shows the SEM micrographs of the fracture surface of the WP composites made from PP. Unirradiated sample (a) and irradiated onces (b and c) show similar behaviour, since irradiation does not influence dispersion, but could affect polymer-filler interaction. Fig.1 (a) shows the pull out effect in untreated wood fiber/polymer composite, but from the fig.1 (b) and 1(c) (irradiated samples), there is less pull. It revealed that there is poor adhesion between the wood fiber and polymer matrix in 1(a) and better adhesive force between wood and polymer matrix in the other ones. In the irradiated samples, it can be seen that the effects of radiation depend on the dose applied to the sample. Moreover, for these samples, a loss of weight (less than 1% after 50 KGy irradiation and about 1.25% after the 100 KGy dose) was observed and a visible yellowing appeared already after 50 KGy. These phenomena were most probably signs of degradation. Smirnov and Deyun [13] showed that the average molecular weight of PP, when subjected to gamma radiation, did not show significant variations when the radiation dose was under 10 kGy. The irradiated samples exhibited a slight improvement of more smooth surfaces than those unirradiated, probably due to a reorganization of the matrix chains during irradiation. The most evident changes were achieved already after a dose of 100 KGy. When the irradiation dose increases, the crosslinking density between the resultant copolymer chains increases. This restricts the expansion of the chains of the prepared copolymer and decreases the amorphous phase.

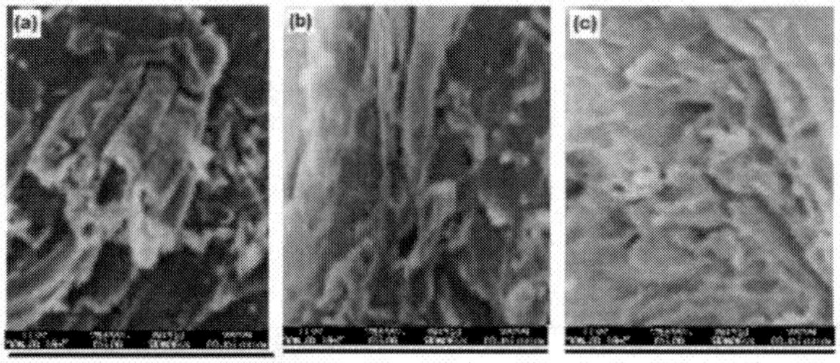

Fig.1. Electron micrographs of WPC: (a) unirradiated, γ -irradiated (b) 50 kGy, (c) 100 kGy.

3.2DSC of the composites

Figure 2 presents the thermograms of the second heating of the PP/wood flour blends subjected to different irradiation doses. The curves were vertically moved for clarity. Only one endothermic peak corresponding to PP can be observed in these figures. As can be seen, these values remain constant at low irradiation doses (< 25 kGy) and then tend to increase as dose increases. This behavior is generated by the oxidative degradation of PP when it is submitted to the irradiation dose, which promotes the scission of the polymer chains. Gamma-radiation at lower doses (0-10) kGy does not give a visible change in the melting behavior of WPC.

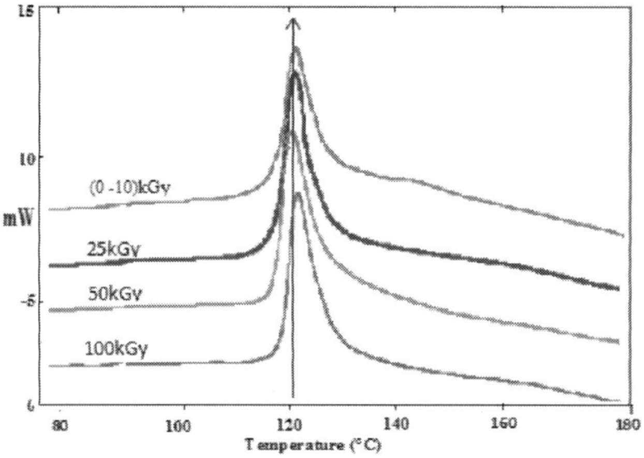

Fig. 2. The effect of γ -irradiation on the DSC thermograms (melting) of WPC with MAPP: unirradiated and irradiated samples.

In figure 3, around 160 ° C, we observe the presence of a peak corresponding to the melting of pure polypropylene. The thermograms of the irradiated composites (25, 50,100kGy) have the same melting peak shifted slightly but these thermograms have an additional shoulder at around 175 ° C. We can provide two possible explanations: the radiation dose is strong enough to change the crystallization

IOP Conf. Series: Materials Science and Engineering **62** (2014) 012025 doi:10.1088/1757-899X/62/1/012025

of PP and the reappearance of a new population of smaller crystals which fuse at a lower temperature. This shoulder comes directly from the reinforcements which are sensitive to gamma rays. The presence of fibers decreases the thermal stability, radiation in turn causes scissions of chains and all these phenomena generate an early fusion and it is more pronounced whit samples without coupling agent. The presence of MAPP does not significantly modify the crystallization temperature but leads to an increase in the degree of crystallinity. It is recognized that wood and MAPP act as nucleating agents [10,] [14], the presence of these two elements generates the formation of more crystals. However, the higher the radiation dose increases, we see that the merger tends to begin at an early stage. The fact that Tc is shifted toward higher temperatures at the maximum dose of (100kGy) can be explained by a higher irradiation dose acted as a slight precursor of crystallization.

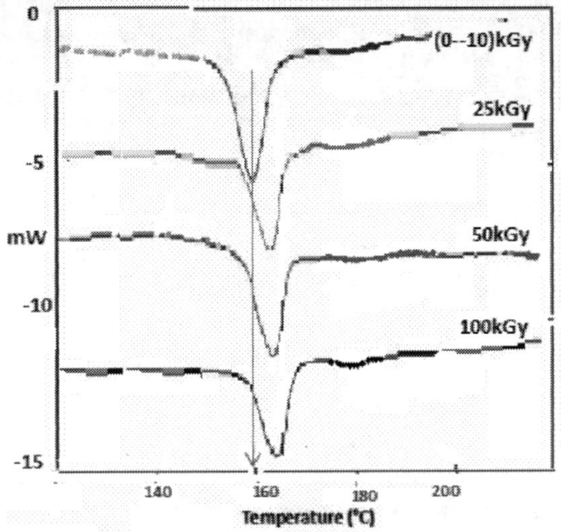

Figure 3. The effect of γ -irradiation on the DSC thermograms (melting) of WPC: unirradiated and irradiated samples.

3.3 TGA of the composites

Thermal gravitational analysis (TGA) provides quantitative information on weight change process while differential thermal gravimetry (DTG) provides the rate of weight loss (dw/dt).

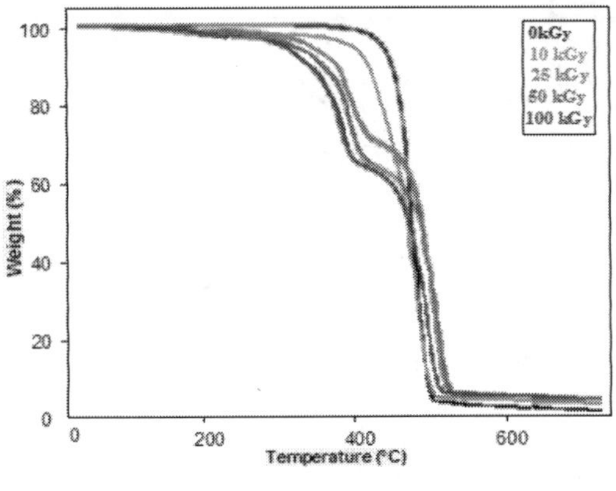

Figure 4: TGA curves of WPC with different gamma doses.

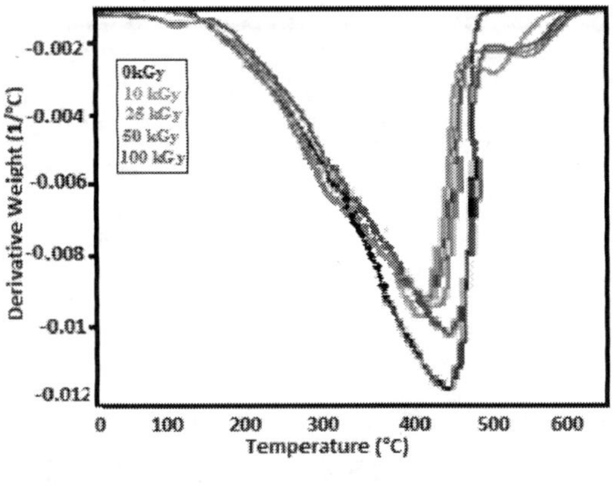

Figure 5: DTG curves of WPC with different gamma doses.

Figures 4 and 5 present respectively the TGA and DTG (the first derivative of TGA) curves of irradiated and unirradiated samples of WPC. The initial mass loss below 100°C is due to the gradual evaporation of absorbed moisture. The initial decomposition temperature for the composite is typical of wood, in which pyrolysis of wood components took place below 250°C. The important degradation process of wood starts at around 200 – 220°C and it concerns the three major constituents of wood (cellulose, hemicellulose and lignin). Therefore, the temperature profile throughout the extruder screw zones was set from 140 °C to 185 °C, from the feed to the metering zone. The mass loss steps of pure PP occurs very slowly under 500°C and above this temperature, this process grows rapidly. As the Gamma dose increased, the thermal stability of the composites decreased and at the same time, the decomposition temperatures of the composites decreased slightly with increasing the gamma doses. Moreover, the DTG curve of the irradiated composite showed degradation of coupling agent around 300°C may be the alcohol product during degradation [15]. An increase in the mass loss rate with gamma dose is also observed in the first stage of decomposition. High irradiation doses generate chains scissions into the material in turn contribute to the decrease in thermal stability of composites. For the irradiated composites, it was verified that the maximum degradation rate was shifted to a lower temperature. This is indicated that the present of the wood flour lowered the thermal stability of the materials and gamma irradiation does not change anything. Because of high inflammability of cellulose, the addition of wood makes the composite less thermally stable. The TGA measurements indicate that when the polymer is heated, its thermal stability is usually appraised from the loss of mass and its molecular weight sharply diminishes due to the degradation of the molecular chains.

3.4FT-IR of the composites

To investigate the effect of gamma-ray irradiation on the mechanical properties of the wood/PP composite, the sample of fifty-fifty was exposed to gamma ray at various doses. To clearly observe any change to the chemical integrity of the composite as a function of radiation dosage, FT-IR was employed and the results are shown in figure 6 below.

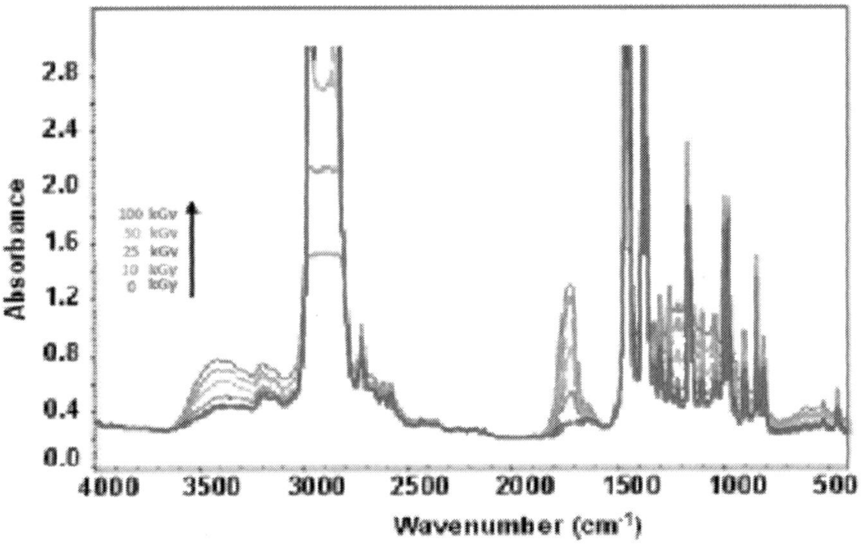

Figure 6: Infra-red spectra of gamma irradiated WPC in presence of oxygen

It is evident that, the signals associated with the aldehyde functional groups became stronger with increasing the radiation dosage. Specifically, the carbonyl stretching at around 1710 cm^{-1} and the C-H absorption in the range of 2700 to 2800 cm^{-1} were identified. In the presence of oxygen, the photooxidation of polypropylene leads to the formation of hydroxyl (mainly hydro peroxides and alcohols) and carbonyl groups (mainly ketones, esters and acids) which are easily detectable by infrared spectroscopy, respectively, in the 3200-3600 cm^{-1} areas and 1500-1900 cm^{-1}. The first mechanism is free radical formation along the polymer backbone via chain scission. It is assumed that when free radical formation is the dominating event during irradiation at a particular total dose. At relatively low total doses, free radical formation dominates the kinetics of polymer degradation. Once these free radicals are formed, the second mechanism, cross-linking, becomes feasible. Cross-linking occurs when free radicals recombine. If the cross-link density becomes higher than that of the original polymer before irradiation, the free radical may also react with oxygen. It is assumed that both the mechanisms of chain scission and cross-linking are occurring randomly and simultaneously [16]. The predominance of one mechanism over the other dictates the increase or decrease of mechanical properties with each total dose. The presence of oxygen (air) during gamma irradiation affected the performance of the sample. Upon initial irradiation in an environment containing oxygen, free radicals are formed within the polymer and readily react with free oxygen (O_2) according to the following reaction to form peroxy radicals: $R + O_2 \rightarrow ROO*$ The participation of oxygen during the beginning of this free radical chain reaction often leads to increased chain scission activity [17, 18]. The formation of these peroxy radicals prevents further cross-linking. The effect of high- energy radiation on organic polymers is to produce ionizing and excitation. The polymer may under cleavage or scission (i.e., the polymer molecules may be broken into smaller fragment). Subsequently rupture of chemical bonds yields fragments of the large polymer molecules. The free radicals thus produced may react to change structure of the polymer and after the physical properties of the materials. It also may undergo cross-linking (i.e., the molecules may be linked together into large molecules) [19, 20]. During degradation, there will be loss in strength due to primary bond breakage in the cellulose constituents and, therefore, be related to changes taking places in the middle lamella, which reduce the ultimate cell [21].

3.5 Tensile properties of the composites

Table 1: Effect of gamma radiation on bending strength of WPC.

Total dose (kGy)	0	10	25	50	100
Bending strength (Mpa)	12,35	13,75	14,15	14,5	12,4

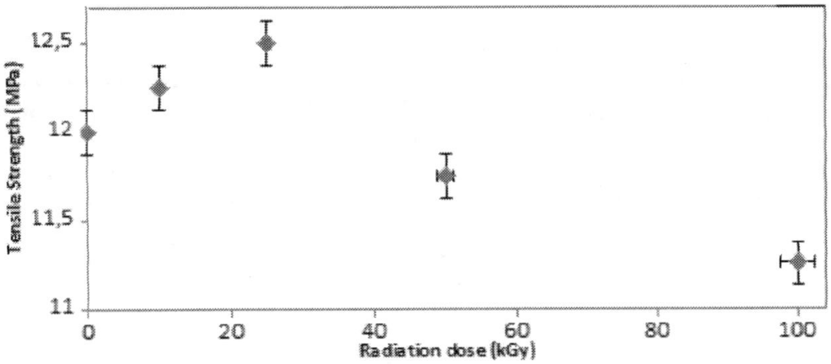

Figure 7: Effect of gamma radiation on tensile strength of WPC.

The behaviour of tensile strength of the unirradiated and irradiated WPC, with MAPP compatibilizer, is shown in Fig. 7. It can be noticed that this property decreases when the dose range increased until 50 kGy and then it seemed constant. Beyond a certain dose, irradiation of the polymer causes the breaking of covalent bonds of the latter which is accompanied by the creation of free radicals. This division is also responsible for a decrease in molecular weight; these chains can bind to the initial split point or rearrange by covalent cross-links with neighboring chains, leading to crosslinking. Gamma irradiation may also remove moisture from the composite, which in turn contributed to better fiber matrix adhesion. With the increase of the radiation dose, degradation through the formation of peroxy and hydroperoxy radicals in the amorphous regions appears to be the primary cause of physical property loss. When sufficient numbers of tie molecules between crystallites are cut through this chain scission process increases the local stress concentration on the crystals. Therefore these mechanical properties (tensile strength and bending strength) are reduced drastically [22]. In presence of MAPP, the tensile strength is also enhanced by forming more chemical bonding between the migration of maleated PP and wood fibers since enough fibers provide more active sites on surface for esterification that is hydroxyl groups. Thus, more ramified structures would be formed as more fibers combined which increase the tensile strength due to load transfer [23, 24]. It is observed that tensile properties increase with Gamma radiation up to a certain dose and then decrease due to the two opposing phenomena, namely, photocross-linking and photodegradation that take place simultaneously under Gamma radiation. At lower doses, free radicals are stabilized by a combination reaction and, as a result, photocross-linking occurs. The higher the number of active sites generated on the polymeric substrate, the greater the grafting efficiency. But at higher radiation, the main chain may be broken down and polymer may degrade into fragments and, as a result, tensile properties will decrease after certain Gamma doses. An intense radiation results in a loss of tensile strength and a reduced degree of polymerization is observed. From this investigation, it is clear that gamma radiation has a strong role on the improvement of the mechanical properties of the composites.

4. Conclusion

The wood polymer composites were irradiated by the variation of gamma irradiation doses. It has been observed from the SEM micrographs that, gamma irradiation improved the interfacial adhesion between the wood fibers and matrix. In thermal behavior, oxidative degradation leads to chain scission, implying that polymer chains fuse at lower temperatures The modifications of the IR spectra of all the films of WPC upon gamma irradiation in the presence of oxygen are comparable to that of pristine polypropylene, and are mainly characterized by an increase of absorbance in the hydroxyl and carbonyl regions. It can be noticed that at a certain limit of gamma irradiation doses improve the mechanical properties of composites such as tensile strength and bending strength. So, gamma irradiation could be an important way to improve the mechanical properties of the WPC and at the same time offer several advantages such as no chemical reagents are required and there is no residual polluting by-product.

References

[1] Diene N , Elisabeth F, Sandrine MT, et al. Durability of wood polymer composites: Part 1. Influence of wood on the photochemical properties. *Compos Sci Technol* 2008;68: 2779–2784.
[2] Karmarkar A, Chauhan SS, Modak JM, et al. Mechanical properties of wood fibre reinforced polypropylene composites: effect of a novel compatibilizer with isocyanate functional group. Compos: Part A 2006; 38: 227-233.
[3] Bledzki AK, Faruk O. Wood fibre reinforced polypropylene composites: effect of fibre geometry and coupling agent on physico-mechanical properties. Appl Compos Mater 2003; 10:365-379.
[4] Lundquist L, Marque B, Hagstrand PO, et al. Novel pulp fiber reinforced thermoplastic composites. Compos Sci Technol 2003; 63:137–152.
[5] Singh, A, Bahari, K. L A. Utraki(Ed) Polymer Blends Handbook, Kluwer Academic publishers, Dordrecht 2001.
[6] Singh, A. Irradiation of polymer blends containing a polyolefin. Radiat Phys Chem 2001; 60: 453- 459.
[7] Chipara MI, Constantinescu B, Reyes Romero J et al. Conferinta Nationala de Fizica Constanta, Romania 1998.
[8] Reyes J, Albano C, Davidson E, et al. Proceedings of 8th Polychar. USA: Texas, 2000.

[9] Adams T., Yasushi W. Gamma-oxidation of linear low-density polyethylene: The dose–rate effect of irradiation on chemical and physical modifications. J Polym Sci Part A: Polym Chem 1995;33 (9):1455–1460.
[10] Diène N, Matuana, LM, Sandrine MT et al. Thermal and Mechanical Properties of Polypropylene/Wood-Flour Composites. J Appl Polym Sci 2011;119: 3321–3328.
[11] Clough R. High energy and polymers: A review of commercial processes and emerging applications. Nucl Instrum Methods B 2001; (185): 8-33.
[12] Valentini, L; Biagiotti, J; Kenny, J M et al. Morphological characterization of single-walled carbon nanotubes-PP composites. Compos Sci Technol 2003; 63: 1149-1153.
[13] Smirnov L P , Deyun EV. Evolution of the molecular mass distribution parameters during gamma radiolysis. Polym Sci Serv A 1999; 41(5):506-511.
[14] Peter Z. Materials of cellulose derivatives and fiber-reinforced cellulose–polypropylene composites: Characterization and application. Pure Appl. Chem. 2006; 78 (10): 1843-1855
[15] Sonia, MBN, Graziela, SC, Simone, MLR. New polymeric coupling agent for polypropylene/wood flour composites. Polym Test 2007; 26: 619-628.
[16] Woods R J, Pikaev A K. Appl Radiat Chem : Radiation Processing. Book, ISBN 0471544523, New-York Wiley & sons 1994.
[17] Seguchi H O, T, Ogawa, K. Localisation of Ca2+ activated adenosine triphosphatase activities in the transitional epithelium of the rabbit urinary bladder. Acta Histochem. Cytochem 1982; 15:76-89.
[18] Andrzej G C, Mohammad HS, Shamshad A, Progress in radiation processing of polymers in Ionizing Radiation & Polymers Proceedings of the 6th International Symposium on Ionizing Radiation and Polymers (IRaP 2004).
[19] Zaman,HU, Khan, MA, Khan, RA. Effect of gamma radiation and bulk monomer on jute fabrics polyethylene/polyvinyl chloride composites. J polym engin 2012; 32: 301-309.
[20] Khan, M A., Shehrzade, S. and Hasan M M. Effect of Pre-treatment with Gamma-radiation on Physical and Mechanical Properties of 1, 6, Hexanediol diacrylate (HDDA) Jute Yarn by UV Radiation. Polym Plast Technol Eng 2003; 42(5): 795 - 810.
[21] C. Albanoa, J. Reyes, M. Ichazo, J. et al. Analysis of the mechanical, thermal and morphological behavior of polypropylene compounds with sisal fiber and wood flour, irradiated with gamma rays. Polym Degrad Stab (2002)76, 191–203.
[22] Sretenovic A, Müller U and Gindl W. Mechanism of stress transfer in a single wood fiber-LDPE composite by means of electronic laser speckle interferometry. Compos Part A 2006; 37(9): 1406–1412.
[23] Osswald TA. Fundamental principles of polymer composites: processing and design. Proceedings of the 5th International Conference of Wood fiber–Plastic Composites, Forest Products Society, Madison, WI, 26–27 May 1999.
[24] Marcovich, N. F., Ostrovsky, A. N., Aranguren, M. I. and Reboredo, M. M. Proceedings from the 3rd International Symposium on Natural Polymers and Composites-ISNaPol/2000, Sao Pedro, SP, Brazil, May 14–17

Physicochemical Properties Of Starch From Sago (*Metroxylon Sagu*) Palm Grown In Mineral Soil At Different Growth Stages

U. Uthumporn*, N. Wahidah, A.A. Karim

Food Technology Division, School of Industrial Technology, Universiti Sains Malaysia, 11800, Penang, Malaysia.

Email: sapina@usm.my

*Corresponding author
Phone number: +604-6532220
Fax number: +604-6573678

Abstract. A study was carried out to determine the physico-chemical properties of sago starch from sago palm grown in mineral soil at different growth stages. Four stages of sago palm, namely, Plawei (P), Bubul (B), Angau Muda (AM) and Angau Tua (AT) were studied. Sago starch granules were observed by using scanning electron microscopy (SEM) while the x-ray diffraction patterns were examined to study the starch crystallinity. The highest starch content was found at Plawei stage (94.2%) and Angau Muda stage (97.9%), respectively. The amylose content varied between 29.4 to 31.2% for each growth stages. The highest swelling power was found at the earliest growth stages (P) late growth stages (AT) which are 13.3 g/g and 13.2 g/g, respectively. Granule size distributions were similar as the palm grows to the later growth stages, where highest mean diameter of sago starches granules was found at AM. Variation of starch, amylose and proximate content was observed for starches derived from sago palm different growth stages were insignificant.

Keywords: Sago, *Metroxylon sagu*, growth stages, starch

1. Introduction

Malaysia is currently the largest world exporter of starch from sago (*Metroxylon sagu*) palm i.e. 47,000 metric ton/year where 96% from the starch was produced in Sarawak. Sago palm is characterized by a crown of compound leaves and terminating in a tall, woody and unbranch stem with non-branching roots go straight down to the soil [1]. It grows well in humid tropical lowlands, up to an altitude of 700 meters. Temperature above 25 °C and relative air humidity of 70 % are favorable. It is extremely hardy plant, thriving in swampy, acidic peat soils, submerged and saline soils where few other crops survive but growing slowly in peat soil than mineral soil [2]. The sago palm is hapaxantic, means it flowers once and dies shortly thereafter. The palm is mostly propagated vegetatively through its suckers in the wild as well as cultivation. Individual sucker firstly grows into a rosette of leaves, then produce a stout trunk. During the vegetative stage, just before flowering, the plant converts its stored nutrients into starch, which accumulates in the trunk [2].

The matured trunk will be harvested, leaving the immature and suckers, thereby sustaining sago production [1]. The trunk consists of a central core of soft pale-pink pith that contains most of the starch stored by growing palm, protected by 2 cm thick of fibrous bark. The bark is progressively denser towards the outside where the surface is covered with a thin shiny reddish-brown skin. This shell provides most of the structural strength of the palm and protects against predatory organisms [2]. The sago starch accumulates in the pith core of the stem of the sago palm. Trunk formation starts in the third and forth year of growth of the palm. The vegetative phase of the sago palm takes about 7 to 15 years, during which time the pith is saturated with starch from the base of the stem upwards[3]. The classification of sago palm stages are tabulated as followed [4].

Local name	Estimated age from planting (years)	Duration of trunk growth (years)	Growth description
Pelawai	10	4.5	75 % trunk growth; trunks are 6 to 8 in length
Bubul	12	6.5	Bolting; appearance of torpedo-shaped flowering structures at the palm terminal. It is characterized by the elongation of the trunk at the top of the crown and frond reduction to bract-like structures
Angau Muda	12.5	7	Flowering; well-developed flowering structure with primary, secondary and tertiary flowering axes spreading out at the terminal. Flowers are in the pre- or post anthesis stage
Angau Tua	14	8.5	Mature fruiting; fruits are mature, of diameter 30 to 40 mm. Seeds (if any) are well developed with dark brown seed coat and bony endosperms. Most fronds are in senescent stage

Quality of sago starches is important when starch is designated for export or when it is sold to large-scale food processors [4]. For example, good quality sago starches give high viscosity during gelatinization [5]. The poor quality of sago starch has been attributed to a number of factors such as poor processing conditions, presence of metal ions during processing, freshness of the raw pith, presence of polyphenol compounds and the consequent activity of polyphenol [6]. Quality problems associated with sago starch are inconsistent viscosity (or variable pasting properties), variable moisture content, distinct odor, low profile viscosity, high level of fiber and dull color. Mature palms are essential for the production of high-quality starch.

The major problem faces by the industry or food manufacturer is the variation in quality of sago starch from batch to batch. The existing data are not sufficient to cater the increasing demand for high quality sago starch from the industry. An understanding of basic properties is required to effectively utilize the starch which data is still lacking for sago palm grown in mineral soil. This study was carried out to determine the effects of different growth stages and on the physicochemical properties of starch granule extracted from sago palm. Four commercial growth stages of sago palm i.e. Plawei (palms at maximum vegetative growth), Bubul (appearance of flowering structure), Angau Muda (flowering) and Angau Tua (fruiting) were studied.

2. Materials And Methods

2.1 Materials

Sago starch was supplied by CRAUN Research SDN BHD, (Kuching, Sarawak, Malaysia). The starch sample was taken from sago plantation in Bau, Sarawak. The starches were taken at different growth stages of sago palm namely Plawei, Bubul, Angau Muda and Angau Tua. All reagents and chemicals were of analytical grades purchased from commercial scientific supplier.

2.2 Proximate Composition of Starch

The moisture content of starch samples were determined by using IR-30 Moisture Analyzer (Denver Instrument, Colorado, USA). Starch (5 g) were spread uniformly on the pan and heated at 105 oC.

The protein content of each starch was determined by Macro-Kjeldahl method [7]. Each starch was analyzed in triplicate and the protein content was calculated by converting the nitrogen content with the factor 6.25 and reported in percent of protein.

The ash content of each starch was determined by using standard AOAC procedure [7]. The analysis was done in triplicate and reported in percent of ash.

The fat content of each starch was determined by using Soxhlet method [7]. Analysis was done in triplicate and reported in percent of fat.

2.3 Determination of Starch Content

Starch content in these samples was determined by using Total Starch Assay Procedure from Megazyme International Ireland Limited (2005). Starch samples (100 mg) were dispersed by using 0.2 ml ethanol (80% v/v) and stirred on a vortex mixer. α-Amylase enzyme (3 ml) was added then incubated in a boiling water bath for 6 minutes. In 50 oC water bath, 4 ml sodium acetate buffer (200 mM, pH 4) was added, followed by 0.1 ml amyloglucosidase (20 U), stirred on a vortex mixer, and then incubated at 50 oC for 30 minutes. The content was transferred into volumetric flask and topped up to 100 ml, followed by centrifuged (Bench Top Centrifuge Kubota 5100, Korea) at 3000 rpm for 10 minutes. The aliquot (0.1 ml) was transferred into a test tube, added with 3 ml Glucose Determination Reagent (GOPOD) then incubated at 50 oC for 20 minutes.

The absorbance was read at 510 nm for each samples while D-glucose control was read against the reagent blank. The starch percent (w/w) on dry basis was calculated by the following equation:

$$\text{Starch \% w/w (dry basis)} = \text{Starch\% w/w as is} \times \frac{100}{100 - \text{Moisture content (\% w/w)}}$$

2.4 Determination of Amylose Content

Amylose content of each sample and raw starch was determined in triplicate according to the procedure described by McGrance et al. [8]..

2.5 Scanning Electron Microscopy (SEM)

The scanning electron microscopy study was carried out on starch samples using Leica Cambridge S360 Scanning Electron Microscope (Leica, Wetzlar, Germany). The starch was sprinkled on double-sided adhesive tapes mounted on the studs. The studs were placed in a container and coated with gold in an Argon gas environment. Next, the coated studs were placed on a sample holder of the Leica SEM to be scanned. The captured images were printed and saved.

2.6 X-Ray Diffraction Analysis

X-ray diffraction patterns of native starch and treated starch were obtained using a D5000 SIEMENS X-ray powder diffractometer equipped with a CuKα target at 40 kV and 25 mA at a diffraction angle ranging from 2θ = 5o to 40o. A step-scan was set at an angle of 0.05o per step with a counting of 2 sec.

2.7 Determination of Swelling Power and Solubility

Swelling power and solubility was determined in triplicate according to the method of Leach *et al.* [9].

$$\text{Swelling (g/g)} = \frac{\text{Weight of the wet sediment (g)}}{\text{Weight of the dry starch (g)}}$$

$$\text{Solubility (\%)} = \frac{\text{Weight of dried supernatant (g)} \times 100}{\text{Weight of dry starch (g)}}$$

2.8 Particle Size and Distribution Analysis

The particle size and distribution study was analyzed by using the Long Bench Mastersizer S (Malvern Instrument) fitted with QSpec Dry Powder Feeder. Sago starch powder (2.0 g, dry basis) was prepared in duplicate for this analysis. The value of D [4, 3] i.e. the volume mean diameter; D (n, 0.1) i.e. 10% cut-off point as ten percent of the distribution is below this point; D (n, 0.5) i.e. volume median diameter and divides the distribution exactly in half; D (n, 0.9) i.e. 90% cut-off point as ninety percent of the distribution is below this point were measured.

2.9 Statistical Analysis

The sago palm growth stages and heights are the factors that are being compared in this research. The comparison was done among the growth stages (i.e., Plawei, Bubul, Angau Muda and Angau Tua) at different heights (base and middle). Statistical analysis was carried out by using one-way ANOVA analysis using the Duncan's multiple range tests to compare means (Steel and Torrie, 1960) in SPSS 11.5 Programs. Triplicate analysis were done for each samples while the significance different was determine at $\alpha = 0.05$ (95 % confidence level).

3. Results And Discussion

3.1 Proximate Composition of Starch

The results of moisture, protein, ash and fat content for sago starches are presented in Table 1. There are no significant values of moisture content of sago starch as the palm grows from Plawei to Angau Tua. It is speculated that sago starch granules becomes more hygroscopic as the palm grows to the latest growth stages. Protein content of sago starch in this study was very low (less than 0.1%) when compared to the value of 0.1– 0.3% [11] and 0.4 – 1.4% [12]. The highest protein content was shown at Bubul stage which is 0.09% and no significant value of protein content observed when the palm grows from Bubul towards Angau Tua.

Table 1 Chemical values of sago starch derived from sago palm at different growth stages at two different heights

Samples	Moisture, %	Protein, %	Ash, %	Fat, %
Plawei	15.62 ± 0.73	0.05 ± 0.007	0.14 ± 0.03	0.17 ± 0.02
Bubul	15.54 ± 0.48	0.09 ± 0.013	0.12 ± 0.01	0.21 ± 0.03
Angau Muda	16.96 ± 0.63	0.07 ± 0.001	0.18 ± 0.02	0.17 ± 0.01
Angau Tua	16.03 ± 0.64	0.08 ± 0.006	0.14 ± 0.02	0.24 ± 0.03

Values are means ± SD (n = 3).

Ash content in this study (0.1 – 0.3%) was in relation to the previous researchers i.e. 0.06 – 0.4% [11] and 0.2 – 0.7% [12]. The value increased significantly as the palm grows from Plawei (0.1%) to Angau Muda stage (0.2%). Ahmad *et al.* [11] reported that ash contents of sago starches were low when compared to potato (0.3%) and corn (0.2%).

The highest value for fat content was found at Angau Tua stage (0.24%) and Bubul stage (0.23%). Commercially starches normally contain trace amounts of fatty acid glycerides usually less than 0.1% which can be removed by using Soxhlet extraction using ether or hexane [11]. The formation of starch-lipid or starch-surfactant complexes improves the textural properties of various foods [13] therefore high fat content value in this study will value-add the benefit of sago starch in various food products.

3.2 Determination of Starch Content

Starch content for sago starches in this study is given in Table 2. From the result, no significant different of starch content values except for the latest growth stages i.e. Angau Muda (94%) and Angau Tua (92%). The result in this study agrees to the previous finding where most sago starch accumulates below 6 m height of sago palm trunk [14]. These specify the initiation of an upward mobilization of starch from the base to the upper portion of the trunk, for the conversion of the starch to other forms of energy for flower and fruits development [15].

Table 2 Starch content values of sago starch derived from sago palm at different growth stages at two different heights

Samples	Starch Content, %	Amylose Content, %
Plawei	94.24 ± 1.56	30.79^{ab} ± 0.82
Bubul	91.69 ± 0.55	30.30^{ab} ± 1.10
Angau Muda	93.91 ± 0.32	31.19^{b} ± 0.92
Angau Tua	91.84 ± 1.61	30.58^{ab} ± 1.01

Values are means ± SD (n = 3).

The starch content was found to be decreased as the palm grows from Angau Muda to Angau Tua for both heights. Nozaki *et al.* [13] reported that starch concentration is related to the activities of starch synthetic enzymes in sago palm. Higher amylase activity in Angau Tua stage increased the starch degradation thus reducing the starch concentration inside the sago palm trunk.

3.3 Determination of Amylose Content

From the results in Table 2, no significant different observed for amylose content at different growth stages, where the values varied between 30.3 to 31.2% The amylose content in this study agreed well with the published values of 24 – 31% [6, 11,16]. As shown in the results, the difference in the amylose content was observed as the outcome of harvesting the sago palm at different growth stages. Amylose content (29 – 31%) of starch from sago palm grown in mineral soil in this study is higher than amylose content (23 – 27%) from palm grown in peat soil [3].

3.4 Scanning Electron Microscopy (SEM)

The scanning electron micrographs of sago starches at different growth stages are presented in Figure 1. The starch granules are oval in shape, with some truncated side. From the micrographs, it is estimated that the granule size ranges from approximately 20 – 60 µm. At Plawei stage, most of the starch granules accumulated at the base of the palm contained larger size granules compared to the middle part of the palm. As the palm grows to Bubul stage, smaller starch granules accumulated at the base of the palm while the distribution of small and large starch granules is similar from the base towards the middle heights of the palm at Angau Muda stage. This result is in accordance with the particle size distribution study (Table 7). At Angau Tua stage most of the starch granules have large starch granules.

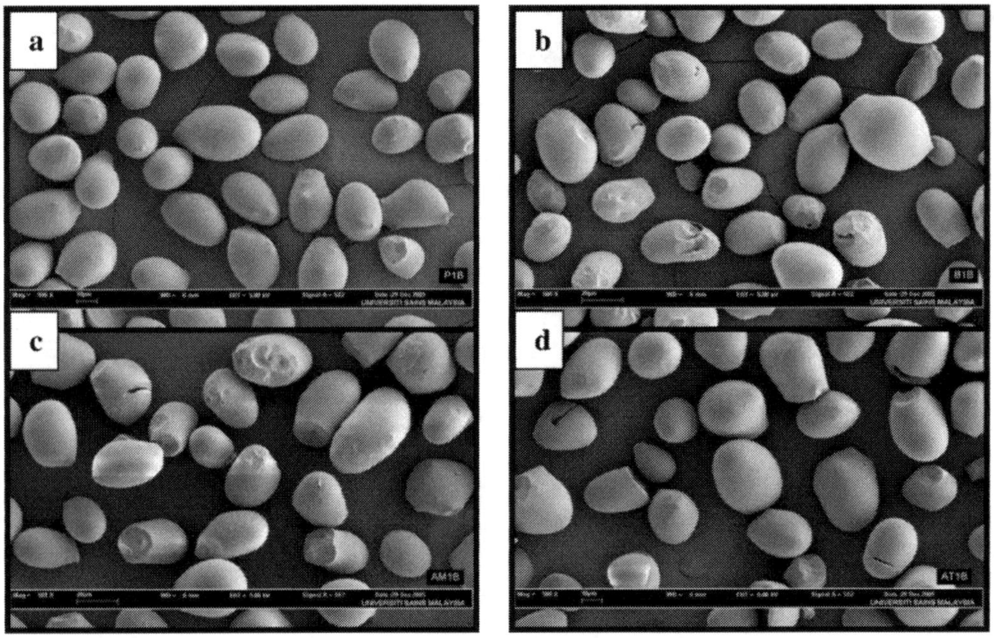

Figure 1 SEM pictures of sago starch at different growth stages Figure a: Plawei stage, Figure b: Bubul stage, Figure c: Angau Muda stage, Figure d: Angau Tua stage (500 ×).

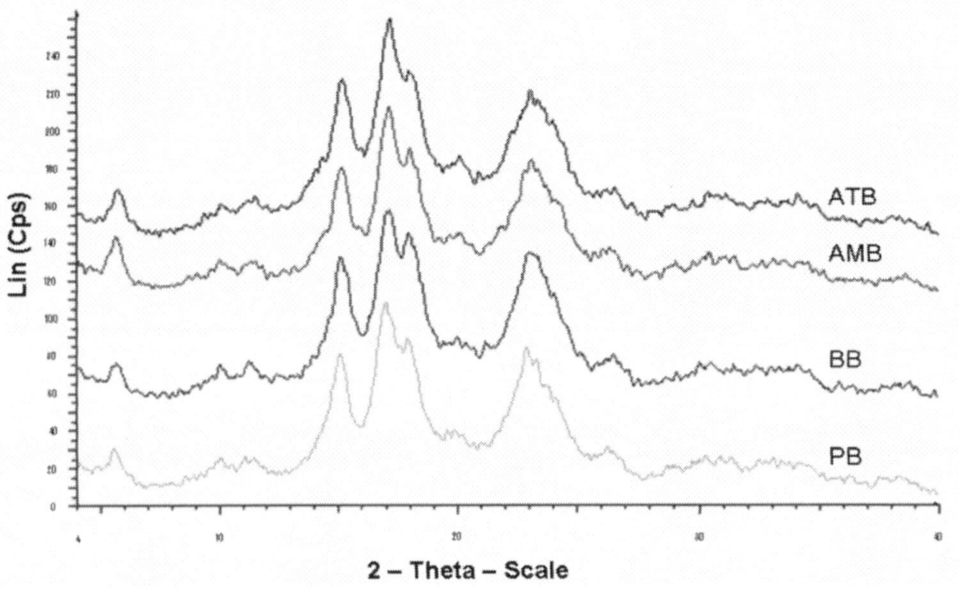

Figure 2: X-ray diffraction pattern for sago starches derived from sago palm at different growth stages (AT = Angau Tua, AM = Angau Muda, B = Bubul, P = Plawei).

3.5 X-Ray Diffraction Analysis

X-ray diffraction pattern for sago starches grown in mineral soil at different growth stages are displayed in Figure 2. The sago starches samples show C-type x-ray diffraction patterns with the reflection intensity at

IOP Conf. Series: Materials Science and Engineering **62** (2014) 012026 doi:10.1088/1757-899X/62/1/012026

5.5°, 10°, 11.1°, 15°, 17.2°, 17.9°, 23.5°, and 26.5° 2θ angles. 2θ values of 17.2° was the characteristic of d B pattern while one peak appeared at 23.5° was the characteristic of the A pattern. Pukkahuta and Varavinit [17] also reported C-type diffraction patterns as reported by for native sago starches. Ahmad *et al.* [11] also reported that the crystalline type of sago starch is intermediate to that of cereal or potato starches.

No clear difference could be seen between the starch samples from different growth stages, meaning that the crystallinity and the polymorphic type did not differ despite some differences in amylose content. Ahmad *et al.* [11] explained that the differences of amylose content are probably not significant enough to induce a change in the crystalline structure. It is well-known that an increase in the amylose content induces an increase in the B-type amount and a decrease in the crystallinity. However, it should be mentioned that C-type patterns are obtained for starches of relatively high amylose content (around 30 %) similar to those of the sago starches studied.

3.6 Determination of Swelling Power and Solubility

The swelling power and solubility of sago starches in this study are presented in Table 3. The results show no specific trend for swelling power and solubility value at different growth stages. The swelling power of starches varies between 11.9 – 13.3g/g. Swelling power provides evidence of non-covalent bonding between starch molecules. Several factors such as amylose-amylopectin ratio, chain length and molecular weight distribution, degree/length of branching and conformation the degree of swelling and solubility inside starch granule [18]. Moorthy [13] pointed the possible relationship between cooking quality and swelling volume, thus the consistent volume of swelling for sago starches in this study suggesting its good cooking quality.

Table 3 Swelling power and solubility of sago starch at different growth stages at two different heights

Sample	Swelling Power, g/g	Solubility, %
Plawei	13.32 ± 0.13	1.80 ± 0.29
Bubul	11.92 ± 0.21	1.48 ± 0.06
Angau Muda	13.01 ± 0.11	2.50 ± 0.04
Angau Tua	12.83 ± 0.83	2.83 ± 0.36

Values are means ± SD (n = 3).

When starch is heated in excess water, the crystallite structure is disrupted and water molecules become linked by hydrogen bonding to the exposed hydroxyl groups of amylose and amylopectin [19]. This causes an increase in granule swelling and solubility. The results showed significant difference for swelling power as the sago palm matured from plawei (P) to bubul (B) stage and from B to angau muda (AM) stage.

Swelling power has been reported to be influenced by strongly bonded micellar networks and amylopectin molecular structure [20]. A low swelling power of starches may be attributed to the presence of a large number of crystallite formed by the association between long amylopectin chains. Crystallite formation increases granular stability, thus reducing the extent of granular swelling [20]. The percent solubility of sago starch granules increases significantly from B to AM stage.

3.7 Particle Size and Distribution Analysis

Table 4 shows the mean particle size and distribution analysis of starch from sago palms from plawei (P) stage to angau tua (AT) stage. The largest mean diameter was observed at angau muda (AM) stage which was 33.3 μm. The calculation of particle size involved the basic assumption that all particles are spherical and of uniform density. In this case, the density was taken to be 1.0 g/cm^3 [15].

The distribution of granule size observed from the D (n, 0.1) and D (n, 0.9) values shows that the starch granule size increased as the palm matured to the latest growth stages. From the results obtained, it was observed that most of the granules has smaller size granules for all growth stages except for bubul stage. This result is different compared to sago palm grown in peat soil at different growth stages [3] where the starch granules distributed more towards higher size at base height.

Table 4 Particle size and distribution of sago starches derived from sago palm at different growth stages at two different heights

Sample	Diameter (μm)		Distribution (μm)		Range
	Mean D [4, 3]	Median D [n,0.5]	D (n, 0.1)	D (n, 0.9)	D(n,0.9) - D(n,0.1)
Plawei	26.8 $\pm$ 0.07	26.7 $\pm$ 0.03	20.9 $\pm$ 0.31	33.6 $\pm$ 0.14	12.7 $\pm$ 0.44
Bubul	29.0 $\pm$ 0.04	28.3^b $\pm$ 0.03	20.9 $\pm$ 0.05	38.5 $\pm$ 0.15	17.6 $\pm$ 0.19
Angau Muda	33.3 $\pm$ 0.03	32.5^f $\pm$ 0.02	24.7 $\pm$ 0.18	43.4 $\pm$ 0.08	18.7 $\pm$ 0.11
Angau Tua	32.2 $\pm$ 0.51	31.8^d $\pm$ 0.22	23.4 $\pm$ 0.23	42.3 $\pm$ 1.09	18.9 $\pm$ 0.86

D [4, 3] is the volume mean diameter; D (n, 0.1) is the 10 % cut-off point as ten percent of the distribution is below this point; D (n, 0.5) is the volume median diameter and divides the distribution exactly in half; D (n, 0.9) is the ninety percent cut-off point as ninety percent of the distribution is below this point. Values are means $\pm$ SD (n = 3).

4. Conclusions

The differences in some physico-chemical properties of sago starches were contributed by the difference in the growth stages. This study revealed that starch at angau muda (AM) stage have significant difference i.e. high starch content, swelling power, solubility and mean diameter compared to the other growth stages. Proximate analysis showed low protein, ash and fat content for all starch samples. Starch content is high with more than 90 % value with the highest value during AM stage. SEM micrographs showed that sago starch granules varied between 20 – 60 μm. X-ray diffraction study depicted that sago starches exhibit C-type of starch crystallization. The swelling power is significantly high during B and AM stage while the solubility increased significantly towards AM stage. Starches at B stage were found to retrograde easily as shown by the highest setback value. AM stage showed the highest mean diameter of starch granules while the particle distribution is similar for most of growth stages. When comparing to sago starches grown in peat soil, sago starches grown in mineral soil was found to have lower swelling power and enthalpy of gelatinization, and larger mean size diameter of starch granules. Both sago palm from peat soil and mineral soil showed the highest starch content during AM stage.

Acknowledgements

This work was also supported by University Sains Malaysia Short Term Grant Research.

References

[1] Cabalerro B, Trugo L C and Finglas P M, Encyclopedia of Food Sciences and Nutrition, 2nd edition. Academic Press 2003 **8** 5035 – 5039.

[2] Rekha S S, Kennedy, J F, Sajilata, M G, Kaczmarek, A, Knill C J and Putri F A 2008 Carbohydrate Polymers **72**, 1 – 20.

[3] Tie A P l, Karim A A and Mohamed A M D 2008 Physicochemical Properties of Starch in Sago Palms (*Metroxylon sagu*) at Different Growth Stages Starch/Starke **60** 408 – 416.

[4] Oates C G, Hicks A, Sago Starch Production in Asia and the Pacific – Problems and Prospects. In: Kainuma K., Okazaki M., Toyoda Y., Cecil J.E., editors. Proceedings of the International Symposium on Sago (Sago 2001) Tokyo, Japan: Universal Academy Press Inc., 27 – 36.

[5] Azudin M N and Lim E T K An Evaluation of the Quality of Sago Starch Produced in Sarawak. In: Ng T-T., Tie Y-L., Kueh H-S., editors. Proceedings of the Fourth International Sago Symposium, Sarawak, Malaysia: Ministry of Agriculture and Community Development and Department of Agriculture, 1991, 149 – 152.

[6] Sim L L, Oates C G, Wong H A 1991 Starch/Starke **43** 459 – 466.

[7] AOAC. (1990). Official Methods of Analysis (15th ed.) Washington: Association of Official Analytical Chemists.

[8] McGrance S J, Cornell H J, Rix C J 1998 Starch/Starke **50** 4 158 – 163.

[9] Leach H W, Mccowen L D and Schoch T J 1959 Cereal Chemistry **36** 534 – 544.

[10] Siau C L, Karim A A, Norziah M H and Wan Rosli W D 2004 Journal of the Science Food and Agriculture **84** 1722 – 1730.

[11] Ahmad F B, Williams P A, Doublier J L, Durand S and Buleon A 1999 Carbohydrate Polymers **38** 361 – 370.

[12] Sim E S and Ahmed M I 1978 Malaysian Agriculture Journal **51** 4 351 – 358.

[13] Moorthy S N 2002 Starch/Starke **54** 559 – 592.

[14] Nozaki K, Nuyim T, Shinano T, Hamada S, Ito H, Matsui H and Osaki M 2004 Plant Foods for Human Nutrition **59** 85 – 92.

[15] Pei Lang A T, Mohamed A M D and Karim A A 2006 Carbohydrate Polymers **63**, 283 – 286.

[16] Azukawa S and Kawabata A 2003 The Japanese Society of Applied Glycoscience **50** 121 – 126.

[17] Pukkahuta C and Varavinit S 2007 Starch/Starke **59** 624 – 631.

[18] Rickard J E, Asaoka M and Blanshard J M V 1991 Tropical Science **31** 189 – 207.

[19] Lee J S, Kumar R N, Rozman H D and Azemi B M N 2005 Food Chemistry **91** 2, 203 – 211.

[20] Shujun W, Wenyuan G, Hongyan L, Haixia C, Jiugao Y and Peigen X 2006 Food Chemistry **99** 38 – 44.

[21] Liu Q, Weber E, Currie V and Yada R 2003 Carbohydrate Polymers **51** 213 – 221.
[22] Huang C C, Lin M C and Wang C C R 2006 Carbohydrate Polymers **64** 524 – 531.
[23] Tian S J, Rickard J E and Blanshard J M V 1991 Journal of the Science Food and Agriculture **57** 459 – 491.
[24] Cooke D and Gidley M J 1992 Carbohydrate Research **227** 103 – 112.
[25] Abera S and Rakshit S K 2003 Starch/Starke **55** 287 – 296.

IOP Conf. Series: Materials Science and Engineering **62** (2014) 012027 doi:10.1088/1757-899X/62/1/012027

Marangoni instability and nonlinear development of perturbations in fluid films

L.A. Prokudina

South-Ural State University, Chelyabinsk, Russia

E-mail: prokudina-la@mail.ru

Abstract. A mathematical model is presented for the non-isothermal flow of a fluid film. In conditions of Marangoni instability is calculated the area of unstable vertical flow of film, its wave characteristics. The influence of thermocapillary forces and surface viscosity on the nonlinear development of perturbations during the time is investigated when the possibility of rupture of the fluid film and stabilization of its course.

1. Introduction

The investigation of a thin layer of a viscous fluid (a fluid film) that combines thin and large contacted surfaces connected with the implementation of the trends in heat and mass transfer apparatus, for example, thermal power, chemical, food, pharmaceutical industry. The manufactured products are sensitive to high temperatures, is able to burn, to polymerization with prolonged contact with the heating surface (certain types of plastics, dyes, juices, herbal extracts, dairy products, hormones, antibiotics, and others). Application of film apparatus allows are solved these problems. Due to the short-term contact with the heating surface to prevent the thermal decomposition, are provided with the necessary properties and ecological purity of the manufactured products. On regimes of the flow of a fluid film are influenced of various physical and chemical factors. The use of additives of surface-active substances qualitatively changes the character of the flow of a fluid film. The gradients of temperature and concentration cause the heterogeneity of the surface tension and the appearance of the surface tension gradients, resulting in the interphase surface arise force (thermocapillary), the nature of which is determined by the real physical process.

Various heat-mass exchange processes occur in conditions of interfacial instability of the interface, caused by a variety of effects, factors. Special attention attracts researchers interfacial instability associated with the Marangoni effect. The Marangoni effect related to the change at the interface of surface tension from point to point. Emerging Marangoni instability manifests itself in changes of the wave characteristics [1] and the formation in the surface layers of dissipative structures. Marangoni instability, depending on the sign of the gradient interfacial tension determines the orientation of the Marangoni effect on the strengthening of perturbations in fluid films or their extinction.

1.1 Mathematical model of non-isothermal liquid film

Consider for a thin layer of viscous incompressible fluid (fluid film) thick under the action of gravity on solid inclined plane in a rectangular coordinate system OXYZ (Fig. 1).

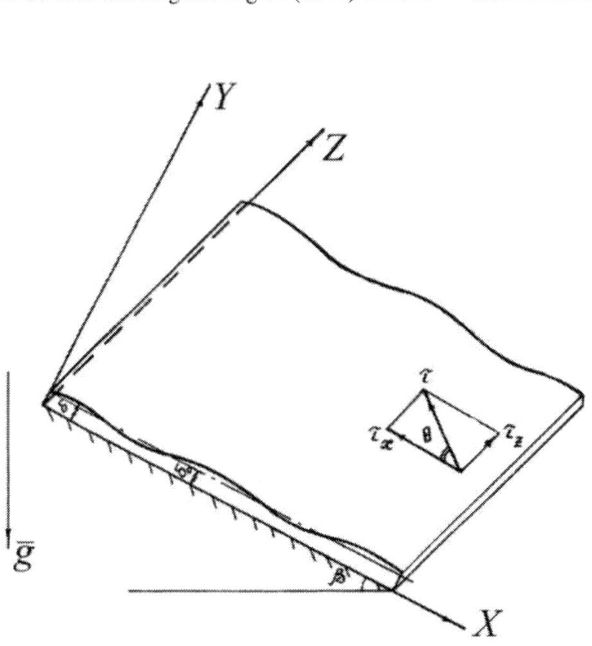

Fig.1. Scheme of the flow of a fluid film

The mathematical model of the flow of a three-dimensional non-isothermal fluid film of thickness δ is a system of Navier–Stokes equations and continuity equations with boundary conditions, taking into account the effect of heat – mass – transfer processes [1]:

$$\begin{cases} \dfrac{\partial u_+}{\partial t_+} + u_+\dfrac{\partial u_+}{\partial x_+} + v_+\dfrac{\partial u_+}{\partial y_+} + w_+\dfrac{\partial u_+}{\partial z_+} = -\dfrac{\partial P_+}{\partial x_+} + F_x + \dfrac{1}{\mathrm{Re}}\left(\dfrac{\partial^2 u_+}{\partial x_+^2} + \dfrac{\partial^2 u_+}{\partial y_+^2} + \dfrac{\partial^2 u_+}{\partial z_+^2}\right), \\ \dfrac{\partial v_+}{\partial t_+} + u_+\dfrac{\partial v_+}{\partial x_+} + v_+\dfrac{\partial v_+}{\partial y_+} + w_+\dfrac{\partial v_+}{\partial z_+} = -\dfrac{\partial P_+}{\partial y_+} + F_y + \dfrac{1}{\mathrm{Re}}\left(\dfrac{\partial^2 v_+}{\partial x_+^2} + \dfrac{\partial^2 v_+}{\partial y_+^2} + \dfrac{\partial^2 v_+}{\partial z_+^2}\right), \\ \dfrac{\partial w_+}{\partial t_+} + u_+\dfrac{\partial w_+}{\partial x_+} + v_+\dfrac{\partial w_+}{\partial y_+} + w_+\dfrac{\partial w_+}{\partial z_+} = -\dfrac{\partial P_+}{\partial z_+} + F_z + \dfrac{1}{\mathrm{Re}}\left(\dfrac{\partial^2 w_+}{\partial x_+^2} + \dfrac{\partial^2 w_+}{\partial y_+^2} + \dfrac{\partial^2 w_+}{\partial z_+^2}\right), \\ \dfrac{\partial u_+}{\partial x_+} + \dfrac{\partial v_+}{\partial y_+} + \dfrac{\partial w_+}{\partial z_+} = 0, \end{cases} \tag{1}$$

$$y = 0: \quad u_+ = w_+ = 0, \qquad v_+ = V_0; \tag{2}$$

$$y = \delta: \quad \frac{1}{\mathrm{Re}}\left[2\frac{\partial u_+}{\partial x_+}\frac{\partial \delta_+}{\partial x_+} - 2\frac{\partial v_+}{\partial y_+}\frac{\partial \delta_+}{\partial x_+} - \left(\frac{\partial v_+}{\partial x_+} + \frac{\partial u_+}{\partial y_+}\right) + \frac{\partial \delta_+}{\partial z_+}\left(\frac{\partial u_+}{\partial z_+} + \frac{\partial w_+}{\partial x_+}\right)\right] + \\ +M\frac{\partial \delta_+}{\partial x_+} + N\left(\frac{\partial^2 u_+}{\partial x_+^2} + \frac{\partial^2 w_+}{\partial x_+ \partial z_+}\right) + \tau_x = 0 \tag{3}$$

$$\frac{1}{\mathrm{Re}}\left[2\frac{\partial w_+}{\partial z_+}\frac{\partial \delta_+}{\partial z_+} - 2\frac{\partial v_+}{\partial y_+}\frac{\partial \delta_+}{\partial z_+} - \left(\frac{\partial w_+}{\partial y_+} + \frac{\partial v_+}{\partial z_+}\right) + \frac{\partial \delta_+}{\partial x_+}\left(\frac{\partial u_+}{\partial z_+} + \frac{\partial w_+}{\partial x_+}\right)\right] + \tag{4}$$

$$+M\frac{\partial \delta_+}{\partial z_+}+N\left(\frac{\partial^2 w_+}{\partial z_+^2}+\frac{\partial^2 u_+}{\partial x_+ \partial z_+}\right)+\tau_z=0$$

$$P_+=\frac{2}{\mathrm{Re}}\left[\frac{\partial v_+}{\partial y_+}-\frac{\partial \delta_+}{\partial x_+}\left(\frac{\partial u_+}{\partial y_+}+\frac{\partial v_+}{\partial x_+}\right)-\frac{\partial \delta_+}{\partial z_+}\left(\frac{\partial w_+}{\partial y_+}+\frac{\partial v_+}{\partial z_+}\right)\right]-\sigma_+\left(\frac{\partial^2 \delta_+}{\partial x_+^2}+\frac{\partial^2 \delta_+}{\partial z_+^2}\right)+ \tag{5}$$
$$+P_0-\operatorname{sgn}\Delta T\frac{\frac{\rho_1}{\rho_2}-1}{(\mathrm{Re}\cdot \mathrm{Pr}\cdot \mathrm{Ku})^2}\frac{1}{\delta^2}$$

$$\frac{\partial \delta}{\partial t_+}=v_+-u_+\frac{\partial \delta}{\partial x_+}-w_+\frac{\partial \delta}{\partial z_+}+\frac{1}{\mathrm{Re}\cdot \mathrm{Pr}\cdot \mathrm{Ku}}\frac{1}{\delta} \tag{6}$$

In formulas (1–6), dimensionless quantities: $u_+=\frac{u}{u_0}$, $v_+=\frac{v}{u_0}$, $w_+=\frac{w}{u_0}$ are projections of velocity onto the corresponding coordinate axes; $t_+=\frac{tu_0}{\delta_0}$ is time; $x_+=\frac{x}{\delta_0}$, $y_+=\frac{y}{\delta_0}$, $z_+=\frac{z}{\delta_0}$ are variables; $\delta_+=\frac{\delta}{\delta_0}$ is the thickness of the fluid film; $\mathrm{Re}=\frac{u_0\delta_0\rho}{\mu}$ is the Reynolds number; $F_x=\frac{g_x\delta_0}{u_0^2}$, $F_y=-\frac{g_y\delta_0}{u_0^2}$, $F_z=\frac{g_z\delta_0}{u_0^2}$ are the projections of the Froude number onto the corresponding coordinate axes; $P_+=\frac{P}{\rho u_0^2}$ is pressure; $N=\frac{k+e}{\rho\delta_0^2 u_0}$ is the surface viscosity parameter; $\overline{\tau}_x=\frac{\tau_x}{\rho u_0^2}$, $\overline{\tau}_z=\frac{\tau_z}{\rho u_0^2}$ are the projections of tangential stress; $M=M_T+M_K$ is the Marangoni parameter, where $M_T=\frac{\partial \sigma}{\partial T^0}\left(\frac{\partial T^0}{\partial y}\right)_{y=\delta}\frac{1}{\rho u_0^2}$, $M_K=\frac{\partial \sigma}{\partial K}\frac{\partial K}{\partial \delta}\frac{1}{\rho u_0^2}$; $\mathrm{Pr}=\frac{\rho c_p \upsilon}{\lambda}$; is the Prandtl number; $\mathrm{Ku}=\frac{r'}{c_p\Delta T}$ is the number of a phase transition; $\sigma_+=\frac{\sigma}{\rho u_0^2\delta_0}$ is the surface tension parameter, where u_0 is the average velocity of the main flow of the fluid film, δ_0 is the thickness of the film in the unperturbed state. During condensation, $\operatorname{sgn}\Delta T=1$, whereas, during evaporation, $\operatorname{sgn}\Delta T=-1$.

On the surface of a fluid film developing disturbances belonging continuous band of wave numbers (wave packet) [2]:

$$\psi = \int\limits_{k_{0x}-\Delta k_x}^{k_{0x}+\Delta k_x} \int\limits_{k_{0z}-\Delta k_z}^{k_{0z}+\Delta k_z} F(k_\Sigma)\exp\{i(k_x x + k_z z - \omega t)\} dk_x dk_z = A\exp\{i(k_{0x}x + k_{0z}z - \omega(k_0)t)\},$$

where k_0 is the center of the wave packet, $\omega = \omega_r + i\omega_i$,

$$A = \int\limits_{-\Delta k_x}^{\Delta k_x} \int\limits_{-\Delta k_z}^{\Delta k_z} F(k_{0\Sigma} + \delta k_\Sigma)\exp\left\{ i\left(\delta k_x x + \delta k_z z - \left(\frac{\partial \omega_r}{\partial k_x}\right)_{k_0} \delta k_x t - \left(\frac{\partial \omega_r}{\partial k_z}\right)_{k_0} \delta k_z t - \right.\right.$$

$$\left.\left. -\frac{1}{2}\left(\frac{\partial^2 \omega_r}{\partial k_x^2}\right)_{k_0} (\delta k_x)^2 t - \frac{1}{2}\left(\frac{\partial^2 \omega_r}{\partial k_z^2}\right)_{k_0} (\delta k_z)^2 t - \left(\frac{\partial^2 \omega_r}{\partial k_x \partial k_z}\right)_{k_0} \delta k_x \delta k_z t - \varepsilon^2 \bar{\omega}_i t \right)\right\} d\delta k_x d\delta k_z +$$

$$+ o(\varepsilon^3),$$

$F(k_\Sigma)$ is the Fourier component of the expansion, k_Σ is the wave vector of perturbations, under the following assumptions: the wave number band is $\Delta k_\Sigma = o(\varepsilon)$,and the increment is $\omega_i = \varepsilon^2 \bar{\omega}_i = o(\varepsilon^2)$,where ε is a small parameter.

The equation for the amplitude A envelope of a wave packet is a nonlinear parabolic equation of the form:

$$\frac{\partial A}{\partial t_2} + \frac{i}{\varepsilon}\frac{\partial \omega_i}{\partial k_x}\frac{\partial A}{\partial x_1} + \frac{i}{\varepsilon}\frac{\partial \omega_i}{\partial k_z}\frac{\partial A}{\partial z_1} - \frac{i}{2}\left(\frac{\partial^2 \omega_r}{\partial k_x^2} + i\frac{\partial^2 \omega_i}{\partial k_x^2}\right)\frac{\partial^2 A}{\partial x_1^2} -$$

$$-\frac{i}{2}\left(\frac{\partial^2 \omega_r}{\partial k_z^2} + i\frac{\partial^2 \omega_i}{\partial k_z^2}\right)\frac{\partial^2 A}{\partial z_1^2} - i\left(\frac{\partial^2 \omega_r}{\partial k_x \partial k_z} + i\frac{\partial^2 \omega_i}{\partial k_x \partial k_z}\right)\frac{\partial^2 A}{\partial x_1 \partial z_1} = \bar{\omega}_i A - \qquad (7$$

$$-(\beta_1 + i\beta_2)|A|^2 A .$$

The coefficients of the nonlinear term in equation (7) characterize the nonlinear damping of perturbations (β_1) and the dependence of phase on the amplitude (β_2). The first- and second-order derivatives in Eq. (7) are calculated using the dispersion equation

$$\omega(a_7 k_x + a_9 k_z + i) + a_1 k_x^4 + a_2 k_x^2 k_z^2 + a_3 k_z^4 - a_4 i k_x^3 - a_5 i k_z^3 - -a_6 k_x^2 - a_8 k_x k_z -$$

$$-a_{10} k_z^2 + a_{11} i k_x + a_{12} i k_z - \frac{1}{\mathrm{Re}\cdot \mathrm{Pr}\cdot \mathrm{Ku}} = 0 , \qquad (8)$$

where $\omega = \omega_r + i\omega_i$, ω_r – frequency, ω_i – increment.

The coefficients of the dispersion equation (8) are of the form:

IOP Conf. Series: Materials Science and Engineering **62** (2014) 012027 doi:10.1088/1757-899X/62/1/012027

$$a_1 = -\frac{\mathrm{Re}\,\sigma}{3},\ a_2 = 2a_1,\ a_3 = a_1,\ a_4 = -\frac{\mathrm{Re}^2 F_x N}{2},\ a_5 = -\frac{\mathrm{Re}^2 F_z N}{2},\ a_6 = a_6^* + a_6^1,$$

$$a_6^* = -\frac{\mathrm{Re}\,F_y}{3} - \frac{\mathrm{Re}\,M}{2} + \frac{3}{40}\mathrm{Re}^3 F_x(\tau_x + F_x),\ a_6^1 = \frac{2}{3}\mathrm{Re}\cdot\mathrm{sgn}\,\Delta T\frac{C}{(\mathrm{Re}\cdot\mathrm{Pr}\cdot\mathrm{Ku})^2},$$

$$a_7 = \frac{5}{24}\mathrm{Re}^2 F_x,\ a_8 = \frac{3}{20}\mathrm{Re}^3 F_x F_z + \frac{3}{40}\mathrm{Re}^3(F_z\tau_x + F_x\tau_z),\ a_9 = \frac{5}{24}\mathrm{Re}^2 F_z,$$

$$a_{10} = a_{10}^* + a_{10}^1,\ a_{10}^* = -\frac{\mathrm{Re}\,F_y}{3} - \frac{\mathrm{Re}\,M}{2} + \frac{3}{40}\mathrm{Re}^3 F_z(\tau_z + F_z),$$

$$a_{10}^1 = \frac{2}{3}\mathrm{Re}\cdot\mathrm{sgn}\,\Delta T\frac{C}{(\mathrm{Re}\cdot\mathrm{Pr}\cdot\mathrm{Ku})^2},\ a_{11} = -\mathrm{Re}\,F_x - \mathrm{Re}\,\tau_x,\ a_{12} = -\mathrm{Re}\,F_z - \mathrm{Re}\,\tau_z.$$

The coefficients of equation (8) include physical-chemical factors of the model (1)–(6).

2. The influence of temperature gradients. Computing experiments

Computing experiments are investigated the development of perturbations in fluid films, caused by temperature gradients, for the interval $\mathrm{Re} \le 20$ of Reynolds numbers.

The influence of temperature gradients on the wave characteristics of the fluid film is considered the Marangoni parameter. With the increase of the magnitude of the temperature gradients (parameter M>0) fluid film becomes more unstable. Qualitatively change its wave characteristics. Figure 2 shows the changes of the wave characteristics of vertical film of water for Reynolds number Re=5. With growth of temperature gradients increases the increment, and the value of the phase velocity decreases.

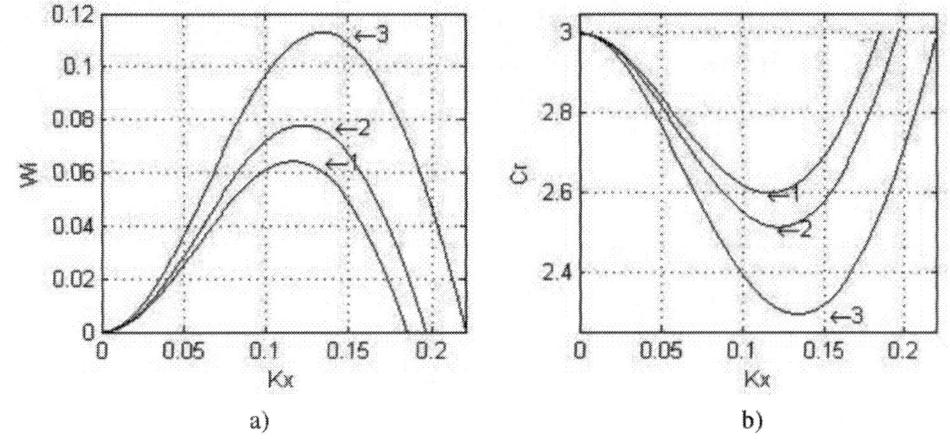

Fig.2. Increment (a) and the phase velocity (b) as a function of wave number: 1 - M=0; 2 - M=0.3; 3 - M=1

Figure 3 shows the change of instability region and curves of the maximum increment. Note that with the increase of the magnitude of the temperature gradients the instability region of a fluid film expands the range of wave numbers.

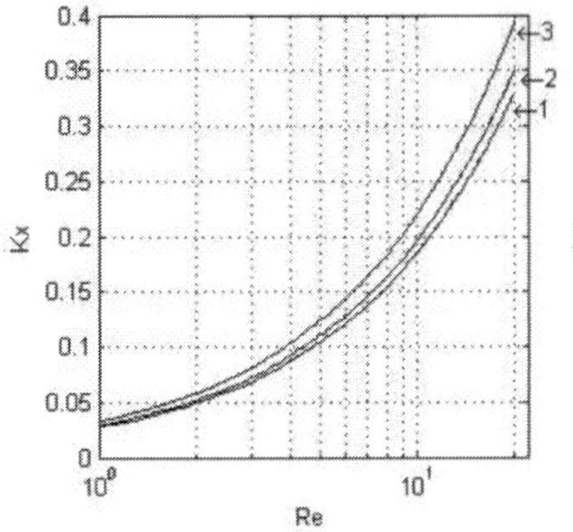

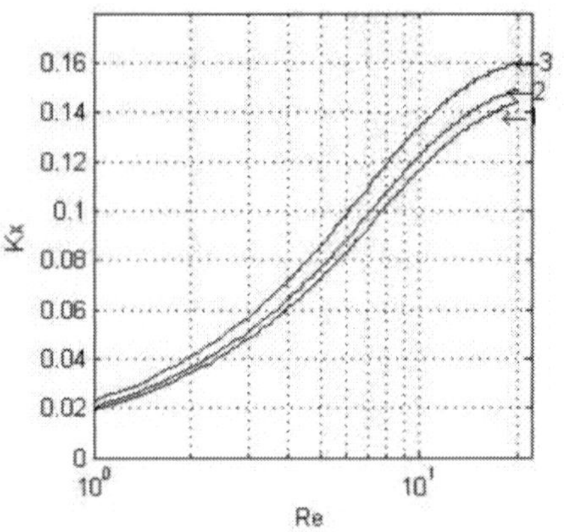

a) b)

Fig.3. Instability region of a vertical film (a) and curves of the maximum increment (b): 1 – M=0; 2 – M=0.3; 3 – M=1

Near the neutral curve (Fig. 3 a) there is an area in which the effect of directed energy transfer along the spectrum of the wave packet and its bias towards harmonic maximum increment [3]. In the surrounding area (Fig. 3 b) there are stationary monochromatic modes of the wave flow of fluid films, as was also mentioned in the experiments.

3. The surface condition of non-isothermal liquid film

The surface tension depends on temperature and the concentration of matter on the free surface of a film. The temperature and concentration gradients lead to inhomogeneity of surface tension and rise to the gradients of surface tension.

3.1 The impact of negative temperature gradients

Negative temperature gradients occur in the flow of a fluid film on a heated surface. The course is accompanied by an increase in the growth rate of perturbations and the expansion of the region of instability [1]. In this case significance $M > 0$ is for the Marangoni parameter.

Under large temperature gradients thermocapillary forces can destroy a fluid film that can cause local overheating cooled film wall and create an emergency situation when working heat and mass–transfer apparatus. The process of development of disturbances in time simulating the destruction of a film (with significant temperature gradients) presented in Fig. 4.

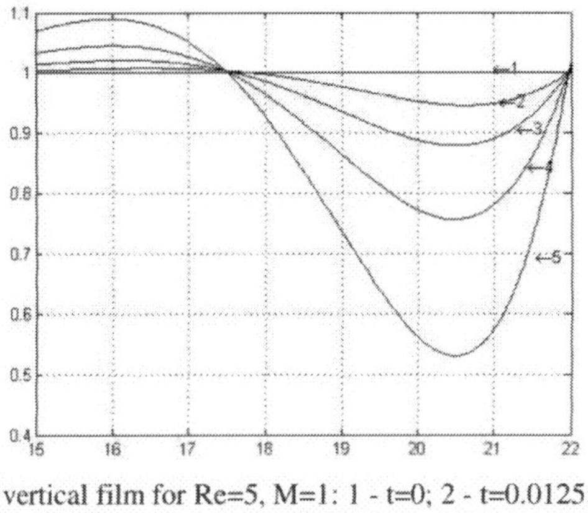

Fig.4. Surface conditions of a vertical film for Re=5, M=1: 1 - t=0; 2 - t=0.0125; 3 - t=0.025; 4 - t=0.0375; 5 - t=0.05

The presence in fluid film insoluble, surface-active substances, such as oils, fats (the parameter N), the viscosity of which is higher than the viscosity of water and increases with their molecular masses, the motion of the fluid layer leads to the appearance of forces of a superficial viscosity, qualitatively affect wave characteristics of the flow of fluid films, the stability boundary. Film insoluble, surface-active substances, having a large internal friction, the motion of a thin layer of viscous liquid pick up a significant part of its energy, which leads to a reduction in the rate perturbed current and smoothing profile waves (Fig. 5).

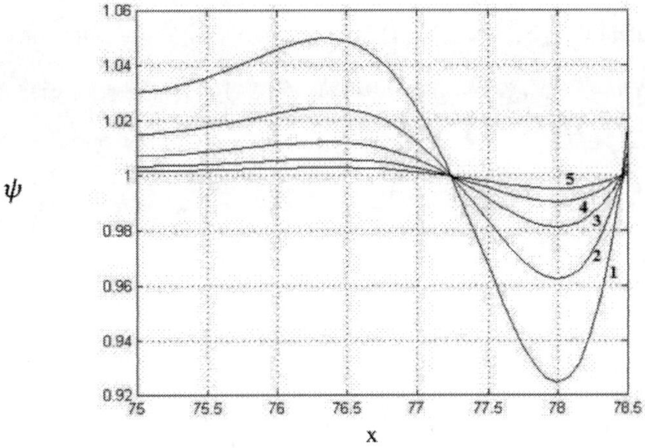

Fig.5. Development of perturbations on the free surface of a fluid film: M=1, N=1: 1 - t=0; 2 - t=0.0125; 3 - t=0.025; 4 - t=0.0375; 5 - t=0.05

Insoluble surface-active substances are damping effects, reduce the intensity of Marangoni instability that prevent the rupture of the fluid film and ensures trouble-free operation mode of tape devices.

3.2 *The impact of positive temperature gradients*

Significance $M < 0$ is the Marangoni parameter. When implemented in the fluid film positive temperature gradients, for example, when blowing of the film surface of high-temperature gas flow, thermocapillary forces stabilize its course, reducing the growth rate of perturbations. The blowout films of high-temperature gas flow leads to the alignment of the profile of the film in place perturbation (Fig. 6).

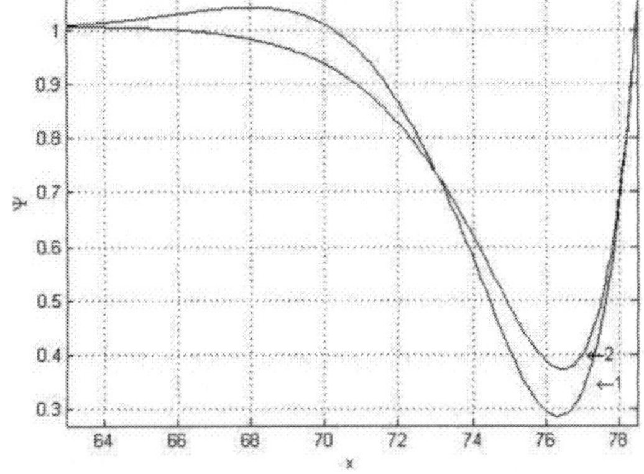

Fig.6. Development of perturbations on the free surface of a fluid film M=-1: 1 - t=0; 2 - t=2

Numerical simulation of non-linear development in time of perturbations on the free surface of a fluid film taking into account the heterogeneity of its surface tension, the surface viscosity promotes the study of the influence of these factors on the development of perturbations and the development of methods of control and management of different regimes of flow of fluid films.

References

[1] Prokudina L A and Vyatkin G P 1998 Instability of a nonisothermal liquid film Doklady Physics. 43(10) 652

[2] Prokudina L A and Vyatkin G P 2011 Self-organization of perturbations in fluid films Doklady Physics. 56(8) 444

[3] Prokudina L A 2014 Nonlinear Evolution of Perturbations in a Thin Fluid Layer during Wave Formation J. Experimental and Theoretical Physics. 118(3) 480

Tensile deformation mechanisms of ABS/PMMA/EMA blends

S H Wang[1], J Gao[1]*, S X Lin[1], P Zhang[2], J Huang[2] and L L Xu[2]

[1]School of Mechanical, Electrical & Information Engineering, Shandong University at Weihai, Weihai 264209, China
[2]Qingdao Haier Mold Co., Ltd, Qingdao 266103, China

E-mail: wesharewshp@gmail.com

Abstract. The tensile deformation mechanisms of acrylonitrile – butadiene - styrene (ABS) / polymethyl methacrylate (PMMA) blends toughened by ethylene methacrylate (EMA) copolymer was investigated by analysing the fracture morphology. ABS/PMMA was blended with EMA copolymer by melt mixing technique using co-rotating twin extruder. Tensile tests show that the elongation at break of ABS/PMMA blends can be efficiently improved with the increase in EMA content. Fracture morphology of ABS/PMMA/EMA blends reveals that the material yield induced by hollowing-out of EMA particles and its propagation into yield zone is the main toughening mechanism. Moreover, the appearance that EMA particles in the central area are given priority to hollowing-out may be related to the skin-core structure of the injection moulded parts caused by the different cooling rate between surface and inside in the process of injection moulding.

1. Introduction

Blending of polymers to achieve better comprehensive properties than those of individual components is of practical importance. Acrylonitrile-butadiene-styrene (ABS) resin itself is a kind of multiphase polymer, and it is very blendable with many other polymers like nylon, polypropylene and styrene-co-maleic anhydride (SMA) [1-3]. ABS blends with poly (methyl methacrylate) (PMMA) can provide ABS with improved surface characteristic, and ABS/PMMA blends have been used where good surface appearance is desired.

However, the addition of PMMA to ABS leads to a low toughness of ABS/PMMA blends, which is a certain limitation on its practical use [4]. In the past decades, the researches about ABS/PMMA blends were mostly focused on its composition and the performance of itself [1, 5]. Very recently, there appeared researches on the modification of ABS/PMMA blends. Zhang [6] and Tayebe [7] investigated the modification of nano-calcium and organoclays on ABS/PMMA blends, respectively. In general, brittle polymers are blended with elastomer particles to improve their fracture toughness. The elastomer particles act as stress concentrators under the action of external force, thereby induce crazes and shear deformation [8]. Poomalai [9] used Ethylene methacrylate (EMA) copolymer toughened PMMA, and conducted studies on physical-mechanical and optical properties. The toughness of the blends was improved significantly with the addition of EMA content, without affecting the optical properties significantly. It was also revealed that the EMA phase dispersed as domains in the continuous PMMA matrix at 5 and 10 wt% [9]. The same methyl methacrylate (MMA)

* J Gao, School of Mechanical, Electrical & Information Engineering, Shandong University at Weihai, Weihai 264209, China, Email: wesharewshp@gmail.com

structure may contribute to the good miscibility of PMMA/EMA blends. In addition, EMA and ABS may also have anticipated compatibility for the vinyl structure in EMA and butadiene composition in ABS.

The objective of this article is to obtain information concerning the nature of toughening mechanisms of ABS/PMMA blends toughened by EMA. The toughening mechanisms will be revealed according to the fracture morphology of specimens that fully deform in a low strain rate.

2. Experimental

2.1 Materials

The ABS matrix polymer (757K) was obtained from Zhenjiang Chimei Chemical Co., Ltd, with Vicat softening temperature of 105°C and flow coefficient of 1.8g/10min. PMMA resin, CM-211, with flow coefficient of 16g/10min and Vicat softening temperature of 102°C, were purchased from Taiwan Chimei Chemical Co., Ltd. EMA was provided by Dongguan AnChen plastic technology co., with Vicat softening temperature of 59°C.

2.2 Preparation

The ternary blends of ABS, PMMA and EMA were prepared by melt mixing technique using co-rotating twin extruder (L/D ratio 40:1) with a screw diameter of 21.7 mm. The ABS/PMMA blends composition was 80/20, with the EMA content of 0, 4wt%, 6wt% and 8wt% of ABS/PMMA/EMA blends. Prior to mixing, ABS and PMMA were pre-dried in an air circulating oven at 80°C for 12 h, as well as EMA, which was dried at 50°C for 4 h. The pre-dried granules of ABS, PMMA and EMA were mixed homogeneously and fed into the extruder. The processing temperature range of 200°C-225°C and screw speed of 21 rpm were used. The extruded strands were quenched subsequently in water and cut into pellets, and then dried again in the hot air oven at 80°C for 12 h for further study.

2.3 Measurements

The dumbbell-shaped test specimens (150 mm in gauge length, 10 mm in width, and 4 mm in thickness) for tensile tests were prepared by a 68 tons injection molding machine with a 6 MPa injection pressure and a 5.5 MPa package pressure. The tensile tests were performed on a CMT 5305 electrical testing machine at room temperature with a crosshead speed of 4 mm/min.

The specimens were broken in the tensile tests, and the fracture surface was coated with a thin layer of gold. They were then examined in a scanning electron microscope (VEGA 3 SBH).

3. Results and discussion

3.1 Tensile deformation

The measured tensile strength and elongation at break of ABS/PMMA blends and EMA toughened ABS/PMMA are presented in figure 1. The percentage elongation at break for ABS/PMMA blends is found to be substantially increased with the addition of EMA. Meanwhile, it is seen that as the EMA content increases, the elongation at break gradually increases. When EMA content of the blends increases from 0 to 8wt%, a large variation of elongation at break from 11.38% to 28.38% are observed, with an increase of 149.4%. On the contrary, tensile strength decreases as EMA content increases, but the drops are not big, as shown in figure 1.

The elongation at break is one of the most important parameters to estimate the toughening effect of rubber particles in modified plastics. Therefore, the addition of EMA toughened ABS/PMMA blends efficiently. This may be due to the good adhesion of EMA with ABS and PMMA, which may improve the compatibility of ABS and PMMA, and improved interface adhesion may prevent the emergence and development of cracks to a certain extent. The linear increase in figure 1 may also involve the optimum composition, as 4wt% to 8wt% EMA is such a limited range that can hardly express the nature and amount of the elastomeric phase. More importantly, the rubber particles generally act as stress concentrators in the tensile process and hence promote the development of crazes at the periphery of rubber domains [10]. It is worth noting that neck-down is observed and stress whitening propagated through the entire gauge portion during the necking process. To obtain the details of the toughening mechanism, SEM images were used for further study.

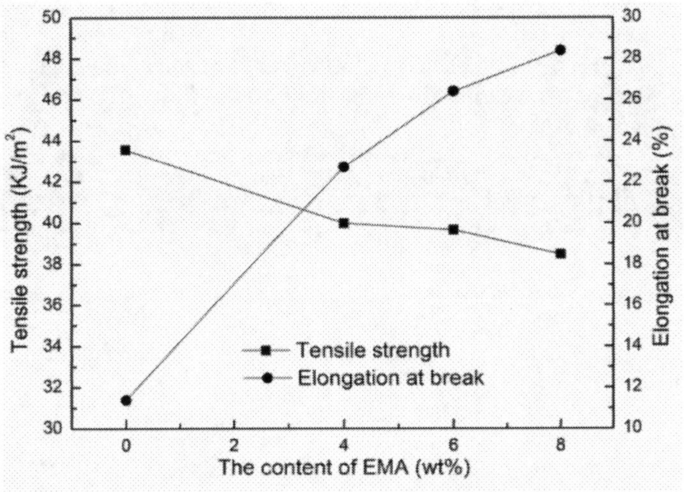

Figure 1. Effect of EMA content on the tensile strength and elongation at break of the blends.

3.2 *Fracture morphology of the blends*

Figure 2 shows the tensile fracture morphology of ABS/PMMA/EMA blends with EMA content of 6wt%. It can be seen from the overall morphology presented by figure 2(a) that the blends exhibited very complicated localized strain behaviour in low-speed stretching with the addition of EMA. Figure 2(b) shows the typical hollowing-out phenomenon of EMA particles under continuous stretch. The article considers that the hollowing-out of EMA induces a yield band that radiates to the surrounding gradually until being blocked by the crazing, as showed in figure 2(c). Figure 2(d) illustrates that the continuous crazing will cause fracture eventually. The hollowing-out of elastomer particles, the yield band, and the crazing are believed to participate in energy dissipation, thus improving the ductility of the blends.

From figure 2(a) and (b) it can be noticed that a set number of hollowing-outs will be induced in the interior of the specimens under the effect of tensile force. EMA particles act as the stress concentrators when ABS/PMMA/EMA blends are subject to external forces and gradually generate cavity inside. The continuous extend of the cavities in EMA particles largely depends on its energy, which is mainly from the additional internal stress of specimens that caused by the different cooling rates between surface and inside in the process of injection moulding. As injection mouldings are known to exhibit anisotropic skin-core morphological behaviour [11], EMA particles in different regions will suffer from different internal stresses. The internal regions will have a set amount of shrinkage during the slow cooling process, and the internal stress of the tensile direction caused by shrinkage makes EMA particles in the core a higher cavitation energy, especially in the central area. Therefore, the EMA particles in the central area are given priority to hollowing-out.

The material deformation resistance is reduced by the hollowing-out of EMA particles, and hence allowing matrix resin yield. The yield tends to be localized and formation of yield zone is the result of strain softening. Furthermore, the strain softening of elastic deformed materials in the surroundings of each peneplain zone causes the plane expansion of the yield zone, and finally forms the oval flat yield zone around the central b point, as shown in figure 2(a). Meanwhile, from figure 2(c) it can be noticed that EMA particles and the rubber composition of ABS resin in the skin area induce crazing deformation under the action of external force. The strain hardening produced by orientation of crazing chain eventually limits the degree of strain, embodied in the crazing voluntary termination and terminating the yield zone. Finally, the end and the fibre of crazes give rise to strain softening of polymer surrounded, and thus can efficiently improve the resin volume that involve into absorbing energy. As discussed above, the transition of strain softening to stain hardening happens in the highly tensile area when it reaches critical elastic deformation determines the toughness of the blends. Moreover, yield may occur a little earlier than crazing in ABS/PMMA/EMA blends, which allows the blends to seem too prone to yield.

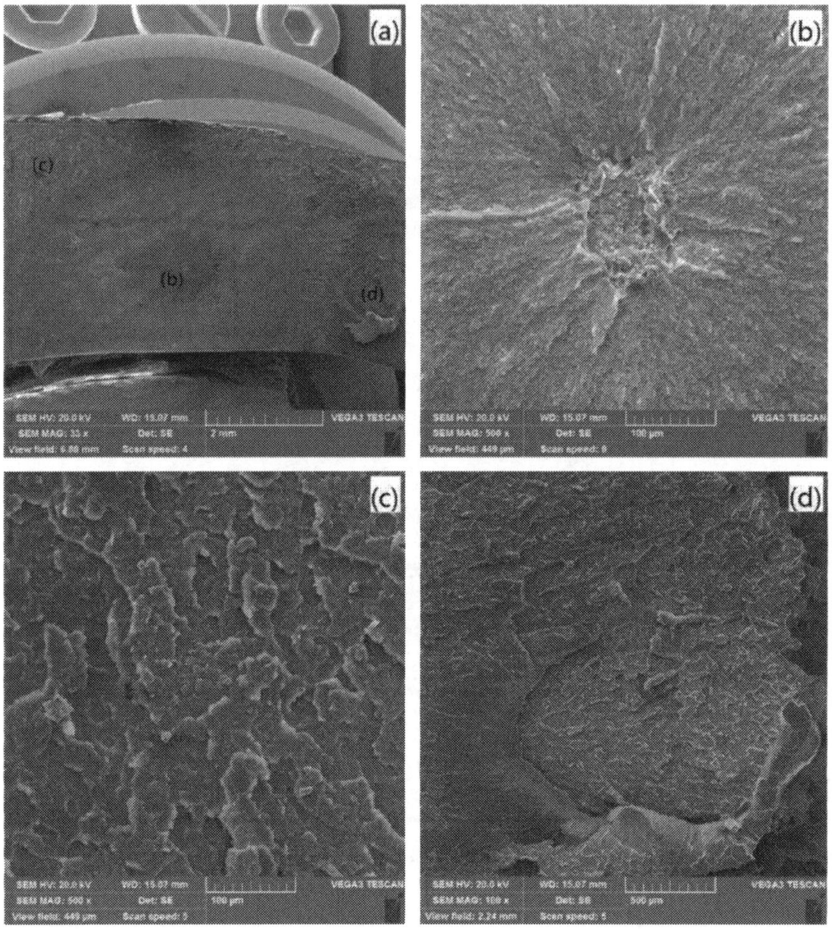

Figure 2. Fracture morphology of ABS/PMMA/EMA (EMA=6wt%) blends.

Figure 3 shows the tensile strength at yield caused by low-rate stretching as a function of EMA content. Virgin ABS/PMMA has the highest tensile strength at yield in comparison with the toughened blends by EMA. It is observed that with the increase in EMA content, the tensile strength at yield gradually decreases, especially when the EMA content is more than 8wt%. Polymer blends are easier to be induced to yield when in a lower tensile strength at yield, which may relate to the phenomenon that the toughened ABS/PMMA blends appear sufficient yield at a low strain rate.

IOP Conf. Series: Materials Science and Engineering **62** (2014) 012028 doi:10.1088/1757-899X/62/1/012028

From the discussion above, it can be observed that ABS/PMMA/EMA blends show better elongation at break compared to the virgin ABS/PMMA. The toughen mechanism is mainly due to the tensile yield, which can dissipate a large amount of energies while it is induced and in the process of propagation. The characteristics of being prone to yield may be associated with the reduction of tensile strength at yield caused by the addition of EMA content. Therefore, EMA particles in ABS/PMMA blends may play the role of stress concentrators that can induce yield through the hollowing-out instead of the rubber particles in ABS resin.

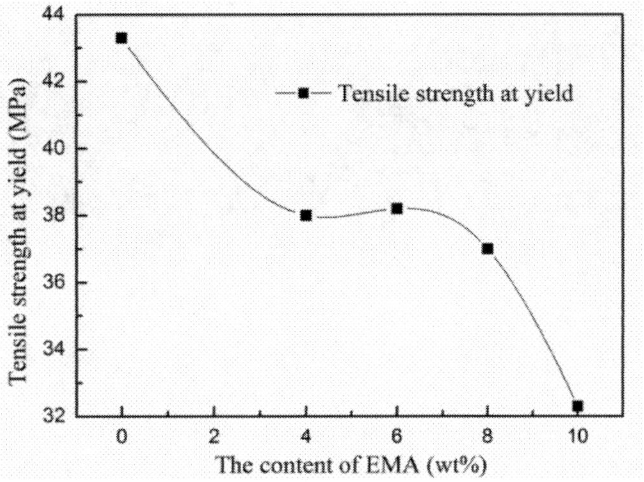

Figure 3. Effect of EMA content on the tensile strength at yield of ABS/PMMA blends.

4. Conclusions

The tensile deformation of ABS/PMMA blends toughened by EMA copolymer and its toughening mechanisms were obtained in this investigation. The elongation at break of ABS/PMMA blends was found to substantially increase with the addition of EMA. When EMA content of the blends increased from 0 to 8wt%, the elongation at break increased by 149.4%. The toughening mechanism for ABS/PMMA by EMA might be due to the reduction in tensile strength at yield, thus the material was easier to be induced to yield by the hollowing-out of EMA particles, especially in the central area. Meanwhile, a large amount of energies would be dissipated in the process of inducement and the formation of yield zone.

Acknowledgments

The work was supported by the National Science and Technology Support Project (Grant No. 2011BAF16B01).

References

[1] Kim Y J, Shin G S, Lee I T and Kim B K 1993 *J. Appl. Polym. Sci.* **47** 295
[2] Araujo E M, Hage JR E and Carvalho A J F 2003 *J. Mater. Sci.* **38** 3515
[3] Machado J M and Lee C S 1994 *Polym. Eng. Sci.* **34** 59
[4] Zhang A M, Zhao G Q and Gao J 2010 *Polym-Plast. Tech. Eng.* **49** 296
[5] Han C D and Yang HH 1987 *J. Appl. Polym. Sci.* **33** 1221
[6] Zhang A M, Zhao G Q and Guan Y J 2013 *J. Appl. Polym. Sci.* **127** 2520
[7] Tayebe N, Hamid G and Ahmad A 2012 *J. Appl. Polym. Sci.* **126** 1637
[8] Xu S A and Tjong S C 1998 *Eur. Polym. J.* **34** 1143

[9] Poomalai P, Varghese T O and Siddaramaiah 2009 *Polym-Plast. Tech. Eng.* **48** 958
[10] Bucknall C B and Clayton D J 1972 *J. Mater. Sci.* **7** 202
[11] Ogazi-Onyemaechi B C, Leong Y W and Hamada H 2010 *J. Appl. Polym. Sci.* **116** 132

IOP Conf. Series: Materials Science and Engineering **62** (2014) 012029 doi:10.1088/1757-899X/62/1/012029

Ionic Liquids as Catalysts for the Radical Acrylate Polymerization Co-initiated by Imine Bases

I Polenz[1] **and S Spange**[2]

[1] Max Planck Institute for Dynamics and Self-organization, Am Fassberg 17, D-37077 Goettingen

[2] Chemnitz University of Technology, Strasse der Nationen 62, D-09111 Chemnitz

E-mail: ingmar.polenz@ds.mpg.de

Abstract. The catalysis of the imine base acrylate (IBA) polymerization by Ionic Liquids (ILs) is reported. Addition of IL traces (~10–50 mM) to an imine base / acrylate mixture leads to both a significant decrease of the activation temperature (40 °C) required for the IBA polymerization process and an increase in the polymerization rate by a factor of 5–40 depending on the IL species. The radical character of the polymerization is proved by copolymerization experiments using methyl methacrylate (MMA) and methacrylonitrile (MAN) and comparison with literature known values of copolymerization parameters r_{MMA} and r_{MAN} of these co-monomers. The influence of the IL on the polymerization kinetics is quantified by the polymerization rate law; the order referring to the IL is 1 indicating its crucial impact on the monomer activation. The IBA activation properties are strongly dependent on the IL interaction strengths with the IBA components verified by the KAMELT-TAFT hydrogen bond donating ability α. The stronger the interaction (higher α) is, the less the IBA polymerization activation. The temperature dependence of four different IL catalysed IBA polymerization is investigated, allows a classification and anomalous non-ARRHENIUS regimes are discussed. Activation energies $E_{A,P}$ span over 20 and 50 kJ·mol^{-1}, which is between the values of thermal- (~80 kJ·mol^{-1}) and photo-initiation (~20 kJ·mol^{-1}).

1. Introduction

Recently, we reported on a novel radical polymerization of acrylates such as methyl methacrylate (MMA), methacrylonitrile (MAN), and methyl acrylate which occurs in the presence of imine bases lacking a protic NH group such as 2-ethyl-2-oxazoline, 2-phenyl-2-oxazoline (POX) and 2-(methylthio)-2-thiazoline (MMT), respectively, at elevated temperatures. [1] A general scheme of the IBA polymerization is shown below (Scheme 1).

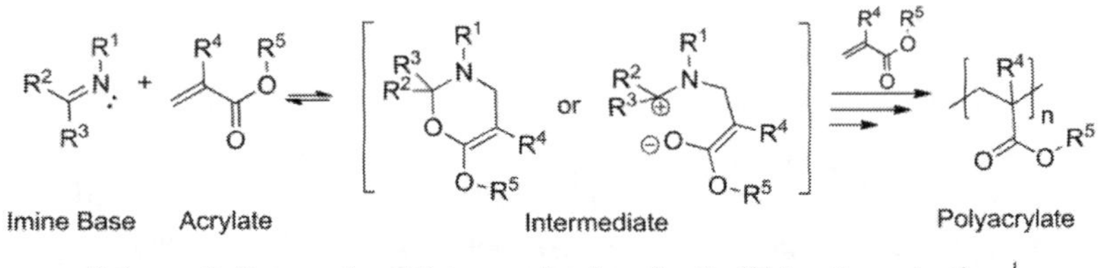

Scheme 1. Proposed multistep mechanism for the IBA polymerization.[1]

Homo- and co-polymerizations of acrylates have been experimentally detected yielding high molecular weight polymers with dispersity around 1.90. We isolated the imine base after polymerization almost quantitatively and it can be multiply re-used. Either a six-membered ring or a zwitterion composed of the acrylate and the imine base is still hypothetically assumed to be an intermediate at the initiation process and, hence, this polymerization has been denoted as imine base / acrylate (IBA) polymerization. However, drawbacks of these systems are the high activation temperature of 100 °C and the high molar amounts of imine required for the polymerization.

Herein, we report on the impact of Ionic Liquids (IL) on the radical IBA polymerization. The use of room temperature Ionic Liquids have gained high interest in academia and industry in the last decades due to their use as co-reactants and catalysts in organic and inorganic chemistry as well as in polymer and material science. [2-13] In these works, most likely, ILs are used in high molar amounts or as solvents to gain notable effects on the reaction path. In the here reported systems, tremendous changes in the polymerization kinetics at IL concentrations even at catalytic amounts of 10–50 mM are detected. The use of Ionic Liquid traces at the IBA polymerization leads to both a decrease in activation temperature and an increase in reactivity of the IBA mixture; both effects are quantified in this work by measuring the polymerization kinetics. It is shown that the polymerization rate is a function of the hydrogen bond donating ability α of the Ionic Liquid and can be adjusted precisely in a range between 5- up to 40-fold the value of the non-IL containing IBA mixture. Scheme 2 shows the imine bases and Ionic Liquids classified by their α-values that are used in this study.

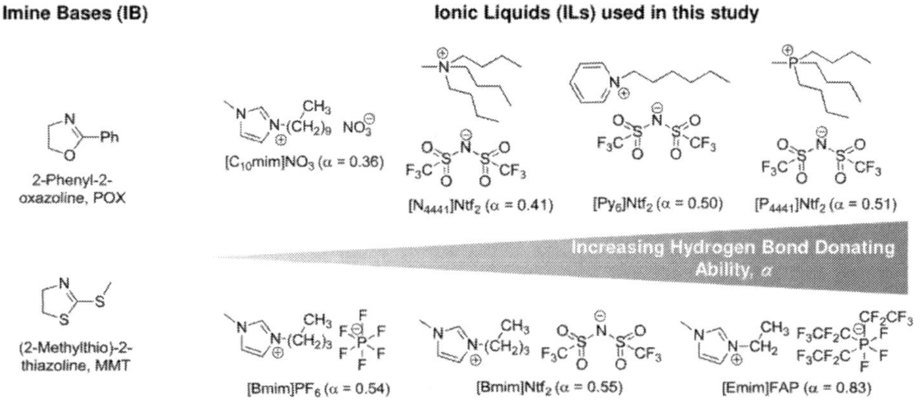

Scheme 2. Imine Bases and Ionic Liquids used in this study. The ILs are systematically arranged by their KAMLET-TAFT hydrogen bond donating abilities α.

Furthermore, we investigated the behavior of IL catalysed IBA polymerization at elevated temperatures. Four different polymerization mixtures were examined. Besides the detection of linear ARRHENIUS dependencies at the *ln k* versus T^{-1} plots ([Emim]FAP) anomalous 'volcano' curves for $[Cmim]NO_3$ and $[N_{4441}]Ntf_2$ were observed and discussed. By knowing the factors that dominate the kinetics of this novel reaction type, we propose polymerization systems with tunable properties for applications in polymer and material science.

2. Results and Discussion

2.1 Qualitative observations

Polymerizations were carried out as described earlier.[1] MMA was distilled prior to use and the amine bases and ILs were used as received. A standard polymerization procedure is as follows; to a solution of 4.2 millimole (mmol) imine base in 2.65 mL (25 mmol) MMA, 10–100 mg of IL is added under inert conditions. The reaction mixture is heated to 65 °C for 4 h. Then, the viscous solution is cooled to room temperature, diluted with 10 mL of chloroform and gently dripped into 400 mL of methanol. After purification and drying of the precipitate, poly(methyl methacrylate) (PMMA) is obtained as a colorless solid. Table 1 shows specific results of IL catalysed IBA polymerizations using MMT and POX.

Table 1. Reaction conditions of IL catalysed IBA polymerizations as well as PMMA analysis results recorded by gel permeation chromatography. The MMA concentration is 4.7 M, the IB concentration is 0.8 M and that of the IL 50 mM.

IB	IL	α	T [°C]	t [h]	M_N [g·mol^{-1}]	M_W / M_N	X_{MMA} [%]
MMT	[Emim]FAP	0.83	65	3	46.700	1.54	25
			77	3	56.200	1.49	49
			95	3	139.000	1.85	60
POX	[Emim]FAP	0.83	65	10	62.400	1.87	61
			85	8	77.200	1.60	67
	$[Bmim]Ntf_2$	0.55	85	10	56.800	1.70	99
	$[Bmim]PF_6$	0.54	85	10	60.000	1.79	99
	$[Py_6]Ntf_2$	0.50	85	5	266.800	1.78	30
	$[C_{10}mim]NO_3$	0.36	85	5	54.800	1.52	66

Both imine bases are suitable for the IL catalysed IBA polymerization; however, we detected significant differences in their reactivity, qualitatively. Thus, MMT has a higher reactivity and yields PMMA with a slightly reduced average molecular weight compared to the POX systems. We used a broad variety of Ionic Liquid species in the experiments and for each IL catalysis of the IBA polymerization was measured, considerably. More drastic changes in the reactivity of certain IBA systems were found while changing the IL type; a detailed quantification of this acceleration is given in chapter 2.2.

Interestingly, we re-obtained the imine base / IL mixture out of the methanol residue and it can be re-used for multiple times. No IL-groups were detected in the product PMMA, which was proven by ^{1}H and ^{13}C NMR (Figure one) studies underlining the fact that the Ionic Liquid acts as a catalyst in the IBA process.

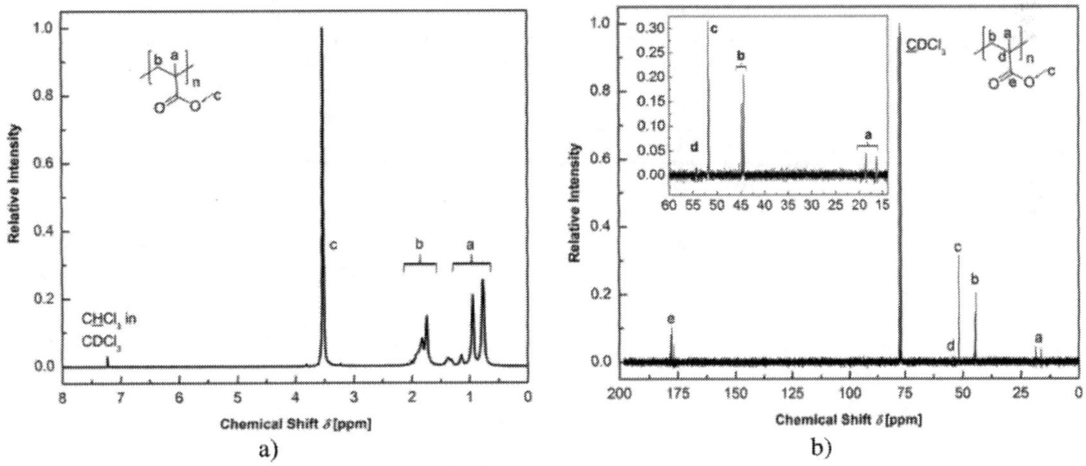

Figure 1. ^{1}H (a) and ^{13}C NMR of PMMA (250 MHz and 62.5 MHz, in $CDCl_3$). It was polymerized using the IBA combination POX / MMA and the IL [Emim]FAP. No Signals of the IL can be detected in neither of the spectra.

The radical character of the IL catalysed IBA polymerization is proved by copolymerization experiments of MMA and MAN. Experiments yield the corresponding copolymerization parameters

r_{MMA} and r_{MAN} that represent the molar ratios of the co-monomers in the copolymer as given by the MAYO-LEWIS equation. [14] Therefore, in an anionic polymerization the molar ratio of the stronger electrophile MAN is much higher than that of MMA (statistical non-azeotropic copolymerization) and in a radical copolymerization an azeotropic copolymerization is found. [15] The MMA/MAN copolymerization diagram using the combination POX / [Emim]FAP and literature known plots of an anionic and radical copolymerization of this co-monomer pair is shown in Figure 2 and the corresponding *r*-values are given in Table 2. Comparison of the results with literature known data of these co-monomers admits to confirm the propagating species at the IL catalysed IBA polymerization to be of a radical nature.

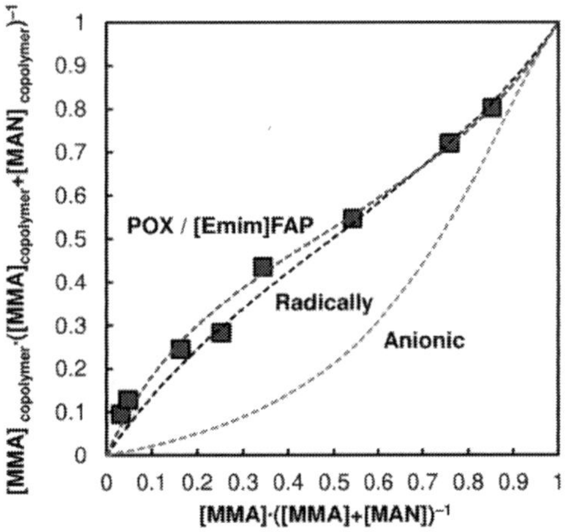

Figure 2. Copolymerization diagrams of MMA / MAN using POX / [Emim]FAP (■) and fitted curve (---) as well as comparison with literature known plots for radical (---) and anionic (---) copolymerizations of these co-monomers.[15]

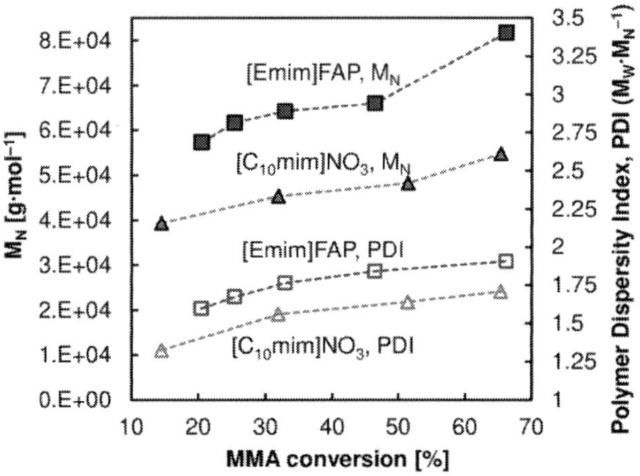

Figure 3. PMMA average numbermolecular weights M_N (filled symbols) and polymer dispersity indices (hollow symbols) of IL catalysed IBA polymerizations using the combination POX (0.8 M) / MMA (4.7 M) as a function of the monomer conversion. It was catalysed with 50 mM [Emim]FAP (■) and [C_{10}mim]NO_3 (▲) at 85 °C.

Table 2. Copolymerization parameters r_{MMA} and r_{MAN} using POX / [Emim]FAP as well as literature known *r*-values of radical and anionic copolymerizations of these co-monomers.[15]

Initiation	r_{MMA}	r_{MAN}
POX/[Emim]FAP	0.57	0.58
MMT/[Emim]FAP	0.54	0.53
Radically[15]	0.67	0.65
Anionic[15]	0.67	5.20

At the IL catalysed IBA homopolymerization of acrylates, we found that with increasing monomer conversion the macromolecule chain length increased. The results of MMA polymerizations using POX and Ionic Liquids [Emim]FAP and [C_{10}mim]NO_3 are shown in Figure 3. Besides the increase in M_N a slight increase in the polymer chain length dispersity (PDI) is notable. In common living polymerizations the PDI value decreases with increasing conversion; [16] the character of the IL catalysed IBA polymerization, hence, cannot be classified as a purely living polymerization. As found earlier for the non-IL containing IBA polymerization in bulk the measured behavior belongs to a quasi-living process. The observations manifest the fact that the Ionic Liquid dominantly influences the initiation process *via* an activation of the monomer and co-initiator; quantification of the effects by means of the polymerization kinetics is the next step for a detailed description of the IL catalysis.

2.2 Kinetic studies

We used standard approaches of polymer physics for determining the polymerization rate law.[16] In a previous study we quantified the shape of the IBA polymerization rate law that reads[1]

$$v_{P,0} = k_{P,0} \cdot [\text{POX}]^{0.5} \cdot [\text{MMA}]^{1.5}, \tag{1}$$

where, $v_{P,0}$ is the net polymerization rate, $k_{P,0}$ the polymerization rate constant and [POX] and [MMA] the concentrations of POX and MMA. Inclusion of the Ionic Liquid in the polymerization rate law gives

$$v_{P,0} = k_{P,0}{}^{*} \cdot [\text{POX}]^{0.5} \cdot [\text{MMA}]^{1.5}, \tag{2}$$

$$k_{P,0}{}^{*} = k_{P,0} \cdot [\text{IL}]^{x}, \tag{3}$$

where [IL] is the Ionic Liquid concentration and x equals the order referring to the IL. For an entire description of $v_{P,0}$ in the next step the order x referring to the IL is defined. Therefore, we transferred equation 2 into a logarithmic straight line equation

$$ln\ v_{P,0} = ln\ k_{K,0} + x \cdot ln\ [\text{IL}] + 0.5 \cdot ln\ [\text{POX}] + 1.5 \cdot ln\ [\text{MMA}]. \tag{4}$$

A set of IL catalysed IBA polymerizations was performed; $v_{Br,0}$ was determined from the initial slopes of the [MMA] versus time plots and concentrations of POX and MMA were held constant whereas [IL] was modified.[16] The slope of the correlation $ln\ v_{P,0}$ over ln [IL] yields x. In Figure 4 the experimental data set as well as the x-determination of the IL catalysed IBA polymerization using the combination POX / [Emim]FAP / MMA are shown. The experiments yield a value of one for the order referring to the Ionic Liquid that indicates a crucial impact of the IL on the whole kinetics of the IBA polymerization. A full description of the IBA polymerization rate law allows for a detailed quantification of the influence of ILs on the polymerization kinetics.

In previous studies, investigations on the interaction of ILs with acrylates and imine bases by means of near field attenuated total reflection infrared spectroscopy and correlation with KAMLET-TAFT parameters have been done. [17, 18] There, we detected a significant shift of the MMA ν(C=O) stretching vibrational mode to lower wavenumbers in the presence of ILs. This observation indicates an increase in the electrophilicity of the monomer that contributes to the nucleophilic imine addition rate. Consequently, an increase in the IBA reactivity results (Scheme 3). The outcomes of this study encourages us to use the hydrogen bond donating ability α of Ionic Liquids as an indicator for the interplay strengths of ILs with the IBA components. In this work, we used Ionic Liquids that drastically vary in their hydrogen bond donating abilities; α-values span over the range between 0.36 (weak H-bond donor) and 0.83 (strong H-bond donor).

IL catalysed IBA polymerizations were carried out using the combination POX / MMA; the IL species was changed. Polymerization rate constants $k_{P,0}$ were determined and brought into correlation to the corresponding hydrogen bond donating ability of the Ionic Liquid; the results are given in Figure 5.

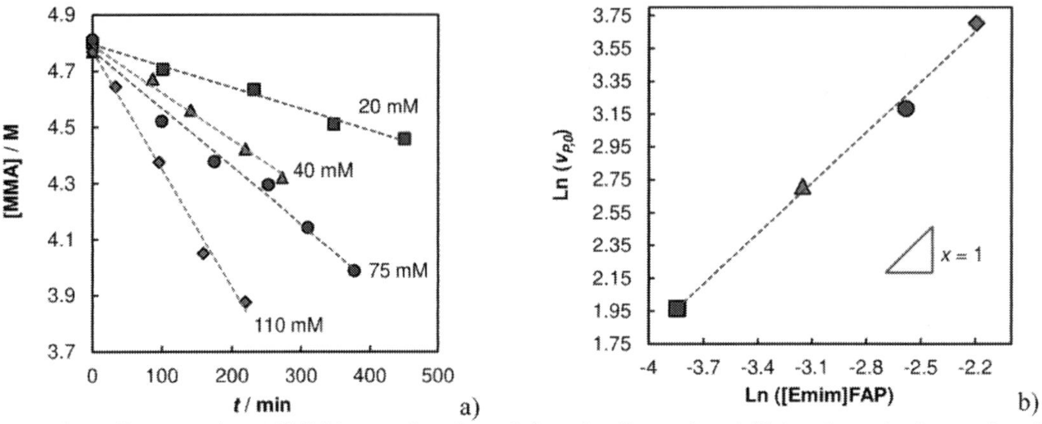

Figure 4. a) Concentration of MMA as a function of time for IL catalysed IBA polymerizations using the combination POX / [Emim]FAP / MMA at 65 °C at [Emim]FAP concentrations of 20 mM (■), 40 mM(▲), 75 mM (●), 110 mM (■). The linear fits of the initial stages are $[MMA]_{20mM} = -7.2 \cdot 10^{-4} \cdot t + 4.77$; $[MMA]_{40mM} = -15.2 \cdot 10^{-4} \cdot t + 4.77$; $[MMA]_{75mM} = -24.2 \cdot 10^{-4} \cdot t + 4.78$; $[MMA]_{110mM} = -40.9 \cdot 10^{-4} \cdot t + 4.76$. b) Double-logarithmic plot of the polymerization rate $v_{P,0}$ as a function of the [Emim]FAP concentration. The linear fit is $ln\ v_{P,0} = 1.02 \cdot ln$ ([Emim]FAP) + 5.9, $R^2 = 0.99$.

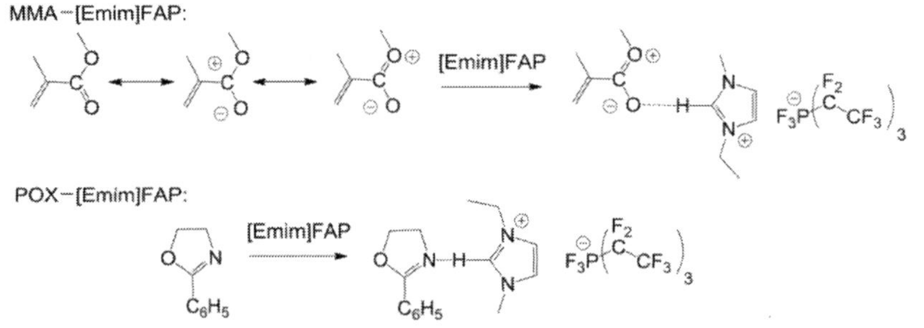

Scheme 3. Interactions of [Emim]FAP with MMA and POX.[17,18]

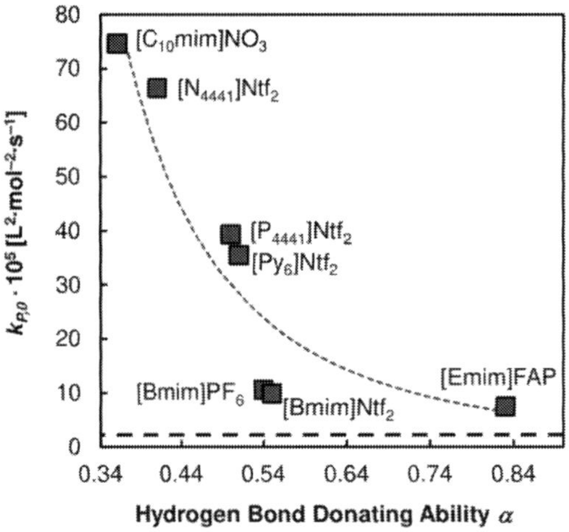

Figure 5. IBA polymerization rate constants $k_{P,0}$ at IL catalysis as a function of the IL hydrogen bond donating abilities α. The power law fit is $k_{P,0} \cdot 10^5 = 3.7 \cdot \alpha^{-3}$, $R^2 = 0.83$. The dotted black line indicates the $k_{P,0}$ value at the non-IL containing IBA polymerization.

Each $k_{P,0}$-value of the IL catalysed IBA polymerization is higher than that of the non-IL containing IBA polymerization ($1.72 \cdot 10^{-5}$, Fig. 4, dotted black line). We observed that with increasing IL hydrogen bond donating ability the $k_{P,0}$-values decreased. Besides the MMA activation by the IL as described above, a complexation of the imine bases has to be taken into consideration, which is the dominating equilibrium, obviously. With increasing hydrogen bond donating ability of the IL the imine bases is stronger complexed which results in a reduction in $k_{P,0}$. The IBA polymerization kinetics can be fine-tuned facilely in a range between 9.5 and $76.5 \cdot 10^{-5}$ $L^2 \cdot mol^{-2} \cdot s^{-1}$ by varying the catalyst IL. Hence, we are able to control the monomer conversion, enabling the potential use of these systems in various fields of material fabrication.

2.3 Behavior at varying thermal treatment

It is practicable to determine the thermal properties of the system described in this work for its establishment in the field of material science. To quantify the behavior of the IBA polymerization with IL catalysis at varying thermal treatment we use the ARRHENIUS approach:

$$k_{P,0} = A \cdot \exp(-E_{A,P} \cdot (R \cdot T)^{-1}), \tag{5}$$

$$ln\ k_{P,0} = ln\ A - E_{A,P} \cdot (R \cdot T)^{-1,} \tag{6}$$

where A is the pre-exponential factor, R the ideal gas constant, T the absolute temperature and $E_{A,P}$ the activation energy of the polymerization rate constant. The activation energy $E_{A,P}$ is taken as a measure for the activation barrier for the net polymerization process including the initiation, propagation and termination. The double logarithmic form of equation five given in equation 6 enables the verification of $-E_{A,P} \cdot R^{-1}$ by determining the slope of the plot $ln\ k_{P,0}$ over T^{-1}. In Figure 6, experimental results of [Emim]FAP catalysed MMA IBA polymerizations using the imine bases POX and MMT are shown.

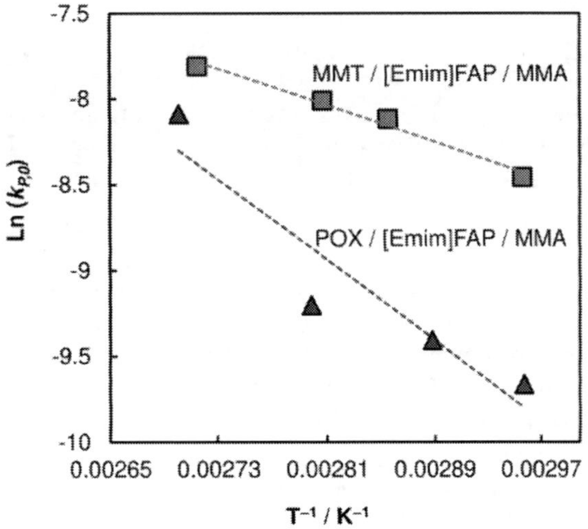

Figure 6. ARRHENIUS plots of IL catalysed IBA polymerizations using MMT / [Emim]FAP / MMA (■) and POX / [Emim]FAP / MMA (▲). The linear fits are $ln\ k_{P,0}(MMT) = 2683.1 \cdot T^{-1} - 0.48$; $R^2 = 0.97$ and $ln\ k_{P,0}(POX) = 5955.3 \cdot T^{-1} + 7.81$; $R^2 = 0.88$ and the corresponding activation energies are 22.3 $kJ \cdot mol^{-1}$ (MMT) and 49.5 $kJ \cdot mol^{-1}$ (POX).

The experimental data strictly follow ARRHENIUS's law; we measured activation barriers of 22.3 (MMT) and 49.5 kJ mol^{-1} (POX). Clearly it is found, the IBA polymerization kinetics using POX are more inhibited by [Emim]FAP due to a stronger complexation. The $E_{A,P}$ values enable a classification of the IL catalysed IBA polymerization. Thus, it can be ranged between radical polymerization initiated by photo-initiators ($E_{A,P} \sim 20$ $kJ \cdot mol^{-1}$) and radical polymerizations initiated by thermal initiators ($E_{A,P} \sim 80$ $kJ \cdot mol^{-1}$).[15]

A detailed look on the POX-mediated IBA polymerizations catalysed by [C_{10}mim]NO_3 and [N_{4441}]Ntf_2 under thermal treatment reveals an anomaly; the appropriate correlations $ln\ k_{P,0}$ over T^{-1} more rather equal 'volcano'-like curves (Figure seven).

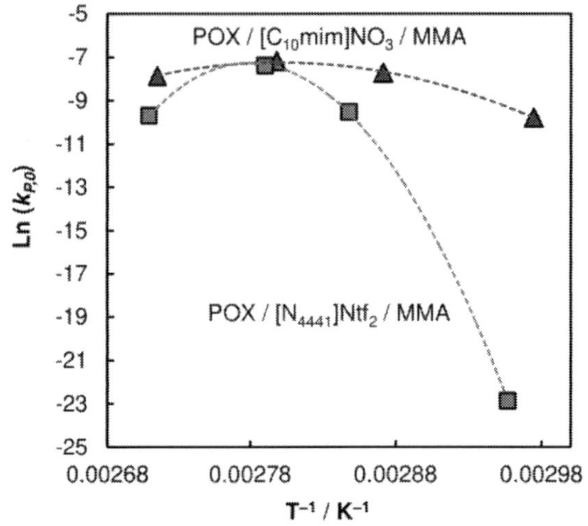

Figure 7. ARRHENIUS plots of IL catalysed IBA polymerizations using POX / [N_{4441}]Ntf_2 / MMA (■) and POX / [C_{10}mim]NO_3 / MMA (▲). The polynomial fits are $ln\ k_{P,0}([N_{4441}]Ntf_2) = -4.96\cdot10^8\cdot T^{-2} + 2.76\cdot10^6\cdot T^{-1} - 3841.7$; $R^2 = 0.99$ and $ln\ k_{P,0}$ ([C_{10}mim]NO_3) $= 4.69\cdot10^5\cdot T^{-2} - 8.38\cdot10^7\cdot T^{-1} - 664.7$; $R^2 = 0.99$.

Experimental data cannot be fitted linearly. The system reaches a certain optimum temperature with a maximum $k_{P,0}$; above that temperature the polymerization rate decreases. A drastic increase in the IL–POX interaction for [N_{4441}]Ntf_2 and [C_{10}mim]NO_3 at higher temperatures is most propably to be responsible for this trend. Thereof, at a certain threshold temperature the POX–IL interaction exeeds the sum of interactions that lead to the formation of the initiating species which, consequently, because the initiation is the rate determining step,[16] results in a reduction in $k_{P,0}$. The data manifest the versatility of interactions at the IL catalysis of IBA polymerizations.

3. Conclusion

The addition of Ionic Liquid traces to the binary imine base / acrylate (IBA) polymerization mixture has notable effects on the whole polymerization process. Most considerably, IBA polymerization activation temperatures are decreased and kinetics is accelerated in the presence of ILs. The activation properties are strongly dependent on the IL–IBA component interplay strengths that are elucidated here by the hydrogen bond donating ability α of the IL. Therefore, the stronger the IL–IBA component interaction (the higher α) is, the less are the IL activation properties. The quasi-living behaviour of the IL catalysed IBA polymerization is indicated by the molecular weight increase of acrylate polymers at increasing monomer conversions, which is achieved solely by catalytic IL amounts. The form of the polymerization law of the IL catalysed IBA polymerization is determined allowing a formal classification of the catalysis effect of particular IL / imine base / acrylate mixtures. The reaction rate at IL catalysis can be precisely controlled and reach values up to 40-fold the reactivity of the non-IL IBA mixture. Our investigations of IL catalysed IBA polymerization at thermal treatment yield the specific polymerization activation energies. The polymerization is ranged between thermal- and photo-initiation. The results reveal novel approaches on the way to designing acrylate coatings based on IBA polymerization mixtures.

References

[1] Polenz I, Schmidt F G, Tchernook I, Friedrich J, Spange S 2013, *Macromol. Chem. Phys.* **214** 1473

[2] Abbott A P, Capper G, Davies D L, Rasheed R K, Tambyrajah G 2002 *Green Chem.* **4** 24
[3] Early M J, McCormac P B, Seddon K R 1999 *Green Chem.* **1** 23
[4] Lourenco,N M T, Afonso C A M 2003 *Tetrahedron* **59** 789
[5] Allen G D, Buezzo M C, Davies I G, Villagran C, Hardacre C, Compton R C 2004 *J. Phys. Chem. B* **108** 16322
[6] Conte V, Floris B, Galloni P, Silvagni A 2005 *Pure Appl. Chem.* **77** 1575
[7] Lancaster N L, Welton T 2004 *J. Org. Chem.* **69** 5986
[8] Brinichi L, Germani R, Savelli G 2003 *Tetrahedron Lett.* **44** 2027
[9] Debreul J F, Bazureau J P 2000 *Tetrahedron Lett.* **41** 7351
[10] Wheeler C, West K N, Liotta C L, Eckert C A 2001 *Chem. Commun.* **10** 887
[11] Chiappe C, Conte V, Pieraccini D 2001 *Org. Lett.* **3** 1061
[12] [12] Carmichael A J, Haddleton D M 2002 *Polymer Synthesis in Ionic Liquids* (Ionic Liquids in Synthesis) ed Wasserscheid P, Welton T (Weinheim: Wiley-VCH Verlag)
[13] Schmidt-Naake G, Woecht I, Schmalfuss A, Glueck T 2009 *Macromol. Symp.* **275–276** 204
[14] Mayo F R, Lewis F M 1944 *J. Am. Chem. Soc.* **66**(9) 1594
[15] Brandrup J, Immergut E H, Grulke E A, Abe A, Bloch D R 1999 *Polymer Handbook*, (New York: Wiley-Interscience)
[16] Matyjaszewski K, Davis T B 2002 *Handbook of Radical Polymerization* (Hoboken: Wiley Interscience)
[17] Polenz I, Schmidt F G, Spange S 2012 *J. Polym. Sci. A* **51** 687
[18] Polenz I, Spange S 2014 *Ionic Liquids as Catalysts for the Imine Base/Isocyanate-Mediated (IBI) Radical Polymerization: Mechanistic Studies and Polymerization Kinetics Control* (Ionic Liquids: Synthesis, Characterization and Applications) ed Brooks A (Hauppauge, NY: Nova Science Publishers)

Effect of compatibilizer on the strength of ABS/PMMA blends

J C Zhou[1], J Gao[1]*, S H Wang[1], S X Lin[2], J Huang[2] and L L Xu[2]

[1] School of Mechanical, Electrical & Information Engineering, Shandong University at Weihai, Weihai 264209, China
[2] Qingdao Haier Mold Co., Ltd, Qingdao 266103, China

E-mail: winter003@163.com

Abstract. To explore the influence of different compatibilizers on the strength of ABS/PMMA, two groups of tests were performed. Polymers were mixed by melt blending, and then the ternary blends were tested. SMA makes the tensile strength and impact strength of ABS/PMMA blends decrease obviously, while SBS improves the two properties except the addition of 2wt%. The results of different compatibilizers on ABS/PMMA blends were unlike.

Keyword. ABS; PMMA; compatibilizer; tensile strength; impact strength

1. Introduction

Blending modification is a convenient and effective method that makes materials possess the excellent properties of each part and satisfy different demands from different domains. As engineering plastic, acrylonitrile-butadiene-styrene (ABS) has nice toughness, hardness and rigidity. Good mechanical and processing properties make ABS widely used in many fields [1]. Poly (methyl methacrylate) (PMMA) has good optical property and light aging performance. Although the blends of ABS and PMMA increase heat resistance of ABS and notch impact strength of PMMA, different mixed degrees have great impact on the nature of the blend as shown by studies in the past decades [2-4].

The compatibility between ABS and PMMA will affect the nature of the blends clearly [5, 6]. In this case, the increase of compatibility has been studied in the past few years [7]. According to the principle of dissolution in the similar material structure, addition of compatibilizer can improve the performance of polymer [8]. In this study, two compatibilizers were used to explore the effect of mechanical properties on ABS/PMMA blend to explore the different power on polymer by different compatibilizers.

2. Experimental

2.1 Materials

A commercial ABS resin (757K) and PMMA resin (CM-211) from Zhenjiang Chimei Chemical Co., Ltd were used for the matrix. Two different compatibilizers were used in the study: styrene maleic

*To whom any correspondence should be addressed.

IOP Conf. Series: Materials Science and Engineering **62** (2014) 012030 doi:10.1088/1757-899X/62/1/012030

anhydride (SMA) and styrene-butadiene-styrene (SBS), all from Dongguan AnChen plastic technology Co., Ltd.

2.2 Preparation

Prior to use, ABS, PMMA and SMA were dried at 80°C, while SBS was dried at 60°C for 12 h in an air circulating oven. Then, ABS/PMMA was mixed with SMA and SBS, respectively, in a co-rotating twin extruder with a screw diameter of 21.7 mm and a length to diameter ratio of 40. Temperature was set range from 200°C to 225°C from hopper to die and the speed was 150 rpm. The extruded was injected into molding machine after granulation and drying at 80°C for 12 h again. The dumbbell-shaped test piece for tensile test was 150 mm in gauge length, 10 mm in width and 4 mm in thickness; the rectangular test simple was 80 mm long, 10 mm wide and 4 mm thick for impact tests.

2.3 Characterization

The tensile and Izod impact tests were performed on CMT 5305 electrical testing machine at a crosshead speed of 4 mm/min and XC-5.5D impact tester, respectively.

3. Results and discussion

3.1 Effect of SMA content

The styrene-acrylonitrile copolymer (SAN) of ABS has good compatibility with SMA, and SMA can be incorporated into PMMA, because the phenyl of SMA has strong interaction with carbonyl of PMMA. Hence, SMA can increase the compatibility between ABS and PMMA. From previous experience, the blends of ABS/PMMA, at 80/20, have the best comprehensive mechanic performance, by adding SMA into the blends to improve its performance.

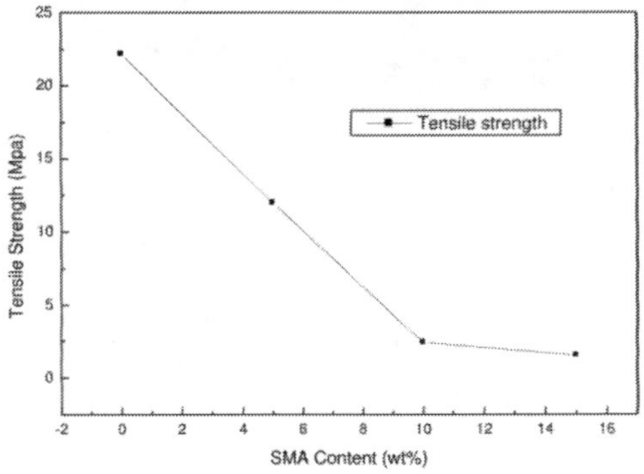

Fig.1. Effect of SMA on tensile strength of ABS/PMMA blends

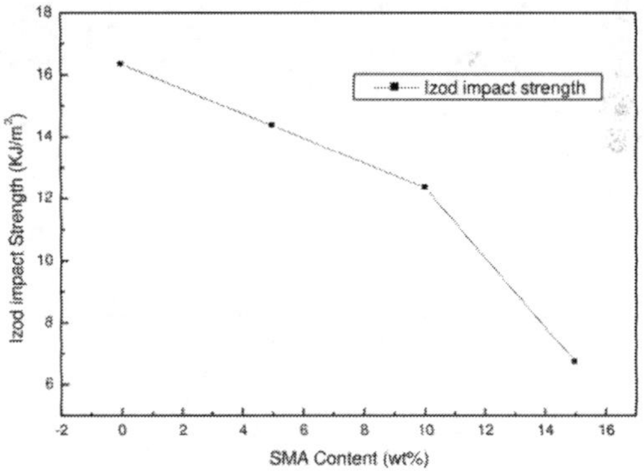

Fig.2. Effect of SMA on Izod impact strength of ABS/PMMA blends

As Fig.1 shows, the blends' tensile strength decreases with the increase of SMA, so does its impact property, as seen in Fig.2.

The ternary mixture was ABS and PMMA at a ratio of 80 to 20, with the SMA content of 0.5wt%, 10wt% or 15wt%. The two figures above indicate that SMA did not toughen ABS/PMMA blends, but obviously lower its performance. This may be caused by the rigidity of maleic anhydride in SMA. Maleic anhydride possesses cyclic structure, and this made the tensile strength and impact property decrease [9]. The macroscopic phenomena indicated that as the compatibilizer of ABS and PMMA, SMA cannot increase their compatibility.

3.2 Effect of SBS content

The butadiene of SBS presents rubber elasticity, and this can promote the impact strength of polymer. The styrene part owns high temperature liquid, and this nature makes the machinability of polymer easier. In addition, butadiene and styrene are also parts of ABS. According to this, SBS can be the compatibilizer of ABS and PMMA [10]. Under the ratio of 80 to 20 of ABS and PMMA, SBS was added at 0, 2wt%, 4wt%, and 6wt%.

After a series of performance tests, the result can be seen as follows.

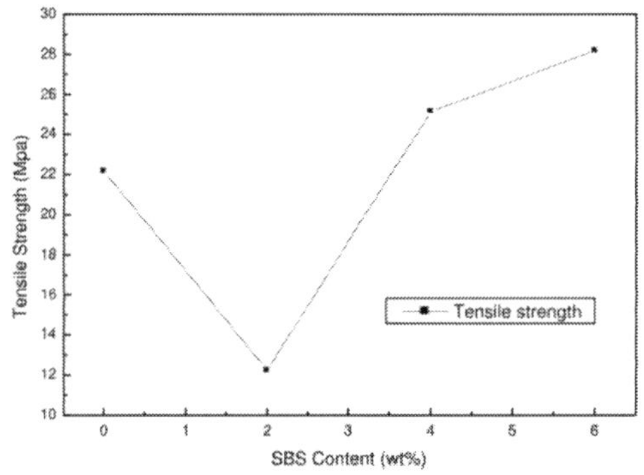

Fig.3. Effect of SBS on tensile strength of ABS/PMMA blends

The tensile strength of ABS/PMMA blends dropped with SBS added before the SBS content reached 2wt%, and then kept increasing as SBS increased, while the impact strength of the blends kept increasing with the increase of SBS, as showed in Fig.4, except the addition of 2wt%.

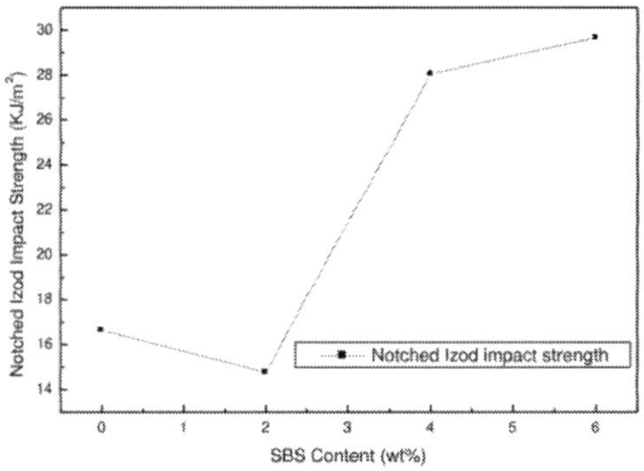

Fig.4. Effect oof SBS on notched impact strength of ABS/PMMA blends

As seen above, SBS differs from SMA clearly. The phenomenon may be that SBS does not work as compatibilizer but toughener. The parts of SBS, Styrene and butadiene, enhance the strength of the ABS/PMMA blends. When the volume addition of SBS is little, SBS may affect the compatibility of the blends, and this leads to decline of tensile and impact strengths. With the content increasing, the function of toughening is in a dominant position, hence, the polyline of the strength goes up.

4. Conclusions

The two compatibilizers, SMA and SBS, play very differently in the blends; SMA makes tensile strength and impact strength of ABS/PMMA blends decrease obviously, while SBS improves the two properties except the addition of 2wt%. It is to be explored whether SMA and SBS promote the compatibility between ABS and PMMA or not.

Acknowledgments

The work was supported by the National Science and Technology Support Project (Grant No. 2011BAF16B01).

References

[1] Tayebe N, Hamid G and Ahmad A2012 *J. Appl. Polym. Sci.* **126** 1637
[2] Eniko S, Marton O and Ferenc R 2011 *J. Anal. Appl. Pyrol.* **92** 19
[3] Zhang A M, Zhao G Q and Guan Y J 2013 *J. Appl. Polym. Sci.* **127** 2520
[4] Chang Dae Han and Heng-Huey Yang 1987*J.Appl.Polym.Sci.***33** 1221
[5] Jan R, Ralf L and Wolfgang G 2012 *Polym.***57** 87
[6] S M Lee, C H Choi and B K Kim 1994 *J. Appl. Polym. Sci.* **51** 1765
[7] Basseri G, Mazidi M M, Hosseini F and Aghjeh M K R 2014*Polym.Bull.***71** 465
[8] YJ Kim, G S Shin, I T Lee and B K Kim 1993 *J. Appl. Polym. Sci.* **46** 295
[9] J S Wu, B H Guo, Chi-Ming Chan, J X Li, Hoi-Shuen Tang 2001 *Els. Sci.* **42**8857
[10] R W Wang and W Wang 2003 *J.Appl.Polym.Sci.* **90**2260

Fabrication and Properties of sc-PLA/SiO_2 Composites

Huan Ge[1], Zhiguo Zhu[1], Huihui Yin[2], Xiuqin Zhang[1], Rui Wang [1,2]

[1]Department of Materials Science and Engineering, Beijing Institute of Fashion Technology, Beijing 100029
[2]Department of Polymer Science and Materials, State Key Laboratory of Polymer Mat erials Engineering,

Sichuan University, Chengdu, China 610065

Abstract. Stereocomplex poly(lactic acid) /SiO_2 (sc-PLA/SiO_2) composites were prepared through solution mixing followed by vacuum drying. In the composites, good dispersion of SiO_2 particles could be obtained when the loading was less than 3wt%. Crystallization properties of sc -PLA were improved due to the addition of SiO_2. The maximum of both melt crystallization temperature (T_{mc}) and crystallinity (X_c) simultaneously occurred when the loading of SiO_2 was 1wt%. Furthermore, thermal stability of the composites was not affected greatly by agglomeration in a certain range. The initial decomposition temperature increased by 75℃ (5wt% loading of SiO_2) compared with the crude.

Keywords. sc-PLA; SiO_2; Crystallization; Thermal-stability

1. Introduction

As a plant-derived biodegradable polyester, poly(lactic acid) (PLA) is widely used as engineering plastics, pharmaceutical, wrapping films, etc[1]. The enhancement of crystallization rate and heat-resistance of semicrystalline PLA is crucial to the broadening of the scope of application. Stereocomplexe PLA (sc-PLA) prepared from PLLA and PDLA has been investigated as advanced PLA to some special applications. The most valuable attraction of sc-PLA is the melting point (up to about 230℃), which is about 50℃ higher than that of the homocrystals [2]. However, the crystallization rate and heat-stability of sc-PLA still need great improvement for some plastic application.

One of the most effective methods to improve the properties of sc-PLA is nano-modification. For instance, Brzezinski [3] prepared PLA stereocomplexes from equimolar mixtures of PLLA and PDLA in the presence of MWCNT-g-PLA. In this case, crystallization of stereocomplexes from molten state was completely reversible, without formation of any homocrystallites. Yukiko Furuhashi [4] mixed equimolar PLLA and PDLA with addition of organo-modified montmorillonite in solution. Homocrystallization and stereocomplexation of PLAs both enhanced after annealing of the blends.

As one of the most commonly used modifier, nano-SiO_2 has been widely used in plastic and paint. Liu [5] prepared PLA/nano-SiO_2 composites by in-situ polycondensation. Thermal stability and tensile strength increased compared with the crude. However, the effect of nano-SiO_2 on other properties of sc-PLA has not been investigated to our best knowledge. This paper explores in detail the effect of nano-SiO_2 on the crystallization and thermal stability of sc-PLA, in order to find an effective method to obtain high thermal-stability PLA composites to optimize the application of PLA materials. .

2. Experimental

2.1 Materials

L-lactic acid (90%) and D-lactic acid (90%) were purchased from Musashino (Jiangxi Province, China). Stannous dichloride dehydrate ($SnCl_2 \cdot 2H_2O$), *p*-toluenesulfonic acid monohdrate (TSA) were obtained from Tianjin Guangfu Technology Company. Dichloromethane and alcohol were from Beijing Chemical Works. Nano-SiO_2 was supplied by Nanjing Haitai nano Material Co.

2.2 Polymerization

2.2.1 Preparation of PLA

The linear PLLA and PDLA enantiomeric polymers were synthesized according to known procedure [6]. Take PLLA as an example, mixture of OLLA (50 g) with $SnCl_2.2H_2O$ (0.05 g) and TSA (0.05 g) in a three-necked flask was heated at 180℃ under reduced pressure of 13.3×10^2 Pa for 3-5 h. The crude product was purified through two dissolve-precipitation cycles in chloroform and ethanol, respectively. The relative weight-averaged molecular weight (M_w) of PLLA in white solid powder ranged between 2 and 3×10^4. PDLA was prepared with the same procedure.

2.2.2 Preparation of sc-PLA/SiO2 composites

SiO_2 was pre-dispersed in DMF through high speed ball-milling procedure lasting for about 7 hours. PLLA and PDLA were respectively dissolved in dichloromethane at a concentration of 1 g/dl. SiO_2 dispersion solution was added drop by drop into the solution mixture of PLLA and PDLA under vigorously stirring. The solvent was evaporated off under ambient temperature or vacuum oven.

2.3 Measurements

A DSC-6200 (Seiko Instruments Inc) analyzer was used for analyzing the thermal properties of polymer samples under a nitrogen atmosphere. The samples were heated from 30 to 240℃ at a heating rate of 80℃/min (the 1st heating scan). Then the samples stayed at 240℃ for 2 min and then cooled at 10℃/min (cooling scan), followed by heated from 30 to 240℃ at a rate of 10℃/min (the 2nd heating scan). TG209F1 (Netzsch)was used for analyzing the thermal stability. In the procedure, the temperature rised form 40 to 400℃ at a heating rate 10℃/min. FE-TEM (Tecnai G^2 F20) was used to investigate dispersion of SiO_2 at an acceleration voltage of 200KV.

3. Result and discussion

3.1 Morphology

As shown in Figure 1, the microstructures of the composites were characterized by TEM. Nanoparticles can be well dispersed in DMF through milling. However, there exists some hard agglomeration when the content of SiO_2 reaches to1wt%. Hard agglomeration phenomenon becomes more obvious with increasing loadings. As a result, the crystallization and other properties of the composites decreased. In other words, solution-casting was one of several possibilities, but not the optimum one. Moreover, such phenomenon might be greatly improved by in-situ polymerization in our follow-up research.

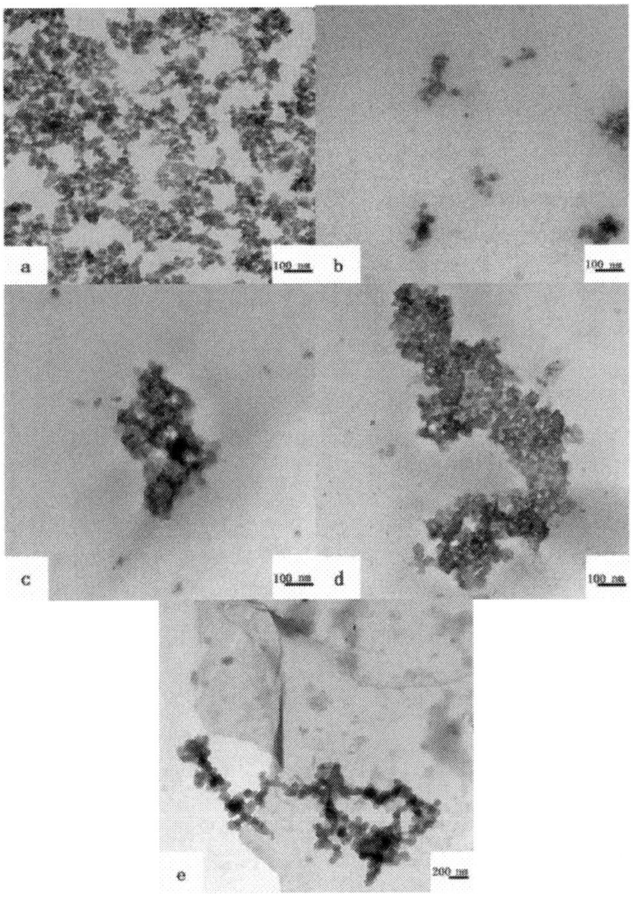

Figure 1 TEM image of SiO_2 dispersed in DMF (a) and in sc-PLA with different contents 0.5wt%(b), 1wt%(c), 3wt%(d) and 5wt%(e)

3.2 Crystallization

As demonstrated in the data in Figure 2 and Table 1, nano SiO_2 obviously played a role of heterogeneous nucleating agent in sc-PLA, which can accelerate the crystallization rate of PLA. The nanoscaled SiO_2 provides large specific surface due to their small size and thus it is reasonable to consider that they acted as efficient nucleating agent. However, it should be noted that the samples appear double-melting peaks in the second heating scan except the pure sc-PLA. The double melting peaks for semicrystallization polymer usually are attributed to melt-reorganization (recrystallization)-remelt mechanism. Both the maximum melt crystallization temperature (T_{mc}) and the maximum crystallinity (X_c in second heating) occurred when the loading of SiO_2 was 1wt%. These should be contributed from good dispersion of nano SiO_2 in the matrix, which offers high efficiency for SiO_2 to be as nucleating agent. When the addition reached to 5wt%, agglomeration of nanoparticles became obvious. Therefore, the advantage of high specific surface and nucleation role declined greatly.

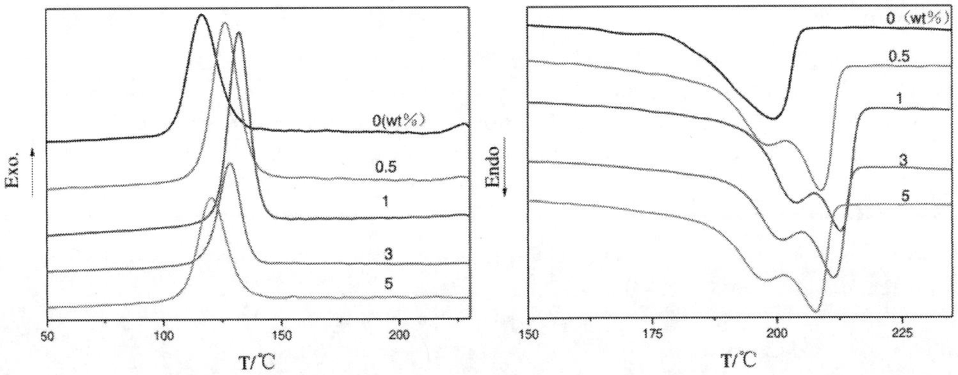

Figure 2.DSC traces of sc-PLA and sc-PLA/SiO_2 composites (Top: cooling scan from the melt; Bottom: the second heating scan)

Table 1. Thermal Properties of sc-PLA and sc-PLA/SiO_2

SiO_2(wt%)	T_{mc}(℃)	T_m(℃)	ΔH_m(J/g)	X_c(%)
0	116	199.7	46.6	30.1
0.5	126	197.5 208.9	68.9	44.5
1	131	203.4 212.8	76.3	49.2
3	128	197.4 207.7	71.3	46
5	120	200.8 211.1	70	45.2

3.3 *Thermal Stability*

Thermal stability of sc-PLA has been improved greatly after SiO_2 being loaded [7]. As shown in Figure 3, both the initial decomposition temperature and the final decomposition temperature increased. Initial decomposition temperature was 75 ℃ higher than sc-PLA at the addition of 5wt% despite of agglomeration. It was also found that such initial decomposition temperature can be as high as PLA with which molecular weight reached to 10^5. Therefore, a simple process for preparing low molecular weight PLA with high thermal-stability polyester by low molecular was achieved. A grid or net structure possibly formed through the hydroxyl function of SiO_2, especially in the presence of high SiO_2 loading. The network might suppress the mobility of the volatile substance, leading to a delay in decomposition. In addition, we need to explore the improvement with high molecular sc-PLA in detail in future study.

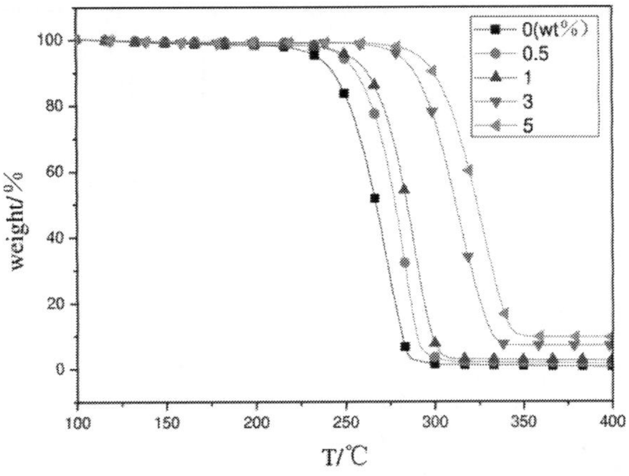

Figure 3. TG curves of sc-PLA and sc-PLA/SiO_2 composites

4. Conclusion

In the present study, well dispersed SiO_2 was obtained by milling in PLAs when the loading was lower. The involved SiO_2 can obviously improve the crystallization of sc-PLA, indicated by the DSC scans. The optimum addition to improve the crystallization behavior was 1wt%. Thermal stability of sc-PLA increased greatly with the addition of SiO_2, despite of some agglomeration in the matrix.

Acknowledgement:

This research was supported by graduate scientific research and innovation projects Beijing Institute of Fashion Technology（No. x2014-089)

References:

[1] Yang Bin,Polylactic acid [M].Beijing：Chemical Industry Press,2007,12-18
[2] Hideto Tsuji,SuongHyu Hyon,Yoshito Ikada.Stereocomplex Formation between Enantiomeric Poly(lactic acid)s. 3. Calorimetric Studies on Blend Films Cast from Dilute Solution[J]. Macromolecules.1991,24：5651-5656
[3] M. Brzezinski,M. Bogus awska,M. Ilcikova.Unusual Thermal Properties of Polylactides and Polylactide Stereocomplexes Containing Polylactide-Functionalized Multi-Walled Carbon Nanotubes[J].Macromolecules 2012, 45：8714–8721
[4] Yukiko Furuhashi,Kousuke Morioka,Hidey uki Tamegai,Naoko Yoshie.Preparation and Some Properties of Stereocomplex-Type Poly (lactic acid)/Layered Silicate Nanocomposites[J] J. Appl. Polymer. Sci. 2013,127：1615–1622,
[5] Liu Lizhu,Preparation and properties of polylactide/nano$_2$ silica in situ composites [J].Journal of Beijing University of Chemical Technology,2008,35(1)：66-69
[6] S.-I. Moon,C.-W. Lee, I. Taniguchi, M. Miyamoto.Melt/solid polycondensation of l-lactic acid: an alternative route to poly(l-lactic acid) with high molecular weight [J].Polymer 2001 (42)：5059–5062
[7] H. Yamane, K. Sasai.Effect of the addition of poly (D -lactic acid) on the thermal property of poly(L -lactic acid)[J] .Polymer 2003,44：2569–2575

IOP Conf. Series: Materials Science and Engineering **62** (2014) 012032 doi:10.1088/1757-899X/62/1/012032

PAN/PS elctrospun fibers for oil spill cleanup

Qiao Ying, Zhao Lili, Sun Haixiang, Li Peng*
State Key Laboratory of Heavy Oil Processing, China University of Petroleum (East China), Qingdao, Shandong266580, China
Email: Lip@upc.edu.cn

Abstract. A high-capacity oil sorbent was fabricated by electrospinning using PS/PAN blend. Morphology, contact angle and oil adsorption of PAN/PS fiber and PP nonwoven fabric were studied. It was found that the PAN/PS fiber had a smaller diameter than PP, and the maximum sorption capacities of the PAN/PS sorbent for pump oil, peanut oil, diesel, and gasoline were 194.85, 131.7, 66.75, and 43.38 g/g, which were far higher than those of PP. The sorbent PS/PAN fiber showed a contact angle of water144.32° and diesel oil 0°. The sorption kinetics of PAN/PS and PP sorbent were also investigated. Compared with the commercial PP fabric, the PAN/PS fiber seems to have the ability to be used in oil-spill cleanup application.

Keywords: electrospinning; PS/PAN; fiber; sorption capacity

1. Introduction

In recent years, with the frequent occurrence of oil spills during oil exploration, transportation and processing, oil absorption materials have attracted extensive attention because of their efficient, economical and easy recovery of oil. Currently, oil adsorption materials can be divided into three categories [1]: inorganic products, organic natural products and organic synthetic products. Among them, organic synthetic products are considered to be potential sorbents for oil spill cleanup for their oleophilic and hydrophobic properties, low cost, and high buoyancy. However, the oil sorption capacities of these sorbents are relatively low, which has limited the application. An ideal sorbent should have high oil sorption capacity, fast uptake capacity, high oil/water selectivity and ready availability. Electrospinning is a method for efficiently producing micro- and nanofibers. There have been some works about polystyrene (PS) fiber [1, 2], which shows high oil sorption capacities. However, PS fiber can be dissolved in gasoline because of its non-polarity. In addition, PS fiber has a poor mechanical behavior and can be dispersed in oil after oil adsorption. PAN [3] fiber has been widely used in industry because of its good conductivity, solvent resistance and high strength. However, PAN fiber shows very poor interaction with non-polarity solvents. To solve the problems of PS fiber, PAN/PS sorbents were fabricated by electrospinning using PS/PAN blend.

2. Experiments

2.1.Materials

Polystyrene (PS), analytical grade, with molecular weight of Mw = 250,000 g/mol was purchased from Energy Chemical. Polyacrylonitrile(PAN), industrial grade, with molecular weight of Mw =80,000 g/mol was purchased from Beijing Yili Fine Chemicals Co, Ltd. N, N-dimethylformamide (DMF) was purchased from Shanghai Chemical Reagents Co., Ltd., China. The oils used for absorption were vacuum pump oil (Beijing Si Fang special oil factory), peanut oil (Shandong Lu Hua Group Co., Ltd.), diesel and gasoline oil (Shandong Qingdao SINOPEC Co., Ltd.). The oil properties investigated are listed in Table 1. The nonwoven polypropylene (PP) was obtained from Shanxi Changqing Oilfield.

2.2 Method

A mixture of PAN (Mw=80000, China) and PS (Mw=250000, USA) was dissolved in DMF by stirring for 4 h to form 18wt% solution. PAN/PS with mass ratio of 1/1 was named as P1. The mixed solution was defoamed standing. All the electrospun fibers were prepared at 25 kV in a relative humidity of 40% at 30℃, at a flow rate of 0.1 mm/s and with the distance between the tip and the collector fixed of 10 cm.

2.3 Characterization

The morphologies of the electrospun fiber and PP nonwoven fabric were examined by field emission scanning electron microscopy (S-4800, Hitachi Ltd., Japan).

Contact angles of electrospun fiber and PP nonwoven fabric were measured in a contact-angle system (SL200B, Kino Inc, USA) at 25 °C. The average contact angle of water and oil was obtained by measuring at three different positions of the same sample.

2.4 Oil absorption

Approximately 0.15 g of oil sorbent was dropped into a glass beaker filled with 100 mL of oil. After 60 min of sorption, the wet sorbent was drained for 60 s. The oil sorption capacities for the sorbents were obtained with the following equation [4]:

$$Q = \frac{m - m_0}{m_0}, \qquad (1)$$

where Q is the oil sorption capacity (g/g), m is the total mass of wet sorbent after the oil was drained for 60 s, and m_0 is the mass of the sorbent before sorption (g). Each sample with a same sorption time was measured three times independently, and the average value and standard deviation were calculated. All of the oil sorption measurements were conducted at 20 °C.

3. Results and discussion

3.1 Surface morphology of the sorbents

The morphologies of P1 and PP sorbents are illustrated in figure 1. As shown in Figure 1a and c, the P1 sorbent has diameters of 1.2-2.7 μ m, while the diameters of PP are in the range of 10-15 μ m. The P1 sorbent has a smaller diameter, contributing to a higher specific surface area and porosity [5], which is conducive to oil sorption. In addition, Figure 1b and 1d show that the surface of P1 sorbent is rougher than that of PP, which is helpful for the adhesion of oil on its surface. Thus, the P1 fiber has a smaller diameter and rougher surface, which will contribute to a higher oil sorption capacity than PP.

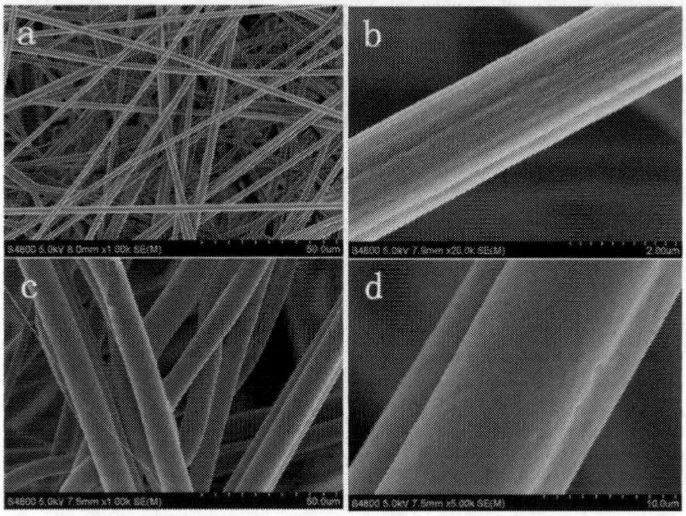

Figure 1. (a,c) Low and (b,d) high SEM images magnification of P1 and PP sorbents.

3.2 The wettability of the sorbents

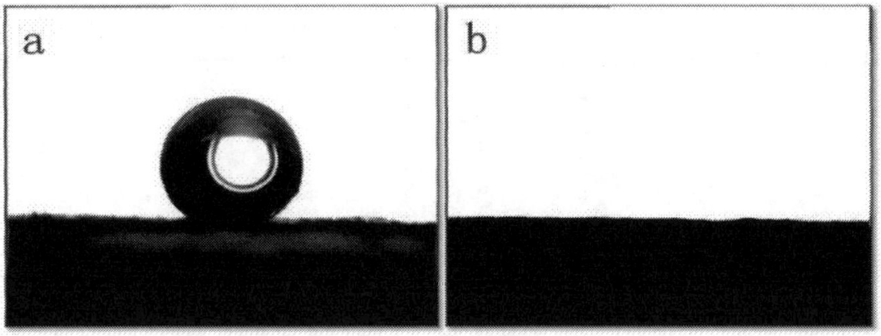

Figure 2. Contact angle photos of (a) Water and (b) diesel oil droplet on the P1 sample with a contact angle of 144.32° and 0°

The wettability of electrospinning fiber was measured by the contact angle of water and diesel oil droplet, as shown in Figure 2. It shows that the water and diesel oil contact angles on P1 sample are 144.32° and 0°, respectively. The wettability of liquid droplet on a surface is determined by two aspects[6]: chemical composition and surface roughness. P1 sorbent consists of PAN and PS, all known as hydrophobic materials. Electrospinning is an efficient method producing fibers with diameters on micro- and nano-scales. Thus, the sorbent P1 have smaller diameter and rougher structures consistent with SEM photos that showed excellent hydrophobicity and oleophilicity.

3.3 Oil absorption and oil sorption mechanisms

Table 1 Characteristics of Studied Oils at Room Temperature (20°C)

Oil	Viscosity(mPa S^{-1})	Density(g cm^{-1})
Vacuum pump oil	162.3	0.8566
Peanut oil	78.45	0.9151
Diesel oil	4.75	0.8449
Gasoline	0.3	0.7724

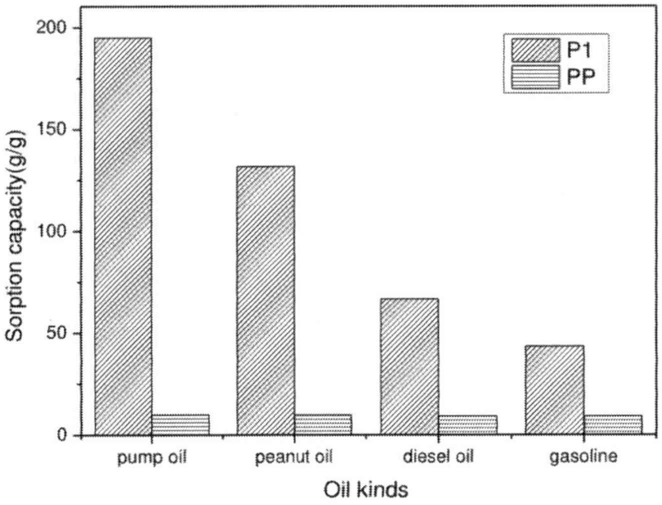

Figure 3. Maximum sorption capacities of P1 and PP for various oils

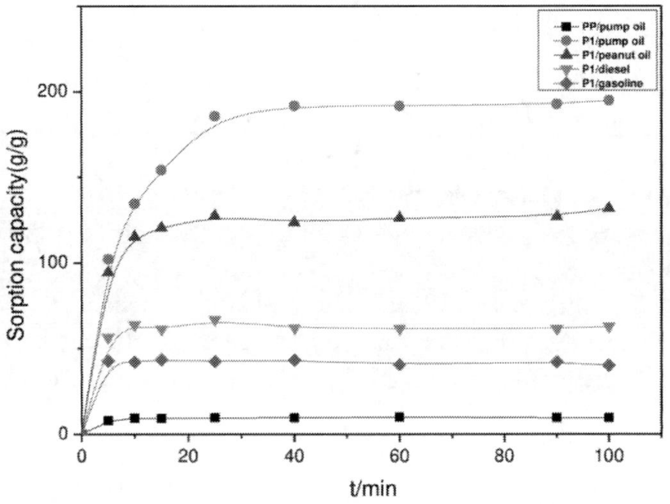

Figure 4. Oil sorption capacities of the sorbents P1 and PP versus contact time with various oils

Oil absorption and oil sorption mechanism are the most important parameters in oil recovery. Figure 3 shows the oil sorption capacities of P1 and PP sorbents. The high sorption capacity of PAN/PS fiber is the combined result of oleophilicity that caused by lower surface tension, smaller diameter and linear grooves, which increase the specific surface area [7].

As can be seen in Figure 4, as absorption time increases, the sorbents' adsorption of the four oils increase rapidly, then slow down, and finally stabilize. At the initial stage of oil sorption, as to oleophilic interaction and the surface capillary force, the oil will be adsorbed on the surface. Then as to van der Waals forces, the oil can be adsorbed into inner fiber voids. Specific surface area of the fibers produced by void fractions, as well as linear grooves would predominantly affect the sorption capacity[8]. Oil sorption capacity is a combination of oil properties and fiber properties. With the increase of oil viscosity, the oil can more likely adhere to the surface of the fibers [8]. As shown in Table1, the oils arranged by their viscosity are pump oil >peanut oil >diesel oil>gasoline, which is consistent with the oil sorption capacities of the fibers. In addition, high viscosity can also inhibit the oil from entering the interior of the fibers, so the sorption capacity of pump oil is the highest, but is balanced by the lowest sorption rate.

Conclusion

PAN/PS fiber is hydrophobic and oleophilic and has a high oil sorption capacity. PAN/PS fiber has high oil sorption capacity and short balance time, which make it suitable to be used in oil spill cleanup.

Acknowledgment

The work was supported by the NSFC (National Natural Science Foundation of China) under Grant 51203186, the Converging Research Center Program funded by Korean Ministry of Education

(2013K000415), the Project of Science and Technology Program for Basic Research of Qingdao (No. 12-1-4-7-(6)-jch).

References

[1] Zhu H, Qiu S, Jiang W, Wu D and Zhang C 2011 *Environ. Sci. Technol.* **45** 4527-4531.
[2] Lin J, Ding B, Yang J, Yu J and Sun G 2012 *Nanoscale.* **4** 176-182
[3] Nam C W, Kim Y H and Ko S W 1999 *J Appl. Poly. Sci.* **4** 2258-2265.
[4] Pourjavadi A, Doulabi M and Soleyman R 2013 *Poly. Int.* **62** 179-183.
[5] Wu J, Wang N, Wang L, Dong H, Zhao Y and Jiang L 2012, *ACS Appl. Mater. Inter.* **4** 3207-3212.
[6] Lim J Y, Yi G R, Moon J H, Heo C J and Yang S M 2007, *Langmuir.* **23** 7981-7989.
[7] Lin J, Shang Y., Ding B, Yang J, Yu J and Al-Deyab S S 2012 *Mar. Pollut. Bull.* **64** 347-352.
[8] Johnson R F, Manjrekar T G and Halligan J E 1973 *Environ. Sci. Technol.* **7** 439–443.

Effect of the Purple carbon black on the properties of NR/BR blend

Zhao Yanfang, Liu Dan, Lin Shengbo, Binjian, Zhao Yinmei, Liao Shuangquan *[1]
Hainan University, Institute of materials and chemical, Hainan Haikou 570228

E-mail:zhaoyanfang818@163.com

Abstract. Purple black is light colored mineral filler mining in recent years in Hainan. The effect of the dosage of the purple carbon black and purple carbon black modificated by Si69 on the vulcanization characteristics, mechanical properties, thermal stability, the damping performance of NR/BR blend rubber were studied, and the blending adhesive tensile sections were analyzed by SEM. Research showed that, with the increasing dosage of the purple carbon black, vulcanization characteristics of NR/BR blend had a little change. Adding the purple carbon black into blending had a reinforcing effect. when the dosage of the purple carbon black was 20, the mechanical properties of blending adhesive was good; Coupling agent Si69 had a modification effect on the purple carbon black. With increasing dosage of Si69, performance of the rubber was improved initially and then decreased; when the mass fraction of Si69 was 8% of the dosage of the purple carbon black, rubber performance was optimal. Purple carbon black had no obvious effect on thermal stability of the rubber, but it improved the damping rubber temperature and damping factor.

Keywords:natural rubber ;BR; purple carbon black; mechanical properties

1. Introduction

Natural rubber (NR) has excellent elasticity, viscoelasticity, the comprehensive mechanical properties and processability, good cold resistance and electrical insulation, but the heat resistance and weather resistance is poor. The modified NR and other analysing blending are used to improve the damping performance [1]. Butadiene rubber (BR) is a crystalline rubber without self-reinforcing effect; it has low elasticity and high strength, good low temperature resistance, wear resistance, small heat generating, and it's commonly used to manufacture tire tread [2].

Purple carbon black is a kind of light color mineral filler recently developed in Hainan [3]. Report about the application of the purple carbon black in rubber products has not been found within China or

[1] * Corresponding author, Liao Shuangquan, male, professor, E-mail:shqliao@263.net
Fund project: Hainan province natural science foundation project (513145); Key science and technology plan projects of Hainan province (ZDXM20120090); Production-teaching-research combination of Hainan province special fund (CXY20130051)
Author introduction: Zhao Yanfang (1973 -), female, associate professor, mainly engaged in polymer materials and processing aspects of the teaching and scientific research work.

abroad. Based on NR, BR as raw materials, this paper studiedthe effect of the dosage of the purple carbon black and the effect of Si69-modified purple carbon black on the vulcanization characteristics, mechanical properties, thermal stability, the damping performance of NR/BR blend rubber, in order to provide the reference for further study of its application in rubber products.

2. Experimental

2.1 Materials

Natural rubber (NR), SCR - 5, Bayi rubber factory in Hainan; Butadiene rubber (BR9000), China petroleum and chemical co., LTD., Beijing yanshan branch; Green purple carbon black (PCB, chemical components are shown in table 1), 1500 mesh, Hainan long sheng mining co., LTD; Stearic acid, zinc oxide, accelerator M, promoting agent, accelerator NOBS, antiager D antiager 264, antiager 2246, naphthenic oil, sulfur, Si69（Bis-(3-thiethoxy silylpropyl）-tetrasulfide) were commonly used in rubber industry.

Table 1 The chemical composition of the purple carbon black

Content	Mass
SiO_2	＞55%
Al_2O_3	＞18%
Fe_2O_3	＜7.5%
K_2O	＜5%
TiO_2	＞0.5%
pH	5.8±0.2
Calcinations of weightlessness	＜6.5%
Oil absorption/10g	2～3.5
The content of water	＜1.0%

Note: The chemical composition analysis according to GB/T4734-1996

2.2 Preparation of compounds

Natural rubber was masticated on a laboratory two-rollmill (JTC - 752 type) and reclaimed rubber was added at different percentages. The blends were prepared as per the formulation given in the below according to the following sequence: NR, BR - stearic acid - zinc oxide-accelerator M - accelerator D - accelerator NOBS - Antiager 264 - filler - naphthenic oil, sulfur and mixed evenly to get the rubber compound. The formulations: NR. 80; BR 20; Stearic acid 1.5; Zinc oxide 5 ;promoter M 0.6;

Accelerator D 0.3; Accelerator NOBS 1; Antiager 264 0.8; Naphthenic oil 5; Sulfur 2.2; Purple carbon black variables.

2.3 Characterization of reclaimed rubber

Thermal gravimetric analyzer was used to conduct vulcanized rubber thermal analysis. With nitrogen as the carrier gas, and a heating rate of 10℃ /min, each sample, weighing about 10 ~ 15mg, was heated from 25 ℃ to 600 ℃. records were obtained as TG curves and DTG curves (mettler Toledo company)

2.4 Rheology

Rheology of compounds was analyzed using rubber processing analyzer

(RPA) (Prescott co., LTD) with strain sweep test at 145 ℃ and 117 Hz (cycle per minute) frequency.

2.5 Curing

Cure rate, scorch time (t_{10}) and cure time (t_{90}) of the rubber compounds were detected with an oscillating disk curemeter at a test temperature of 145 °C. The compounds were compression molded using a 25 tone press (Huzhou star rubber machinery company) at 145°C based on their respective t90 value.

2.6 Mechanical properties

Tensile properties were measured following GB/T528-1998 and GB/T528-91 using a microcomputer control electronic tensometer (Shanghai hualong test instrument co., LTD) at 500mm/min cross head speed. Hardness tests (shore A) were conducted using LX-A (Jiangsu) hardness tester according to GB/ T531-1999. Abrasion loss tests were conducted using WX-76 (Wuxi) resilience and abrasion testers according to GB/T 1689-1998. Fatigue life of vulcanizates was measured using a fatigue to failure tester (401A, Shanghai) at 100°C for 24 hours according to GB 3512-2001.

2.7 Morphology

The failure mechanisms of rubber vulcanizates were investigated using an scanning electron microscope (SEM), (S-3000N, Japan) at 25kv accelerating voltage, examining fractured surfaces of rubber vulcanizates obtained after pulling off.

2.8 Dynamic Mechanical Analyzer (DMA)

Test conditions: stretching mode, the temperature ranges from -100 ℃ to 100 ℃, heating rate of 10℃ / min, test frequency is 5Hz.

Table 2 The effect of purple carbon black content on the properties of NR/BR blends

Properties	Consumption / mass parts				
	0	10	20	30	40
M_H/ (N/m)	0.974	1.094	1.109	1.169	1.185
M_L/(N/m)	0.105	0.075	0.075	0.080	0.080
t_{10}/min	2.35	2.17	2.55	2.33	2.27
t_{90}/min	6.07	5.88	5.45	5.57	5.42
Shore A hardness / Shore A	42	45	46	47	45
Tensile strength /MPa	9.29	10.99	15.56	14.23	13.68
300% Modulus /MPa	1.32	1.9	2.31	2.77	2.84
Elongation at break /%	750	670	740	650	660
Tensile set at break /%	12	12	16	20	20
Tear strength / （kN/m）	19.68	27.43	26.06	25.86	25.20
Abrasion loss /%	1.05	1.27	1.61	1.95	2.39

3. Results and discussion

3.1 Mechanical properties of NR/BR vulcanizates filled with the purple carbon black

The mechanical properties of NR/BR vulcanizates were presented in Table 2. From table 2, the addition of the purple carbon black in NR/BR blend slightly lower M_L, improved M_H, and changed little on t_{10} and t_{90}. Namely, the vulcanization characteristics of rubber had no obvious impacts. Purple carbon black had a certain reinforcing effect on NR/BR blends. This may be due to the surface active group of purple carbon black particle that can combine with rubber reinforcing effect (not quite understand this sentence). With the increasing amount of carbon black, the hardness and tear strength changed little; but the tensile strength and the elongation at break increased initially and then decreased. The 300% Modulus increased. The quality change of the abrasion increased gradually. When the mass fraction of purple carbon black was 20%, the blends had good performance.

3.2 Mechanical properties of NR/BR vulcanizates filled with the purple carbon black modified by coupling agent Si69

The affect of different dosage of Si69 modified purple carbon black on the properties of NR/BR blend vulcanizates was analyzed. The result was shown in table 3.

According to Table 3, with the increasing mass fraction of coupling agent Si69, NR/BR blend of M_H was slightly lower, M_L became slightly higher and t_{10} and t_{90} went down slightly.. This demonstrated that the amount of coupling agent Si69 had no obvious impacts on vulcanization characteristics of NR/BR vulcanizates. The modified purple carbon black could further improve the performance of NR/BR vulcanizates. this is due to the findings that, Si69 transformed to silanol by hydrolyzing which combined with the surface of purple carbon black, and the organic functional group of Si69 reacted with the double bonds of the rubber molecular chain, resulting in a bridge linking up the purple carbon black molecules and rubber molecules, thereby enhancing the effect of purple carbon black. With the increase of the mass fraction of coupling agent Si69, Shore A hardness basically remained unchanged; The tensile strengthincreased initially and then decreased. The tear strength showed a trend of increase; the change percentage of abrasion loss was slightly lower, which showed that the wear resistance had improved. 8% of NR/BR blends that containing 20 phr purple carbon black, which was modified by the mass fraction of coupling agent Si69, exhibited desirable mechanical properties. When the dosage of the coupling agent increased,the reinforcing effect of the purple carbon black on NR/BR blend weakened. This indicated that the coupling agent Si69 may act as diluents which can weaken the interaction between the purple carbon black molecules and the rubber molecules. [5, 6]

Table 3 Mechanical properties of NR/BR vulcanizates filled with the purple carbon black modified by coupling agent Si69

Properties	The mass fraction of Si69 /%				
	0	4	6	8	10
M_H /(N/m)	1.109	0.949	1.00	1.03	0.939
M_L/(N/m)	0.075	0.075	0.095	0.1	0.095
t_{10}/min	2.55	2.12	2.05	1.98	1.88
t_{90}/min	5.45	5.03	5.08	5.12	5.32
Shore A hardness / Shore A	46	48	48	47	48
Tensile strength /MPa	15.56	16.72	17.60	18.02	16.02
300% Modulus /MPa	2.31	2.70	2.84	3.10	3.02
Elongation at break /%	740	650	660	670	630
Tensile set at break /%	16	16	12	16	16
Tear strength / （kN/m）	25.06	26.56	27.66	27.80	30.58
Abrasion loss /%	1.61	1.61	1.71	1.43	1.54

3.3 Thermal degradation

General thermal stability refered to the thermal stability of the polymer chain, namely the main chain of polymer molecular chain after heated the fracture, the decomposition of side chain, oxidation and structured process. [7] The experiment was conducted under nitrogen atmosphere, with a 10 ℃ / min temperature programmed rate. The thermal stabilities of different dosage of filling were determined, and the TG and DTG curve were shown in Fig. 1.

Fig.1 showed that NR/BR vulcanizates of weightlessness was divided into two stages: the first stage was the quickest stage for weightlessness with the temperature range of 300 ℃ ~ 450 ℃. r The NR and BR thermal decomposition of the backbone of the polymer happened in the first stage. The second stage was carbonization process, and the weightlessness curve flattens.

From Fig.1, the initial decomposition temperatures of three kinds of filling system were all about 350 ℃. The final degradation temperature was at 440 ℃, which demonstrated that the purple carbon black had certaineffects on the thermal stability of the vulcanizates.

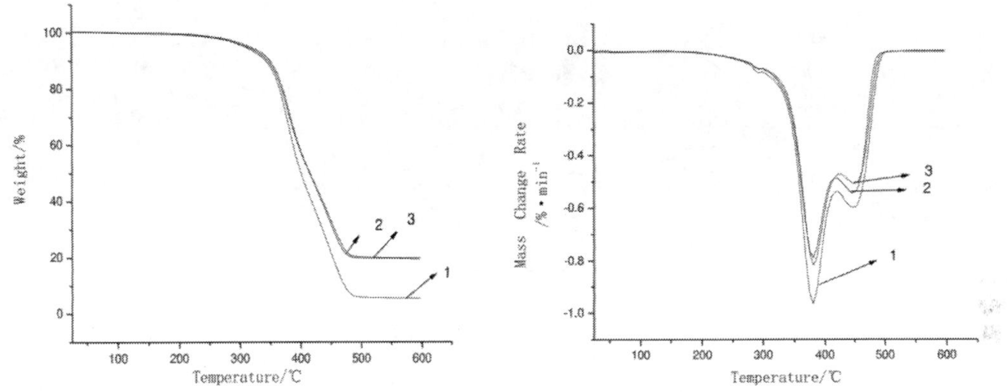

Fig. 1 The TG and DTG analysis of different filling system
Note: 1.0 per the purple carbon black 2.20 per the purple carbon black 3. 20 per the purple carbon black modified by 8% Si69

3.4 SEM studies

Fig.2 showed the SEM photographs of fractured surface of NR/BR vulcanizates. Comparing Fig.2 (a), (b) and (c) showed that the section of NR/BR vulcanizates without purple carbon black was smooth, and showing obvious homogeneous structure. While the blend section of added purple carbon black displayed obvious two phase structure, The particles in 2 (c) were uniform, and the surface showed deep hole. This indicated that the modified purple carbon black had improved the compatibility of NR/BR. this finding has also been demonstrated in the previous mechanical property analysis.

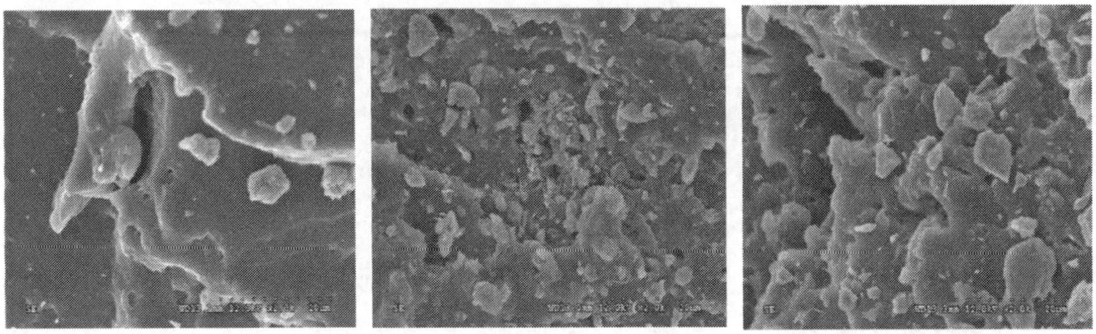

(a) 0 per (b) 20 per (c) modified by 8%Si69

Fig.2 The cross-section of SEM

3.5 Damping properties of NR/BR vulcanizates

When the filler packaged into the rubber matrix, the filler particles could affect the network of the polymer matrix, which also affect the dynamic performance of a filled vulcanizates[8]. The dynamic mechanics performance of the purple carbon black filled vulcanizates was shown in figure 3. According to the graph, the damping factor of vulcanizates increased, the glass transition temperature of vulcanizates filled with modified purple carbon black was higher, and the damping factor was bigger after the purple carbon black was added

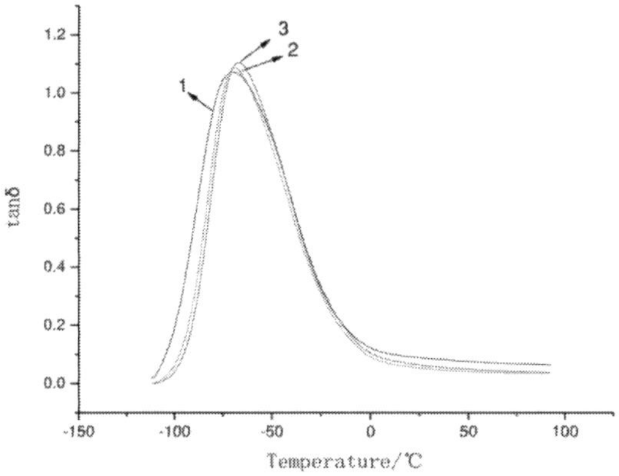

Fig. 3 The DMA curve of vulcanizates

1. without purple carbon black 2.20 per of the purple carbon black
3. 20 per of the purple carbon black modified by 8% Si69

Conclusion

Addition of the purple carbon black had little effect on the vulcanization characteristics of NR/BR vulcanizates, but the addition could increase mechanical properties of the vulcanizates. When the dosage of the purple carbon black was 20, the mechanical properties of the vulcanizates were better. With the increasing dosage of the purple carbon black, abrasion loss increased gradually. When the Si69 mass fraction was 8%, the modification effect of the purple carbon black was optimal. The TG and DTG analysis showed that the purple carbon black had no effect on the thermal stability of vulcanizates. The damping factor of the modified purple carbon black filled NR/BR vulcanizates was higher than the unmodified ones, and the effective damping temperature increased.

References

[1] Chen Kun, Xie Xin. Study on silica segregating in NR/BR blend [J]. Journal of elastomeric, 2010, 20 (2) : 40 - 43.

[2] Zhao Jie,Liu Jjia. Qian Haiyan. Application of precipitated silica in rubber products [J]. Journal of inorganic salt industry, 2009, 9 (7) : 9 ~ 11.

[3] Fang Yuesheng,Liu Yunchun,Ding Jianping.The studies of properties and modification on Sericite/SBR composite material [J]. China Rubber Science and Technology Market, 2011, 12:21-24.

[4] Li Zhijun editor. The analysis of the natural rubber and test [M]. Beijing: China agricultural university press, 2007.174 ~ 200.

[5] Liu Huiyuan, Li Deling, Li Xing. Surface modification of nano-silica by silane coupling agent KH - 560 [J]. Chemical world, 2011, 08: 462 456-458.

[6] Ren Hui, Qu Yixin, Zhao Sugehe, etc., Silica modified by silica coupling agent used for reinforcing styrene-butadiene rubber [J]. Journal of the Chinese ceramic society, 2006,08:951-955.

[7] Ma Lisheng, Liao Xiaoxue, Chen Yongping, etc. Extraction and separation of β-sitosterol from Natural Rubber latex and its influence on rubber aging properties and thermal stability[J]. Journal of materials science and engineering, 2011, 12:766-769.

[8] Zhao Yanfang, Liao Shuangquan, Li Puwang, etc. Effect of carbon black on the damping property of the black filled NR/ENR blends [J]. Chinese Journal of tropical crops, 2008, 03:265-269.

2014 Global Conference on Polymer and Composite Materials (PCM 2014) IOP Publishing
IOP Conf. Series: Materials Science and Engineering **62** (2014) 012034 doi:10.1088/1757-899X/62/1/012034

Application of Sequence Control Injection in Modified Design of Car Front Bumper

WANG Feng, ChenGuodong, BU Jiling, ZHOU Bin

Zhuzhou Times New Material Technology Co., Ltd. Hunan,412007,China

Email: wangfeng4@teg.cn, wangfeng4@csrzic.com

Abstract: In view of lightweight and cost reduction, a bumper has been redesigned, which results in weld lines during the manufacture. For that, a new method of sequence control technology is introduced. The simulation results with the new method and the traditional method were compared. The results showed that the weld lines were removed, and at the same time, the injection balance was better and the injection pressure was decreased by using the sequence control technology. The sequence control technology is of great practical significance to bumper manufacturing.

Key words: Front bumper; Weld line; Sequence control technology; Numerical simulation

1. Introduction

With the development of the automobile industry, bumper, as one of the main components of car body, has increasingly higher requirements, not only pursuing its harmony with the car's style, but also meeting the requirements of lightweight and cost reduction [1]. In this study, a bumper was investigated. The up grille, the down grille and the main body were made together. Meanwhile, the thickness of the bumper was reduced from 3.5mm to 3mm; it not only reduces the weight of the product, but also reduces the number of molds from three sets to one set, which means the molds' cost is reduced at the same time. By using Moldflow software, the modified product was simulated. It was found that the product had good shape molding performance, but the weld line was obvious. In order to fix the weld line defect, the sequence valve control injection system was introduced [2-5]. By contrasting the simulation results with the results using traditional injection system, it was proved that the sequence valve control technique can not only effectively improve the weld line quality, so the surface quality and mechanical properties are improved, but also reduce the mold loss and increase the economic benefits.

2. Structure optimization

In order to reduce the product's weight and the mold's cost, a bumper was optimized, of which the dimensions were length × width × height = 320 × 1595 × 627 mm. The original thickness was 3.5mm.The main body, the up grille and down grille were made independently, thus three sets of molds were needed. After modification, the main body, the up grille and down grille were made

IOP Conf. Series: Materials Science and Engineering **62** (2014) 012034 doi:10.1088/1757-899X/62/1/012034

together, so that the molds were reduced from three sets to one set, and the mold cost was significantly reduced. In consideration of lightweight, the thickness was reduced from 3.5mm to 3.0mm, but the thickness of matching surfaces and bolt connections were not changed. The 3D model of the bumper is shown in Fig.1. The original section is shown by the red lines, and the optimized section is shown by the blue lines in Fig. 1(a).

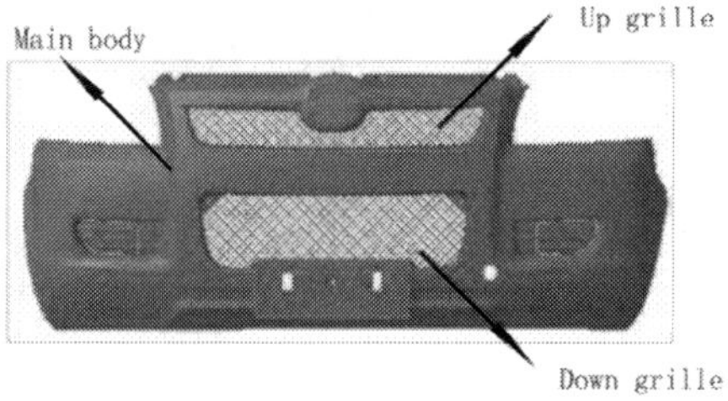

(a)original design

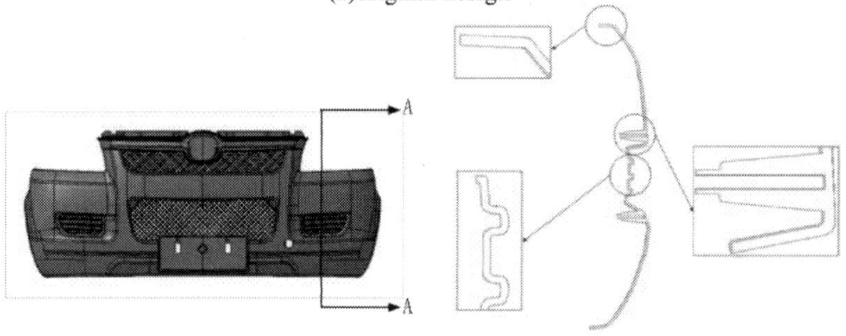

(b)optimized design (c) specific section of optimized design

Fig.1 original and optimized designs

After optimization, the number of molds was reduced from three sets to one set, and the weight was reduced from 4.86Kg to 3.64Kg, which means that the weight was reduced by about 17.6%. Thus, the effectiveness of the optimization is obvious. It greatly reduces the product's cost, and improves the business competitiveness, which meets the strategy of fuel economy and sustainable development. Meanwhile, the new design may cause some injection molding problems.

3. Traditional injection system scheme

3.1 Traditional injection system

Because the up grille, down grille and the main body are made together, the runner cannot be placed in the grille, and can only be arranged at the position of the upper and lower edges. For another reason, the bumper has a large dimension; there are always some injection problems. In order to solve these problems, in traditional injection system, the combination of cold and hot runners and more gates is used [1]. In this paper, the combination of cold and hot runners and five gates is adopted, as illustrated in Fig.2.

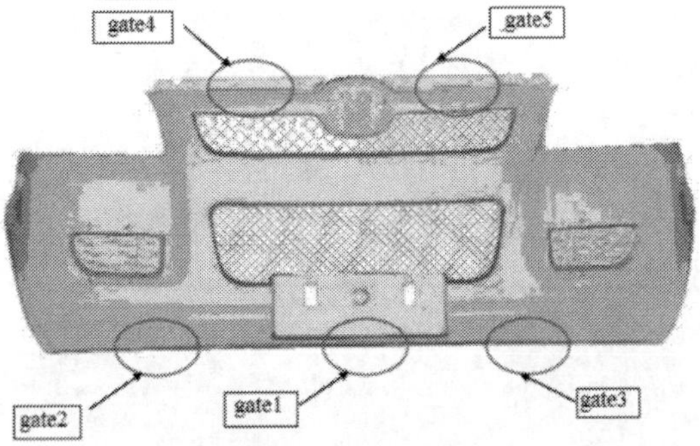

Fig.2 gates setting

3.2 The setting of initial parameters

Modified polypropylene material is used in traditional design. This material has many advantages, such as low cost, impact resistance, weather aging resistance, good high and low temperature performance, high heat-distortion temperature, stiffness and hardness suitability, etc., which are beneficial to the car's lightweight, less energy consumption, and beautiful decoration. Modified polypropylene is an ideal material for manufacturing automobile bumpers [4]. During this simulation, the PF 236 X115 in the Moldflow material database was chosen, and its main physical properties are shown in Table 1.

Table 1. material's physical properties

Melt density/g.cm^3	0.78113
Solid density/g.cm^3	0.95023
Ejection temperature/℃	100
Max shear stress/MPa	0.25
Max shear rate/s^{-1}	10000
Recommended melt temperature/℃	240
Recommended mold temperature/℃	50
Failure temperature/℃	320

The main molding parameters are shown in table1: mold temperature 50℃; melt temperature 240℃; injection time 5S, velocity/pressure control (V/P) conversion 99%, holding pressure 80MPa, and holding time 20s.

3.3 Preliminary analysis

The results of preliminary analysis are shown in Fig.3. Because the five gates inject at the same time, there is a long, deep weld line (marked by red line) that seriously affects the appearance of the products.

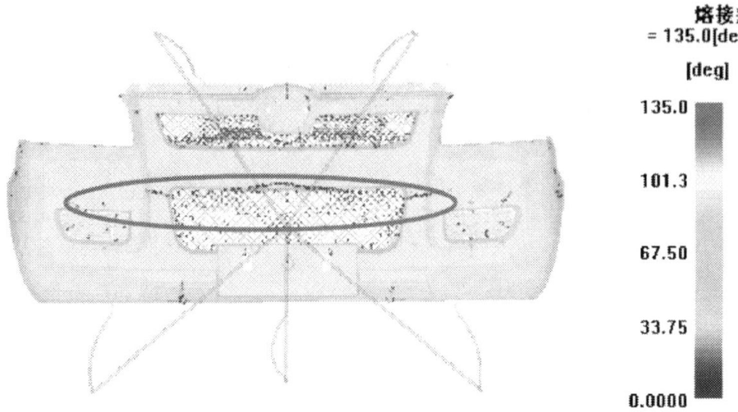

Fig.3 weld line distribution in traditional injection system

4. Optimization of injection system

4.1 Theory of sequence Control Injection System

In order to fix the weld lines defect, the sequence control injection system is introduced. By using the precisely controlled needle-gate injection system, each gate's opening and closing time depend on the plastic melt filling process, and can be controlled individually. Needle-gate injection nozzle adopts the cascade injection control technique. The melt from the firstly opened nozzle is used as the front melt flow. The melt from the subsequently opened nozzle directly flow into the inside of the previous melt, and promote the melt to keep filling, so there is only one flow front in the filling direction, meanwhile the flow front temperature increases. By using needle-gate injection system, the weld line can be eliminated effectively or be limited at less important places of the products, so high-grade products can be made.

4.2 Sequence Control Injection System

The location of the injection system is the same in both programs. The dimensions of the sprue, runner and gates remain unchanged. But at the end of the hot runner there are needle-gates, which can control the gate opening time and closing time.

At the beginning of injection, gate1 opens first, and the melt starts filling. When the flow front arrives at gates2 and gate3, both of them open at the same time. The flow front keeps going and reaches a certain place (in order to control the weld line at a certain position, repeated attempts are required), and then gate4 and gate5 open together. As shown in fig. 4, the five gates' injection time is not equal, and all of them keep open until the pressure releases. This sequence control method is very simple, and demands less pressure on the injection molding machine, so it is widely used in the actual production process.

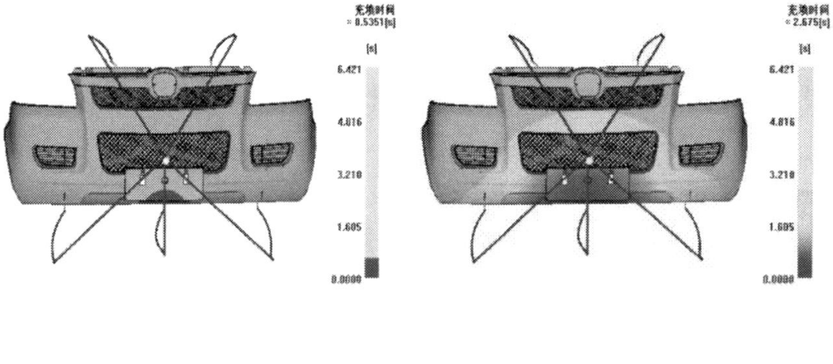

IOP Conf. Series: Materials Science and Engineering **62** (2014) 012034 doi:10.1088/1757-899X/62/1/012034

(a) gate1 opens (b) gate2 and gate3 open

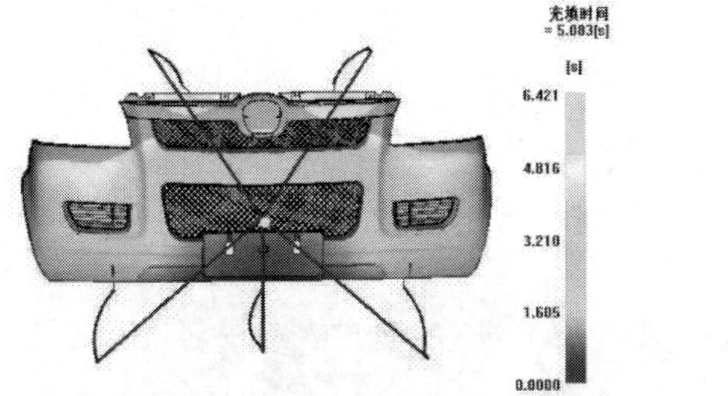

(c) gate4 and gate5 open

Fig.4 gate opening pattern in sequence control technology

4.3 Optimized analysis results

4.3.1 Weld line

The overlay of the weld line, temperature and pressure are shown in Fig.5 (the area of the weld line is marked by blue line). The weld line has a significant influence on the surface quality and mechanical properties of products, so the appearance of it is avoided. If it is not possible, it should be tried to reduce its length, or prevent it from showing at critical locations [6-8]. According to the evaluation standard of the weld line, the quality of weld line is mainly affected by two factors, one is the temperature and pressure when the weld line is forming, and another is the meeting angle and whether there are air traps.

As is shown in Fig.5, compared to the results of the conventional system, the weld line is obviously shorter and only appears at the connection of the up grille and the main body. According to the evaluation standard of the weld line, the temperature difference is small and almost the same with the initial injection temperature when the weld line is forming, at the same time, the pressure is high, and the meeting angle is small, when the sequence control technology is used. Through study, it is fully proved the superiority of sequence valve control injection system in control of weld line defects.

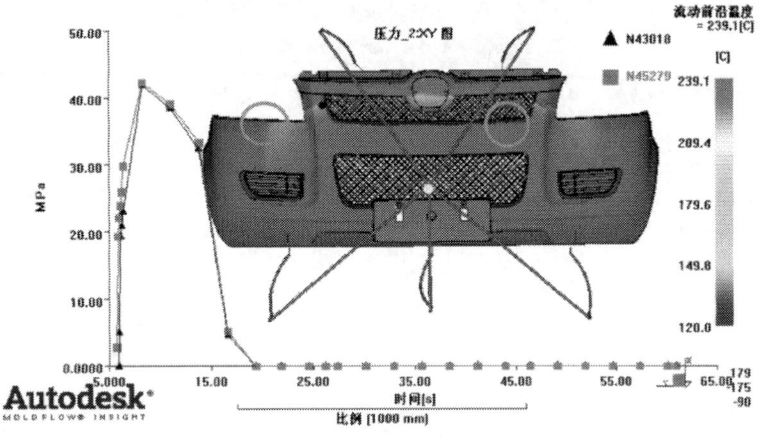

Fig.5 weld line distribution in sequence control technology

4.3.2 Filling time

Figure 6 shows the filling time, where red color indicates the last filling site. By simulation, we can evaluate

IOP Conf. Series: Materials Science and Engineering **62** (2014) 012034 doi:10.1088/1757-899X/62/1/012034

whether the filling is balanced. If the melt reaches the end edges of the product at consistent time, it is said that the filling is balanced. If not, it is easy to cause over pressure and shrinkage mark defects. Meanwhile, it can be decided whether there is short shot. The filling time is 6.307s in traditional injection system and 6.421s in sequence valve control injection system. There are not too many significant differences between two schemes. However, the filling balance is better in valve control injection system than that in traditional injection system (color distribution uniformity is good). By analysis, it can be said that the filling time is not shorter when more gates inject at the same time. Without affecting the production efficiency, sequence valve control injection system can greatly improve the quality of the products.

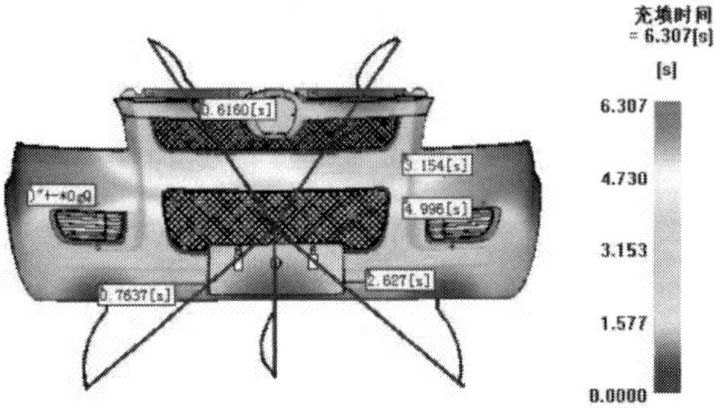

(a)filling time in original injection system

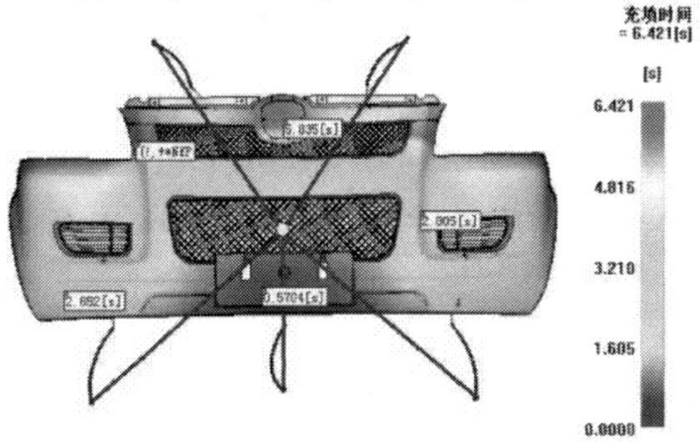

(b)filling time in sequence control technology

Fig.6 Comparison of filling time

4.3.3 Pressure at injection location

In traditional injection system, the cavity pressure continues increasing, and reaches the highest point 94.82 Mpa. The pressure of V/P is 80Mpa. In sequence valve control injection system, the cavity pressure continues increasing similarly, and reaches the highest point 91.50 Mpa. The pressure of V/P is 60MPa. In the sequential injection system, lower injection pressure is required, which shows that there are less resistance in the flow channel and the cavity, so that it can reduce the mold wear, extend the service life and increase economic benefits.

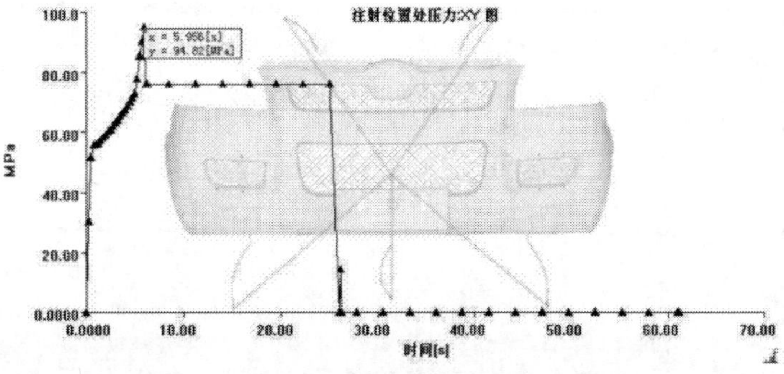

(a)pressure at injection location in original injection system

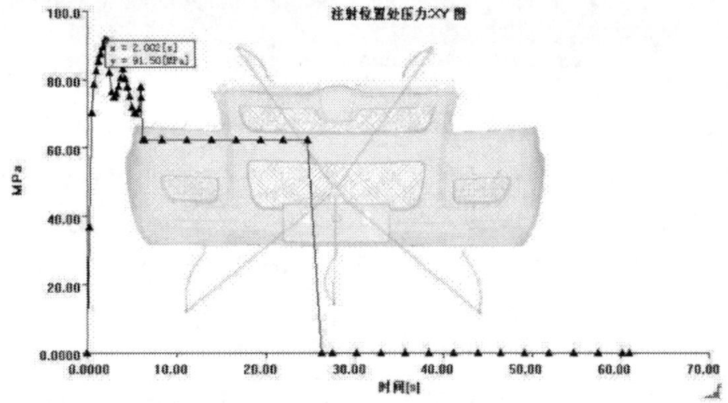

(b)pressure at injection location in sequence control technology

Fig.7 Comparison of pressure at injection location

4.3.4 Actual maneuverability

In order to increase the maneuverability of the sequence valve control injection system in actual production process, the position control mode in simulation analysis is converted to time control mode. The detailed operation guidelines are shown in table 2.

Table 2 optimized gates opening time

gate	**Open time/s**			
	0s-2.17s	2.17s-4.15s	4.15s-6.421s	6.421s-
1	open	open	open	close
2	close	open	open	close
3	close	open	open	close
4	close	close	open	close
5	close	close	open	close

5. Conclusion

(1) Lightweight design and molds integration of the bumper can reduce the cost of products and greatly enhance the market competitiveness, and have an important guiding significance for similar automotive product design.

(2) The introduction of sequence valve control injection system in the automobile bumper production effectively solves the weld line defects, and greatly enhances the quality of the products.

(3) Simulation is a practical method in product development; it can greatly shorten the product development cycle, and improve the quality of the products.

References:

[1] Chen Wanlin. The utility of plastic injection mold design and manufacture of [M]. Beijing: Mechanical Industry Press, 2000:5-8.

[2] Zhou Jie, Lin Li li. Sequential control applications in the injection molding of car front bumper [J]. Engineering plastics application, 2008, 36 (4):37.

[3] Tan Wensheng.Study of fiber orientation in Valve-injection system of injection molding [J]. Light industry machinery, 2006, 24 (2):30- 32.

[4] Luo Yi. Present simulation and prospect of china modified polypropylene particular material used for automobile [J]. Plastic technology, 1998 (10):45-50.

[5] Zhang Yanhui, Diao Hongchao. Research on numerical simulation of controlling weldline in thin wall plastics [J]. Shandong coal science and technology, 2008, (3):144.

[6] Lu Yaping. The formation mechanism and control method of weld line [J]. Mould industry, 2006, 32 (12):61 – 64.

[7] S. C. Malguarnera, A. I. Manisali and D. C. Riggs. Weld line structures and properties in injection molded polypropylene [J]. Polymer Engineering & Science, 2005, 45 (7):1021 – 1030.

[8] Cheng-Hsien Wu and Wan-Jung Liang. Effects of geometry and injection-molding parameters on weld-line strength [J]. Polymer Engineering & Science, Polymer Engineering & Science,1981,21 (17):1149 – 1155.

IOP Conf. Series: Materials Science and Engineering **62** (2014) 012035 doi:10.1088/1757-899X/62/1/012035

HPLC method for the simultaneous quantification of the major organic acids in Angeleno plum fruit

Yanwei Wang[1], Jing Wang[1], Wei Cheng[1], Zhilei Zhao[1*1] and Jiankang Cao[2]

[1] Hebei University, College of Quality and Technical Supervision, Baoding, China
[2] China Agricultural University, College of Food Science and Nutritional Engineering, Beijing, China

zhaozhilei-3208@163.com

Abstract. A method was developed to profile major organic acids in Angeleno fruit by high performance liquid chromatography. Organic acids in plum were extracted by water with ultra-sonication at 50℃ for 30 min. The extracts were chromatographed on Waters Atlantis T_3 C_{18} column (4.6 mm×250 mm, 5 μm) with 0.01mol/L sulfuric acid and water as mobile phase, and flow rate was 0.5 ml/min. The column temperature was 40℃, and chromatography was monitored by a diode array detector at 210 nm. The result showed that malic acid, citric acid, tartaric acid, oxalic acid, pyruvic acid, acetic acid, succinic acid in Angeleno plum, and the malic acid was the major organic acids. The coefficient of determination of the standard calibration curve is $R^2 > 0.999$. The organic acids recovery ranged from 99.11% for Malic acid to 106.70% for Oxalic acid, and CV (n=6) ranged from 0.95% for Malic acid to 6.23% for Oxalic acid, respectively. The method was accurate, sensitive and feasible in analyzing the organic acids in Angeleno plum.

1. Introduction

Plum (*Prunus salicina L)* was flavorful and contains good nutrition, which has been one of the most commonly consumed stone fruits by domestic and foreign consumers [1]. Plum was rich in sugars, amino acids, vitamins, mineral elements [2-3], as well as organic acids. Organic acid is a function of both taste and aroma components, and the content of organic acids dictating the eating quality of the fruit. The sweetness of fruit mainly depends on the acids content [4-5]. The determination of organic acids commonly used titration, spectrophotometry, gas chromatography, fluorescence method, thin layer chromatography, enzymatic, etc. [6-9] Currently, high performance liquid chromatography method has been used in pineapple [10], oranges [11], peaches [12], apricots [13] and other fruit acids composition and content determination, however, the type and the content of organic acids in Angeleno fruit has not been reported at present. In this study, a rapid and effective method of quantitative analysis of organic acids in Angeleno fruit by high performance liquid chromatography was developed, and the content and the type of organic acids in Angeleno plum was analyzed.

[1] *Correspondence author: Zhilei. Zhao
Address:No 180Wusi East Road,Baoding, Hebei,P.R. China

2. Materials and methods

2.1. Reagents and instruments

Standards of oxalic acid, malic acid, tartaric acid, pyruvic acid, acetic acid, citric acid, succinic acid were obtained from Amresco company with purity of 99.0%. Sulfuric acid (GR). Deionized water.

Waters series HPLC system (American Waters companies) equipped with an evaporative light scattering detector (DAD, Waters 600E quaternary pump, 2998 photodiode array detector, Empower chromatography workstation). KQ-500E ultrasonic cleaner with numerical control (the Kunshan Ultrasonic Instrument Co., Ltd). Human ultra-pure water (Ultra Pure Water System). Electronic balance (AR1140, Ohaus Corporation). Flying pigeon centrifuge (TGL-10C, Shanghai An-ting scientific instrument factory). 5 mL disposable syringes, 0.45μm filter needle head aqueous, mortar, 100mL volumetric flask.

2.2. Methods

2.2.1. Chromatographic conditions. The extracts were chromatographed on Waters Atlantis T_3 C_{18} column (4.6 mm×250 mm, 5 μm) with 0.01mol/L sulfuric acid as mobile phase, and flow rate was 0.5 ml/min. The column temperature was 40℃, and chromatography was monitored by the photodiode array detector at 210 nm.

2.2.2. Sample preparation. The extraction of organic acids from plum samples was analyzed according to the method of Wang [14] and Zhang [15] with some modifications. For determining organic acids contents of Angeleno plum, sample was treated as follows, 5 g of fresh sample was grounded into homogenized, followed by transferring to 50 mL volumetric flask and the ultrapure water was added to the scale, organic acids were extracted with ultra-sonication at 50℃ for 30 min. Then 8 mL extracts was transferred to the centrifuge tube, the samples were centrifuged at 10000 r/min for 15 min. The supernatant solution was filtered through a 0.45 μm nylon membrane filter.

2.2.3. Preparation of standard curve. A serial dilution of organic acids (oxalic acid malic acid, succinic acid, tartaric acid, citric acid, pyruvic acid and acetic acid) was made by dissolving the required amount of authentic standard in deionised water (dH2O). All standards were filtered through a 0.22 μm nylon filter before HPLC analysis Each standard was analysed in duplicate for calibration and their concentrations ranged from 1 to 800 mg/L for organic acids.

2.2.4. Determination of sample. Qualitative analysis of extracts was carried out according to the external standard method. The injection volume was 20μL.

3. Results and analysis

3.1. Mobile phase selectivity

3.1.1. The selection of mobile phase component. 0.01 mol/L Phosphoric acid, hydrochloric acid, nitric acid, P-toluenesulfonic acid, sulfuric acid was used as the mobile phase, respectively, the flow rate was at 0.5mL/min, and the column was thermostatic at 40℃. The standard mixture of organic acids and extracst were analyzed by HPLC. The result shows that the Oxalic acid, malic acid, succinic acid and acetic acid can be well separated in the case of phosphoric acid and hydrochloric acid as mobile phase, but poor separation of citric acid and tartaric acid. In terms of nitric acid as the mobile phase, the two peaks of citric acid and tartaric acid overlap with poor separation. Replaced by P-toluenesulfonic acid, the result showed no peaks within 30 min and obtained the worst separation. When taking sulfuric acid as the mobile phase, seven kinds of organic acids had been separated effectively with completed peaks in 20 minutes, was tested to obtain the optimal chromatographic separation.

3.1.2. The selection of sulfuric acid concentration. 0.002, 0.005, 0.01, 0.05, 0.07 mol/L sulfuric acid was used as mobile phase with condition described in section 2.1.1, respectively. Results showed that retention time was declined with the decreased of sulfuric acid, and the separation was poor. Retention time became longer when sulfuric acid increased, and the separation of oxalic acid and citric acid were poor, however, other organic acids were found to give a better separation. When the content of sulfuric acid was changed to 0.07 mol/L, oxalic acid and citric acid can not be separated. Results showed that 0.01 mol/L sulfuric acid as mobile phase could provide a good separation of seven kinds of organic acids. The results were showed in Figure 1, 2.

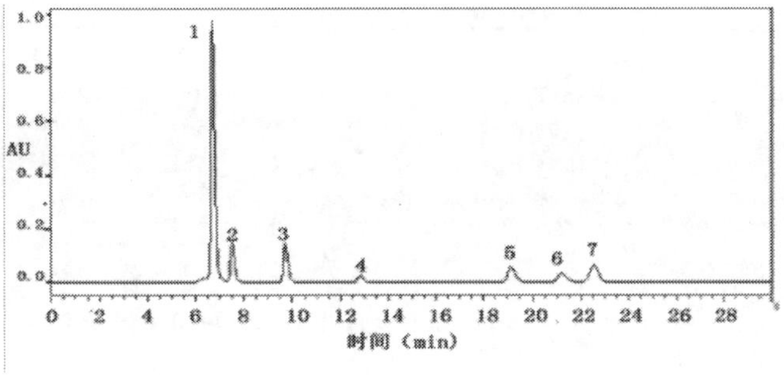

1. Oxalic acid 2. tartaric acid 3. malic acid 4. pyruvic acid 5. acetic acid 6. citric acid 7. succinic acid
Figure 1. Chromatogram of 7 organic acids standards

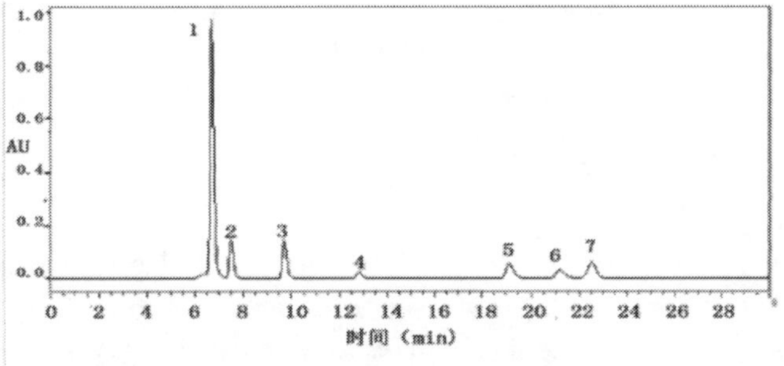

1. Oxalic acid 2. tartaric acid 3. malic acid 4. pyruvic acid 5. acetic acid 6. citric acid 7. succinic acid
Figure 2. Chromatogram of organic acids in plum fruit

3.2. Flow rate selectivity

The analysis were based on 0.01mol/L sulfuric acid as mobile phase, and column temperature was set at 40℃ with different flow rate, ranged from 0.3 to 0.6 mL/min, which were tested to obtain the optimal separation. The results showed that the flow rate was decreased, and the analytical time of separation was not affected markedly. When the flow rate increased, the column pressure increased at the same time and the analytical time was decreased. As a result, the flow rate was set at 0.5 mL/min as the optimal condition.

3.3. Column temperature selection

In order to obtain the optimal separation, we prepared column with 25℃, 30℃, 40℃, 45℃, 50℃, respectively. The results showed that the column temperature was increased, and the analytical time of the organic acids was decreased, but the separation was satisfactory. However, the peak width of acetic acid was broadening when the column temperature was 50℃.The column temperature at 40℃was the optimal condition.

3.4. Linear range investigation

Under the chromatographic conditions described above, the calibration graphs were constructed by plotting peak area vs. concentration. Good linearity was achieved in the range from 0.001 to 1.0mg/ml

IOP Conf. Series: Materials Science and Engineering **62** (2014) 012035 doi:10.1088/1757-899X/62/1/012035

of the standard mixture studied for each organic acid in water with significant correlations. The coefficients of determination was $R^2 > 0.999$ (Tab. 1).

Table 1. The calibration linear and coefficients of determination of organic acids.

Compounds	Linear equations	coefficients(R^2)
Oxalic acid	y = 0.00000006 x + 0.00492105	0.9990
Tartaric acid	y = 0.00000029 x + 0.00462126	0.9995
Malic acid	y = 0.00000051 x + 0.00169815	0.9999
Pyruvic acid	y = 0.00000009 x + 0.00788941	0.9998
Acetic acid	y = 0.00000059 x + 0.00404933	0.9992
Citric acid	y = 0.00000041 x + 0.00437906	0.9997
Succinic acid	y = 0.00000105 x + 0.00503900	0.9995

3.5. Precision inspection

As shown in Table 2, the precision of the method was determined by calculating the coefficient of variation (%CV) of analyte concentration of six repeat mixture standard analysed with the HPLC method. The % CV of the concentration values of organic acids ranged between 0.95% and 6.23%.

Table 2. Precision of the method (n = 6).

Compounds	Concentration/(mg/ml)						Average value	RSD/%
Oxalic acid	0.0134	0.0134	0.0139	0.0138	0.0140	0.0145	0.0138	2.96
Tartaric acid	0.0118	0.0119	0.0120	0.0120	0.0120	0.0121	0.0120	0.81
Malic acid	0.5246	0.5180	0.5266	0.5196	0.5160	0.5282	0.5222	0.95
Pyruvic acid	0.0081	0.0081	0.0081	0.0082	0.0082	0.0081	0.0082	0.78
Acetic acid	0.0052	0.0053	0.0052	0.0055	0.0053	0.0051	0.0053	2.64
Citric acid	0.0231	0.0204	0.0226	0.0208	0.0216	0.0221	0.0218	4.76
Succinic acid	0.0425	0.0389	0.0412	0.0424	0.0412	0.0480	0.0423	7.21

3.6. Repetitive inspection

Approximately 5 groups of the same mixed sample of plums was evalued by the method has been mentioned in 2.2.2. The results shown that the average content of oxalic acid, tartaric acid, malic acid, pyruvic acid, acetic acid, citric acid and succinic acid was 0.085, 0.102, 6.328, 0.081, 0.051, 0.243, 0.255, 0.220 mg/g•Fw, and the CV was 1.83%, 1.79%, 2.31%, 4.26%, 3.79%, 2.38%, 1.97%, respectively. The results show that the repeatability of the method was satisfactory.

3.7. Recovery

The recovery of analytes after extraction from plum spiked with a mixture of standards was analysed in triplicate in order to assess the effectiveness and accuracy of the extraction step. The results confirmed that the separation and analysis conditions were accurate for all compounds (Table 3). The mean of the percentage recovery were 106.70%, 104.44%, 99.11%, 104.61%, 101.18%, 101.55%,105.15%, for oxalic acid, tartaric acid, malic acid, pyruvic acid, acetic acid, citric acid, succinic acid, respectively. For the recovery experiment, plum samples were spiked with all compounds at the same time prior to extraction and HPLC analysis.

Table 3. The recovery of the method.

Compounds	Initial value /(mg/ml)	Added value /(mg/ml)	measured value /(mg/ml)	Average recovery /%
Oxalic acid	0.02	0.03	0.05	106.70

Tartaric acid	0.01	0.02	0.03	104.44
Malic acid	0.49	0.01	0.50	99.11
Pyruvic acid	0.01	0.03	0.04	104.61
Acetic acid	0.01	0.04	0.05	101.18
Citric acid	0.02	0.03	0.06	101.55
Succinic acid	0.05	0.01	0.06	105.15

3.8. Determination of plum fruit samples

The quantitative and qualitative determination of organic acids in plum fruit samples was achieved by the method described in section 2.2.2 and the chromatographic conditions as described in section 2.2.1. The composition and content of organic acids in Angeleno plum were determined. As showed in Table 4, malic acid content was 6.328 mg/g,which showed the highest content in Angeleno plum, citric acid and succinic acid were much lower than malic acid, the contents was 0.243 mg/g and 0.221 mg/g, respectively. Acetic acid content was 0.051 mg/g, which content was the lowest.

Table 4. Organic acids content in Angeleno plum.

Varieties	Content/(mg/g.Fw)						
	Oxalic acid	Tartaric acid	Malic acid	Pyruvic acid	Acetic acid	Citric acid	Succinic acid
Angeleno	0.085± 0.012	0.102± 0.004	6.328± 0.078	0.081± 0.0001	0.051± 0.0004	0.243± 0.001	0.221± 0.006

4. Conclusions

The solvent system chosen for the analysis of organic acids in Angeleno plum exhibited good baseline stability and low background noise. The organic acids profile in Angeleno plum fruit with HPLC system is a difficult task because of several interferences. The proposed method allows the quantification of malic acid, citric acid, tartaric acid, oxalic acid, pyruvic acid, acetic acid, succinic acid in Angeleno plum fruit. The analysis is simple, rapid and does not require any complicated sample preparation, and only one reversed-phase HPLC column is used for the separation in less than 25 min. This HPLC-DAD method is a versatile method that can be used with a simple UV-vis detector. Therefore, the proposed method is set up to analyze up to 7 organic acids in order to control the product quality. The proposed HPLC method is accurate, sensitive and feasible to determine organic acids in Angeleno plum with direct injection after sample extraction.

Acknowledgments

This work was financially supported by Natural Science Foundation of China (31201430); Natural Science Foundation of of HeBei (C2013201113); Special Fund for Agro-scientific Research in the Public Interest (201303075). Correspondence author: Zhao Zhilei.

References

[1] Faragher J D and Chalmers D J 1977 Regulation of an-thocyanin synthesis in apple skin. III. involvement of phenylalanine ammonia-lyase *J. Func. Plan. Bio.* **4** 133-41
[2] Zhang L, Yao S L, and Wang J H 2008 Detection and analysis of nutrients in the fresh fruit of Jiu Qian plum *J. South West Agr. Sci.* **21** 1664-66
[3] Li J, Xu Z, and Zhou L 2011 Determination of nutrients in Wild Cherry Plum in Xinjiang *J. Xinjiang Agr. Sci.* **47** 2145-49
[4] Zhao Y H, and Li X L 2009 Progress of research on the accumulation of sugar and acid in fruit *J. Agr. Sci. and Tech.* **8** 110-12
[5] BERUTER J 2004 Carbohydrate metabolism in two apple genotypes that differ in malate accumulation *J.*

Plan. Phys. **161** 1011-29

[6] Zhang L X, Zhang T F, and Li L Y 1981 *Technical biochemical experiments* (Beijing: Higher Education Press) p 115-118

[7] Zhu X L, Gao Y, and Su Q D 2004 Determination of non-volatile organic, higher fatty acids and low molecular weight carbohydrates in tomatoes by capillary gas chromatography *J. Food Che.* **25** 152-57

[8] Huang Z X, and Dai R Z 1985 Determination of malic acid in wine by fluorescent spectrophotometry *J. Food and Fer. Ind* 26

[9] Lei D F 2006 *Modern biochemistry and molecular biology instruments and equipment* (Beijing: Science Press) p 222

[10] Zhang X M, Du L Q, and Xie J H 2007 HPLC determination of sugars in Pineapple fruit *J. Food Chem.* **28** 450-52

[11] Li Y K, and Pan S Y 2006 Evaluation of orange juice sugar composition by HPLC *J. Food Che.* **27** 190-92

[12] Wu B H, Li S H 2003 The fur and fruit flesh color for the effects and correlation research of sugar and acid content in peach *J. Agr. Sci. of China* **36** 1540-44

[13] Zhang L L, Liu W S 2010 Determination of sugars and acids of five varieties of apricots by HPLC *J. Fruit Sci.* **27** 119-23

[14] Wang Y Y, Hu W Z, and Pang K 2008 HPLC-ELSD determination of soluble sugars content in apples *J. Ana. and Detect.* **34** 129-31

[15] Zhang Y Z, Li P M, and Cheng L L 2010 Developmental changes of carbohydrates, organic acids, amino acids, and phenolic compounds in Honey crisp, apple flesh *J. Food Chem.* **123** 1013-18

AUTHOR INDEX

AUTHOR INDEX